Catalysis by Precious Metals, Past and Future

Catalysis by Precious Metals, Past and Future

Special Issue Editors

Marcela Martinez Tejada
Svetlana Ivanova

MDPI • Basel • Beijing • Wuhan • Barcelona • Belgrade • Manchester • Tokyo • Cluj • Tianjin

Special Issue Editors
Marcela Martinez Tejada
Universidad de Sevilla
Spain

Svetlana Ivanova
Universidad de Sevilla
Spain

Editorial Office
MDPI
St. Alban-Anlage 66
4052 Basel, Switzerland

This is a reprint of articles from the Special Issue published online in the open access journal *Catalysts* (ISSN 2073-4344) (available at: https://www.mdpi.com/journal/catalysts/special_issues/Precious).

For citation purposes, cite each article independently as indicated on the article page online and as indicated below:

LastName, A.A.; LastName, B.B.; LastName, C.C. Article Title. *Journal Name* **Year**, *Article Number*, Page Range.

ISBN 978-3-03928-722-2 (Pbk)
ISBN 978-3-03928-723-9 (PDF)

Cover image courtesy of Jose Luis Santos Muñoz.

© 2020 by the authors. Articles in this book are Open Access and distributed under the Creative Commons Attribution (CC BY) license, which allows users to download, copy and build upon published articles, as long as the author and publisher are properly credited, which ensures maximum dissemination and a wider impact of our publications.

The book as a whole is distributed by MDPI under the terms and conditions of the Creative Commons license CC BY-NC-ND.

Contents

About the Special Issue Editors vii

Preface to "Catalysis by Precious Metals, Past and Future" ix

Svetlana Ivanova and Marcela Martínez Tejada
Editorial: Special Issue Catalysis by Precious Metals, Past and Future
Reprinted from: *Catalysts* **2020**, *10*, 247, doi:10.3390/catal10020247 1

Alberto González-Fernández, Chiara Pischetola and Fernando Cárdenas-Lizana
Gas Phase Catalytic Hydrogenation of C4 Alkynols over Pd/Al$_2$O$_3$
Reprinted from: *Catalysts* **2019**, *9*, 924, doi:10.3390/catal9110924 5

David B. Hobart Jr., Joseph S. Merola, Hannah M. Rogers, Sonia Sahgal, James Mitchell, Jacqueline Florio and Jeffrey W. Merola
Synthesis, Structure, and Catalytic Reactivity of Pd(II) Complexes of Proline and
Proline Homologs
Reprinted from: *Catalysts* **2019**, *9*, 515, doi:10.3390/catal9060515 17

Xiuyun Gao, Lulu He, Juntong Xu, Xueying Chen and Heyong He
Facile Synthesis of P25@Pd Core-Shell Catalyst with Ultrathin Pd Shell and Improved Catalytic
Performance in Heterogeneous Enantioselective Hydrogenation of Acetophenone
Reprinted from: *Catalysts* **2019**, *9*, 513, doi:10.3390/catal9060513 41

Jae-Won Jung, Won-Il Kim, Jeong-Rang Kim, Kyeongseok Oh and Hyoung Lim Koh
Effect of Direct Reduction Treatment on Pt–Sn/Al$_2$O$_3$ Catalyst for Propane Dehydrogenation
Reprinted from: *Catalysts* **2019**, *9*, 446, doi:10.3390/catal9050446 53

Masayasu Nishi, Shih-Yuan Chen and Hideyuki Takagi
Energy Efficient and Intermittently Variable Ammonia Synthesis over Mesoporous
Carbon-Supported Cs-Ru Nanocatalysts
Reprinted from: *Catalysts* **2019**, *9*, 406, doi:10.3390/catal9050406 67

Oscar H. Laguna, Julie J. Murcia, Hugo Rojas, Cesar Jaramillo-Paez, Jose A. Navío and Maria C. Hidalgo
Differences in the Catalytic Behavior of Au-Metalized TiO$_2$ Systems During Phenol
Photo-Degradation and CO Oxidation
Reprinted from: *Catalysts* **2019**, *9*, 331, doi:10.3390/catal9040331 89

Meriem Chenouf, Cristina Megías-Sayago, Fatima Ammari, Svetlana Ivanova, Miguel Angel Centeno and José Antonio Odriozola
Immobilization of Stabilized Gold Nanoparticles on Various Ceria-Based Oxides: Influence of
the Protecting Agent on the Glucose Oxidation Reaction
Reprinted from: *Catalysts* **2019**, *9*, 125, doi:10.3390/catal9020125 105

Alejandra Arevalo-Bastante, Maria Martin-Martinez, M. Ariadna Álvarez-Montero, Juan J. Rodriguez and Luisa M. Gómez-Sainero
Properties of Carbon-supported Precious Metals Catalysts under Reductive Treatment and
Their Influence in the Hydrodechlorination of Dichloromethane
Reprinted from: *Catalysts* **2018**, *8*, 664, doi:10.3390/catal8120664 117

Mengyan Zhu, Lixin Xu, Lin Du, Yue An and Chao Wan
Palladium Supported on Carbon Nanotubes as a High-Performance Catalyst for the Dehydrogenation of Dodecahydro-N-ethylcarbazole
Reprinted from: *Catalysts* **2018**, *8*, 638, doi:10.3390/catal8120638 131

Shanthi Priya Samudrala and Sankar Bhattacharya
Toward the Sustainable Synthesis of Propanols from Renewable Glycerol over MoO_3-Al_2O_3 Supported Palladium Catalysts
Reprinted from: *Catalysts* **2018**, *8*, 385, doi:10.3390/catal8090385 143

Xavier Auvray and Anthony Thuault
Effect of Microwave Drying, Calcination and Aging of Pt/Al_2O_3 on Platinum Dispersion
Reprinted from: *Catalysts* **2018**, *8*, 348, doi:10.3390/catal8090348 161

Guilhermina Ferreira Teixeira, Euripedes Silva Junior, Ramon Vilela, Maria Aparecida Zaghete and Flávio Colmati
Perovskite Structure Associated with Precious Metals: Influence on Heterogenous Catalytic Process
Reprinted from: *Catalysts* **2019**, *9*, 721, doi:10.3390/catal9090721 169

About the Special Issue Editors

Marcela Martinez Tejada was born in Medellin, Colombia. She graduated in chemical engineering from the University of Antioquia, and received her Ph.D. in materials science from the University of Seville in 2008. She obtained a postdoctoral fellowship from the Spanish Ministry of Science and Technology and moved to the Énergie et carburants pour un Environnement Durable research group at the Institute de Chimie et Procédés pour l'Energie l'Environnement et la Santé (ICPEES, CNRS - Université de Strasbourg). She returned to the University of Seville, at the Inorganic Chemistry Department and Institute of Materials Science of Seville (CSIC), within the Química de Superficies y Catálisis reseach group, with a Juan de la Cierva contract in 2012. Since 2016, she has been an associate professor. Her research interests include the synthesis and characterization of heterogeneous catalysts, and catalysts structuration and catalytic reactions for environmental and energetic applications.

Svetlana Ivanova graduated in chemistry (specialty: inorganic and analytical chemistry) from the University of Sofia St. Kliment of Ohrid, Bulgaria, with a Master of Chemical Sciences from the same university and from the University Louis Pasteur, Strasbourg France (University of Strasbourg I), where subsequently she obtained her Ph.D. Her early research interests focus on heterogeneous catalysis based on noble metals (Au, PGM) and their applications to reactions of exhaust gas treatment, CO and VOCs oxidation, and NOx reductions. Later, she was involved in variety of projects including zeolites and silicon carbide application in diverse catalytic reactions, like partial oxidation of methane, production of synthetic fuels (Fischer–Tropsch process, dimethyl ether and olefins production from methanol). In 2008, she joined the Institute of Material Science of Seville, Spain, and shortly after the Inorganic Chemistry Department of the University of Seville, where integrates her teaching and research activities as a professor. Her investigation is centered on the design, synthesis, and application of heterogeneous precious metal catalysts for H2 clean-up processes and reactions for biomass chemical valorization to high-added-value products.

Preface to "Catalysis by Precious Metals, Past and Future"

"Shiny, malleable, and resistant to corrosion" is the obvious definition of precious metals, to which "expensive" and scarce "can" be added. Their use in jewellery, trade, and arts has led to a new era in which metal catalytic potential has been discovered, and precious metals are now key players in the chemical industry. Platinum, alone or in combination with rhodium, was the first precious metal to be catalytically incorporated into the sulfuric and nitric acid production processes. Gold has entered the group of catalytically active metals in the last few decades. The use of all these metals in their bulk form was successively limited due to their high cost and the highly dispersed and supported metal nanoparticles that appeared. The use of supports improves the dispersion of the precious metals, thus reducing their quantity and decreasing the cost of the final catalyst and also preventing metal sintering, loss of catalytically active sites, and deactivation. Both support and precious metals cooperate in the formation of an efficient catalytic machine. The precious metal–support interaction depends on many factors, like precious metal contents, the nature of support and metal, employed preparation methods, and metal nanoparticles morphology. The addition of small amounts of noble metals into the formulation of other transition metals catalysts and the use of bimetallic noble metal catalysts are also attractive, since they can enhance the precious metal–support interaction. Thus, the diversity of supported precious metal catalysts is reflected in their versatility and enlarges their current and future horizons.

Marcela Martinez Tejada, Svetlana Ivanova
Special Issue Editors

Editorial

Editorial: Special Issue Catalysis by Precious Metals, Past and Future

Svetlana Ivanova * and Marcela Martínez Tejada

Departamento de Química Inorgánica-e Instituto de Ciencia de Materiales de Sevilla, Centro mixto Universidad de Sevilla-CSIC, 41092 Sevilla, Spain; leidy@us.es
* Correspondence: sivanova@us.es

Received: 19 December 2019; Accepted: 13 February 2020; Published: 19 February 2020

Precious metal catalysis is often synonymous with diversity and versatility. These metals successfully catalyze oxidation and hydrogenation due to their dissociative behavior towards hydrogen and oxygen, dehydrogenation, isomerization and aromatization, propylene production, etc. The precious metal catalysts, especially platinum-based catalysts, are involved in a variety of industrial processes. Examples include the Pt-Rh gauze for nitric acid production, the Ir and Ru carbonyl complex for acetic acid production, the Pt/Al_2O_3 catalyst for the cyclohexane and propylene production, and Pd/Al_2O_3 catalysts for petrochemical hydropurification reactions etc. A quick search over the number of published articles in the last five years containing a combination of corresponding "metals" (Pt, Pd, Ru, Rh and Au) and "catalysts" as keywords indicates the importance of the Pt catalysts, but also the continuous increase in Pd and Au contribution (Figure 1).

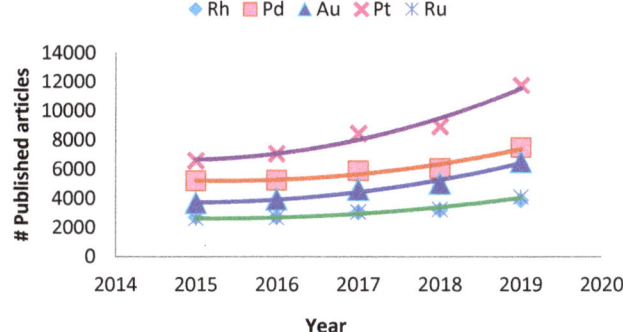

Figure 1. Number of published papers in the last 5 years, search directed on science direct page (www.sciencedirect.com) using combinations of simple keywords relating to corresponding metals (Pt, Pd, Ru, Rh or Au) and catalysts.

An important part of the Pt, Pd and Rh market includes the three-way catalyst (TWC catalyst), although the last research of the last 5 years reflects to a greater extent their participation in more fine chemistry reactions. The growth of the Pd catalyst market is reflected very well in this Special Issue by reports dealing with homogeneous and heterogeneous applications. Hobart, Jr. et al. [1] studied several palladium(II) bis-amino acid chelates for the oxidative coupling of phenylboronic acid with olefins. Despite having modest enantioselectivity, the Pd-complexes present a multiple cross coupling ability of the single substrate, providing a new horizon for the application of palladium organometallic complexes. The heterogeneous palladium catalysis are represented to a greater extent. González-Fernández et al. [2] described Pd/Al_2O_3 catalyst activity in the gas phase hydrogenation of C4 alkynols. They found a special relationship between the hydrogenation rate and C-C bond polarity.

The rate increased following the order primary < secondary < tertiary alkynol. The secondary alkynol transformation rate increased due to the preferable ketone formation via double bond migration. Gao et al. [3] carefully designed Pd heterogeneous asymmetric catalyst for the hydrogenation of acetophenone. This strategy allowed the authors to obtain a highly dispersed high-loading catalyst, resulting in an important increase in the enantioselectivity. The ability of Pd/carbon nanotube catalysts to catalyze the dodecahydro-N-ethylcarbazo dehydrogenation reaction was studied by M. Zhu et al. [4]. The catalyst revealed its potential as a stable and well performing catalyst for hydrogen production—5.6 wt.% of hydrogen was maintained after five catalytic cycles. S. P. Samudrala and S. Bhattacharya [5] addressed the near future of the supported Pd catalysts, towards the sustainable synthesis of added value chemicals, specifically the direct hydrogenolysis of glycerol to 1-propanol the exemplified reaction. This study proposed a possible route to convert the biomass-derived glycerol (rest from the biodiesel industry) into useful chemicals. The optimization of catalyst and reaction parameters resulted in around 80% of total propanol yield.

On the other hand, A. Arevalo-Bastante et al. [6] compared the activity of the carbon-supported Pd catalysts to their Pt and Rh homologues in the hydrodechlorination of dichloromethane. The Pd catalyst in this case was taken over by Pt and Rh catalysts due to their higher stability upon sintering and their ability to maintain the active site unaltered during the treatment prior reaction and therefore. X. Auvray and A. Thuault [7] chose the Pt/Al_2O_3 catalyst to study the effect of microwave pretreatment over precious metal dispersion. The microwave heating was compared to the conventional method of drying and calcination. It was found that microwave heating is only beneficial during drying but the conventional method was necessary to maintain acceptable metal dispersion. J. W. Jung et al. [8] also concentrated on the effect of reduction treatment over bimetallic Pt-Sn catalyst and its behavior in the reaction of propane dehydrogenation. Different Pt-Sn alloys were identified according to the reduction procedure. Well-dispersed Pt_3Sn alloys were found to allow reaction acceleration together with coke migration and active sites preservation.

Ru was also represented in this Special Issue. M. Nishi et al. [9] designed a series of Cs-Ru catalysts supported on mesoporous carbon for ammonia synthesis. The catalytic results show an important dependence on Ru particle size and reduction behaviour, the latter being especially important to obtain the catalytically active phase metallic Ru with adjacent CsOH species. The ammonia synthesis utility of Cs-Ru catalysts was demonstrated for the first time, using CO_2-free hydrogen from renewable energy with intermittent operation in Fukushima Renewable Energy Institute (FREA) of AIST, Japan.

The last group of publication involves different gold catalysts for photo and catalytic purposes. O. H. Laguna et al. [10] used Au/TiO_2 catalyst for photodegradation of phenol and CO oxidation. The gold catalysts prepared by photodeposition presented an important photoactivity due to the inhibited titania anatase–rutile transition. However, the prepared catalysts were less active in the gas phase oxidation of CO due to the sintering of the active phase. The importance of preserving gold nanoparticle size appears also to be a key factor in the study proposed by Chenouf et al. [11] where preformed gold colloids were stabilized by polymeric or solid-state protecting agents and immobilized on various ceria based oxides. The catalyst series was employed in two catalytic reactions, one in the gas phase and other in the liquid phase. In both reactions, the use of montmorillonite as a stabilizing agent resulted in very active catalysts due to different metal electronic state.

The review proposed by G. Ferreira Teixeira et al. [12] crowned the Special Issue and revised the role of precious metals in the perovskite photocatalytic and electrocatalytic processes. Silver and gold are the most employed metals to promote perovskites photoactivity, where the future points to the use of metal/perovskite hybrids for pollutants degradation or even for water splitting.

Let us finish as we start: the future of the precious metals is "shiny and resistant". Although judged expensive and potentially replaceable by transition metal catalysts, precious metal implementation in research and industry shows the opposite. Literally, every year new processes catalyzed by these metals appear, the best example being the important variety of biorefinery reactions or photocatalytic water splitting. Their versatility reflects their diversity and enlarges their current and future horizons of application.

Author Contributions: All authors contribute in a similar manner. All authors have read and agreed to the published version of the manuscript.

Funding: This research received no external funding.

Conflicts of Interest: The authors declare no conflict of interest.

References

1. Hobart, D.B., Jr.; Merola, J.S.; Rogers, H.M.; Sahgal, S.; Mitchell, J.; Florio, J.; Merola, J.W. Synthesis, Structure, and Catalytic Reactivity of Pd (II) Complexes of Proline and Proline Homologs. *Catalysts* **2019**, *9*, 515. [CrossRef]
2. González-Fernández, A.; Pischetola, C.; Cárdenas-Lizana, F. Gas Phase Catalytic Hydrogenation of C4 Alkynols over Pd/Al2O3. *Catalysts* **2019**, *9*, 924. [CrossRef]
3. Gao, X.; He, L.; Xu, J.; Chen, X.; He, H. Facile Synthesis of P25@ Pd Core-Shell Catalyst with Ultrathin Pd Shell and Improved Catalytic Performance in Heterogeneous Enantioselective Hydrogenation of Acetophenone. *Catalysts* **2019**, *9*, 513. [CrossRef]
4. Zhu, M.; Xu, L.; Du, L.; An, Y.; Wan, C. Palladium supported on carbon nanotubes as a high-performance catalyst for the dehydrogenation of dodecahydro-N-ethylcarbazole. *Catalysts* **2018**, *8*, 638. [CrossRef]
5. Samudrala, S.P.; Bhattacharya, S. Toward the sustainable synthesis of propanols from renewable glycerol over MoO3-Al2O3 supported palladium catalysts. *Catalysts* **2018**, *8*, 385. [CrossRef]
6. Arevalo-Bastante, A.; Martin-Martinez, M.; Álvarez-Montero, M.A.; Rodriguez, J.J.; Gómez-Sainero, L.M. Properties of Carbon-supported Precious Metals Catalysts under Reductive Treatment and Their Influence in the Hydrodechlorination of Dichloromethane. *Catalysts* **2018**, *8*, 664. [CrossRef]
7. Auvray, X.; Thuault, A. Effect of microwave drying, calcination and aging of Pt/Al2O3 on platinum dispersion. *Catalysts* **2018**, *8*, 348. [CrossRef]
8. Jung, J.W.; Kim, W.I.; Kim, J.R.; Oh, K.; Koh, H.L. Effect of Direct Reduction Treatment on Pt–Sn/Al2O3 Catalyst for Propane Dehydrogenation. *Catalysts* **2019**, *9*, 446. [CrossRef]
9. Nishi, M.; Chen, S.Y.; Takagi, H. Energy efficient and intermittently variable ammonia synthesis over mesoporous carbon-supported Cs-Ru nanocatalysts. *Catalysts* **2019**, *9*, 406. [CrossRef]
10. Laguna, O.H.; Murcia, J.J.; Rojas, H.; Jaramillo-Paez, C.; Navío, J.A.; Hidalgo, M.C. Differences in the Catalytic Behavior of Au-Metalized TiO2 Systems During Phenol Photo-Degradation and CO Oxidation. *Catalysts* **2019**, *9*, 331. [CrossRef]
11. Chenouf, M.; Megías-Sayago, C.; Ammari, F.; Ivanova, S.; Centeno, M.A.; Odriozola, J.A. Immobilization of stabilized gold nanoparticles on various ceria-based oxides: Influence of the protecting agent on the glucose oxidation reaction. *Catalysts* **2019**, *9*, 125. [CrossRef]
12. Teixeira, G.F.; Silva Junior, E.; Vilela, R.; Zaghete, M.A.; Colmati, F. Perovskite Structure Associated with Precious Metals: Influence on Heterogenous Catalytic Process. *Catalysts* **2019**, *9*, 721. [CrossRef]

© 2020 by the authors. Licensee MDPI, Basel, Switzerland. This article is an open access article distributed under the terms and conditions of the Creative Commons Attribution (CC BY) license (http://creativecommons.org/licenses/by/4.0/).

Article

Gas Phase Catalytic Hydrogenation of C4 Alkynols over Pd/Al$_2$O$_3$

Alberto González-Fernández, Chiara Pischetola and Fernando Cárdenas-Lizana *

Chemical Engineering, School of Engineering and Physical Sciences, Heriot Watt University, Edinburgh EH14 4AS, Scotland, UK; ag33@hw.ac.uk (A.G.-F.); cp44@hw.ac.uk (C.P.)
* Correspondence: F.CardenasLizana@hw.ac.uk; Tel.: +44-(0)-131-451-4115

Received: 28 September 2019; Accepted: 1 November 2019; Published: 6 November 2019

Abstract: Alkenols are commercially important chemicals employed in the pharmaceutical and agro-food industries. The conventional production route via liquid phase (batch) alkynol hydrogenation suffers from the requirement for separation/purification unit operations to extract the target product. We have examined, for the first time, the continuous gas phase hydrogenation ($P = 1$ atm; $T = 373$ K) of primary (3-butyn-1-ol), secondary (3-butyn-2-ol) and tertiary (2-methyl-3-butyn-2-ol) C$_4$ alkynols using a 1.2% wt. Pd/Al$_2$O$_3$ catalyst. *Post*-TPR, the catalyst exhibited a narrow distribution of Pd$^{\delta-}$ (based on XPS) nanoparticles in the size range 1-6 nm (mean size = 3 nm from STEM). Hydrogenation of the primary and secondary alkynols was observed to occur in a stepwise fashion (-C≡C- → -C=C- → -C-C-) while alkanol formation via direct -C≡C- → -C-C- bond transformation was in evidence in the conversion of 2-methyl-3-butyn-2-ol. Ketone formation via double bond migration was promoted to a greater extent in the transformation of secondary (*vs.* primary) alkynol. Hydrogenation rate increased in the order primary < secondary < tertiary. The selectivity and reactivity trends are accounted for in terms of electronic effects.

Keywords: gas phase hydrogenation; alkynols; 3-butyn-1-ol; 3-butyn-2-ol; 2-methyl-3-butyn-2-ol; alkenols; triple bond electron charge; Pd/Al$_2$O$_3$

1. Introduction

The bulk of research on -C≡C- bond hydrogenation has been focused on the transformation of acetylene (to ethylene) over Pd catalysts where the main challenge is to selectively promote semi-hydrogenation with -C=C- formation [1]. Product distribution is influenced by alkyne adsorption/activation mode [2]. Associative adsorption (through a π/σ double bond) on Pd planes [2] follows the Horiuti-Polanyi model, consistent with a stepwise alkyne → alkene → alkane transformation [3,4]. Alternatively, dissociative adsorption via H + three point σ bond [3] or H + π-allyl specie [5] on electron deficient edges/corners of palladium nanoparticles [6] can lead to direct alkyne → alkane hydrogenation [7] or double bond migration [8]. The electronic properties of the palladium phase and the electron density of the -C≡C- bond functionality can influence the alkyne adsorption/activation which, in turn, impact on olefin selectivity. Taking an overview of the published literature, unwanted over-hydrogenation and double migration are prevalent over electron deficient (Pd$^{\delta+}$) nanoparticles that promote strong complexation with the (electron-rich) -C≡C- bond [9,10]. The triple bond charge has also a direct role to play and can be affected by inductive effects (i.e., electron transfer from/to additional groups in poly-functional alkynes). The literature dealing with -C≡C- bond polarisation effects in hydrogenation of multifunctional alkynes is limited. It is, however, worth noting published work that shows increasing activity (over Pd(II) complexes [11] and Pd-Ru catalysts [12]) for hydrogenation of substituted acetylenes with electron donating (e.g., -R=H, -C$_6$H$_5$, -CH$_3$) functional groups [12]). Terasawa and co-workers [11], investigated the catalytic response for a series of functionalised alkynes over polymer bounded Pd(II) complexes catalyst and concluded that -C=C- selectivity is sensitive to the nature of the substituent (i.e., increased olefin selectivity in the presence of electron withdrawing substituents (-Cl, -OH) vs. electron donating (-C_6H_6) functional groups [12]).

Alkenols have found widespread applications in the manufacture of pharmaceutical (e.g., intermediates for vitamins E, A, K [13] and anti-cancer additives [14]) and agro-food (e.g., dimethyloctenol and citral [13,14]) products. Industrial synthesis involves selective hydrogenation of the correspondent substituted alkynol [15]. Alkynols can be categorised with respect to the number of carbons bonded to the carbon bearing the -OH group (C-α in Figure 1), i.e., primary (one C directly attached; labelled C-$β_1$), secondary (C-$β_1$ and C-$β_2$) and tertiary (C-$β_1$, C-$β_2$ and C-$β_3$).

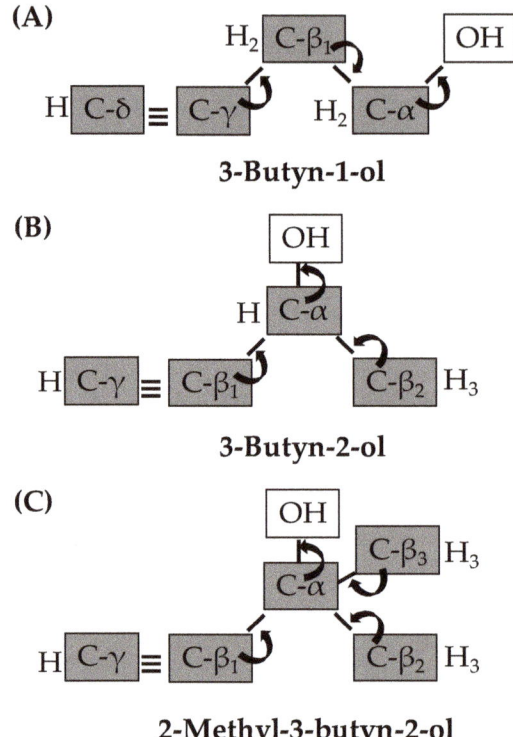

Figure 1. Classification of (**A**) primary, (**B**) secondary and (**C**) tertiary C_4 alkynols. *Note*: Arrows represent associated charge transfer effect.

Work to date has focused on batch liquid mode hydrogenation of saturated (tertiary) alkynols (e.g., 3-methyl-1-pentyn-3-ol [13]) using pressurised (up to 10 atm) reactors [16] with limited research on the selective hydrogenation of primary [17,18] and secondary alkynols [19]. Gas phase continuous operation facilitates control over contact time, which can influence product selectivity [20,21]. We were unable to find any study in the open literature on gas phase hydrogenation of primary or secondary alkynols and only one published paper in the transformation of tertiary alkynols [22]. In this work, we set out to gain an understanding of the mechanism involved in the production of primary alkenols, considering continuous gas phase hydrogenation of 3-butyn-1-ol over a commercial Pd/Al_2O_3 catalyst, as a model system. We extend the catalyst testing to consider secondary and tertiary butynols and prove possible contributions to catalytic performance (i.e., hydrogenation rate and selectivity) due to the position of the hydroxyl group.

2. Results and Discussion

2.1. Catalyst Characterisation

The Pd/Al$_2$O$_3$ catalyst used in this study bore, *post*-H$_2$-temperature programmed reduction (H$_2$-TPR) to 573 K, metal nanoparticles with diameters ranging from ≤1 nm up to 6 nm (see representative scanning transmission electron microscopy (STEM) image (A) and histogram derived from microscopy analysis (B) in Figure 2) and a number weighted mean diameter of 3 nm. An enhanced intrinsic alkenol selectivity for palladium nanoparticles of 3 nm has been reported elsewhere in the liquid (dehydroisophytol over Pd colloids [23]) and gas phase (2-methyl-3-butyn-2-ol using Pd/SiO$_2$ [22]) hydrogenation of alkynols. The STEM images reveal a pseudo-spherical morphology, the most thermodynamically stable configuration [6], indicative of a small area of contact at the interface between the Pd nanocrystals and the Al$_2$O$_3$ support.

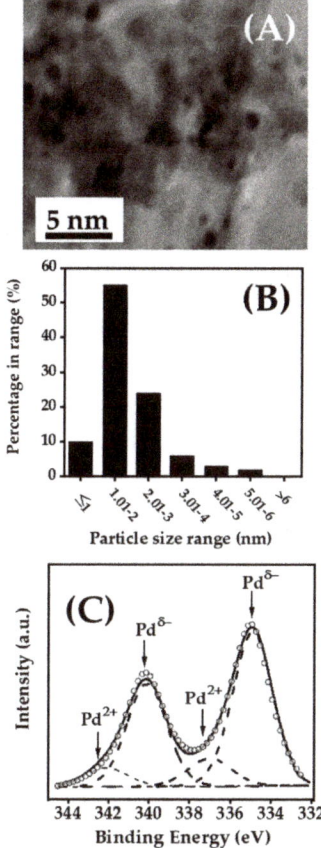

Figure 2. (**A**) Representative scanning transmission electron microscopy (STEM) image with (**B**) associated Pd particle size distribution and (**C**) X-ray photoelectron spectroscopy (XPS) spectrum over the Pd 3*d* region for Pd/Al$_2$O$_3$. *Note:* Raw data is shown as symbols (○) while curve fitted (residual standard deviation = 0.14) and envelope is represented by dashed and solid lines, respectively.

X-ray photoelectron spectroscopy (XPS) measurements were carried out to provide insight into the electronic character of the supported Pd phase. The resulting spectra over the Pd 3*d* binding energy (BE) region is shown in Figure 2C. The XPS profile exhibits a doublet with a main Pd 3$d_{5/2}$ signal at

334.9 eV, that is 0.3 eV lower than that characteristic of metallic Pd (335.2 eV, [24]), a result that suggests partial electron transfer from OH⁻ groups on the alumina support [25]. This is consistent with reported (electron-rich) Pd$^{\delta-}$ (4–5 nm) on Al$_2$O$_3$ [26]. High (94–97%) butene selectivity has been observed in the hydrogenation of butyne over palladium nanoparticles with a partial negative charge [3]. In contrast, the formation of butane and 2-hexene through undesired over-hydrogenation and double bond migration, respectively, has been reported in the hydrogenation of 1-butyne [8] and 1-hexyne [9] ascribed to the presence of (electron-deficient) Pd$^{\delta+}$ nanocrystals. In addition, the profile shows a weak doublet (12%) with curve-fitted values at higher BE (Pd $3d_{5/2}$ = 337.0 eV; Pd $3d_{3/2}$ = 342.2 eV) that can be ascribed to Pd^{2+} as a result of passivation for ex situ characterisation analyses [27]. A similar (10–12%) percentage value was reported by Weissman et al. [28] attributed to oxygen chemisorption on Pd (111) following a passivation step.

2.2. Reaction Thermodynamics

The calculated change in Gibbs free energy of formation at 373 K for each reaction step ($\Delta G_{(I-VII)}$) are included in Figure 3.

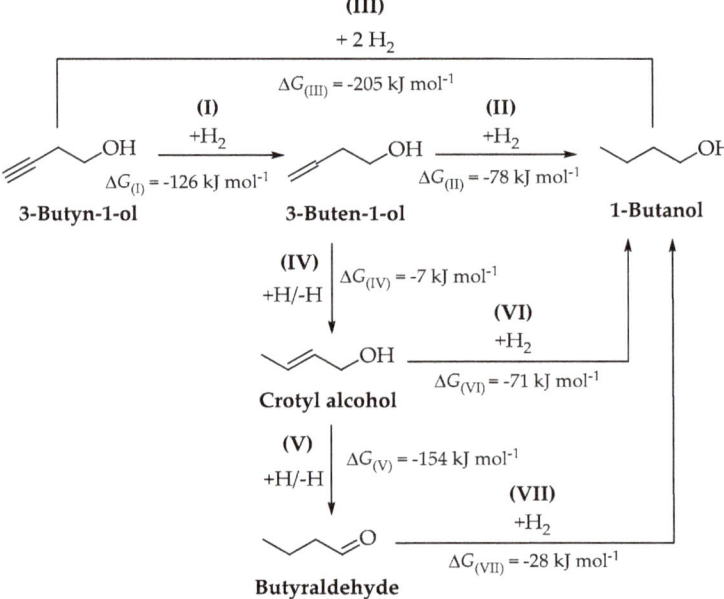

Figure 3. Reaction scheme with Gibbs free energies ($\Delta G_{(I-VII)}$) for each step in the hydrogenation of primary (3-butyn-1-ol) alkynol: *Reaction conditions*: T = 373 K, P = 1 atm.

The $\Delta G_{(I-VII)}$ values serve as criteria in the evaluation of thermodynamic feasibility, where reactions can occur spontaneously when $\Delta G < 0$. Each reaction step exhibits negative ΔG indicating that all products considered are thermodynamically favourable. Under our reaction conditions, a thermodynamic analysis of 3-butyn-1-ol hydrogenation established full conversion predominantly to 1-butanol ($S_{1\text{-butanol}}$ > 99%) with trace amounts of butyraldehyde. Formation of alkanol can result from -C=C- reduction in 3-buten-1-ol (step (**II**) in Figure 3) or direct alkynol hydrogenation via step (**III**). Hydrogenation of the intermediates, that result from alkenol double bond migration (crotyl alcohol (step (**IV**)) and keto-enol tautomerisation (butyraldehyde (step (**V**)), also generates 1-butanol (steps (**VI–VII**)).

2.3. Gas Phase Hydrogenation of 3-Butyn-1-ol

Dependence of hydrogenation path can be effectively proved from a consideration of selectivity as a function of 3-butyn-1-ol conversion; the corresponding data for reaction over Pd/Al$_2$O$_3$ is presented in Figure 4.

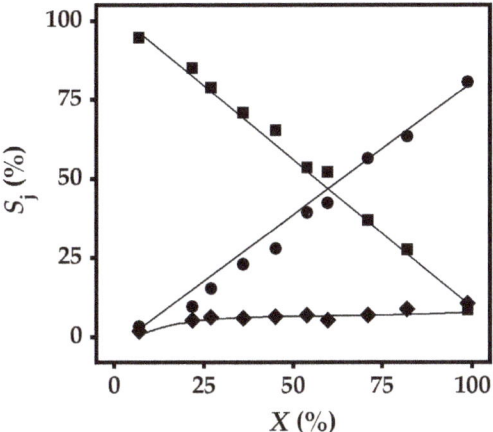

Figure 4. Variation of selectivity (S_j (%), j = 3-buten-1-ol (■), 1-butanol (●), crotyl alcohol + butyraldehyde (◆) with conversion (X (%)) in hydrogenation of 3-butyn-1-ol over Pd/Al$_2$O$_3$. *Note:* solid lines provide a guide to aid visual assessment. *Reaction conditions:* T = 373 K, p = 1 atm.

At low conversions (<25%), product composition deviates from predominant 1-butanol generation under thermodynamic equilibrium, indicative of operation under catalytic control. 3-Buten-1-ol and 1-butanol were the predominant products of partial and full hydrogenation, respectively, but double bond migration (to crotyl alcohol and butyraldehyde) was also detected with a selectivity ≤10%. Formation of 3-buten-1-ol and 1-butanol has been previously reported in the liquid phase (P = 1–6 atm; T = 300–348 K) hydrogenation of 3-butyn-1-ol over MCM-41 [29], Fe$_3$O$_4$ [30] and Fe$_3$O$_4$ coated SiO$_2$ [18] supported Pd catalysts. Production of crotyl alcohol and butyraldehyde observed in this work can be linked to reaction temperature (373 K), where T < 353 K serve to avoid double bond migration [31]. A decrease in 3-buten-1-ol selectivity was accompanied by increased formation of 1-butanol at high conversions, indicative of a sequential hydrogenation route (i.e., Horiuti-Polanyi mechanism) from -C≡C- → -C=C- → -C-C-, typical for gas phase alkyne hydrogenations [32].

2.4. Gas Phase Hydrogenation of 3-Butyn-2-ol and 2-Methyl-3-butyn-2-ol

Reaction pathways in the hydrogenation of secondary (3-butyn-2-ol) and tertiary (2-methyl-3-butyn-2-ol) C$_4$ alkynols are shown in Figure 5.

Both alkynols can undergo sequential (alkynol → alkenol → alkanol, steps **(I-II)**) and direct (alkynol → alkanol, step **(III)**) hydrogenation. Alkenol double bond migration in the transformation of 3-butyn-2-ol generates 2-butanone, (step **(IV)** in Figure 5A) but this step is not possible in the conversion of 2-methyl-3-butyn-2-ol as the C-α (Figure 1) is fully substituted. Alkynol consumption rate at the same degree of conversion (X = 25%) for the three alkynols is presented in Figure 6A where activity decreases in the order: tertiary > secondary > primary. This sequence matches that of decreasing the number of methyl substituents bonded to the C-α (Figure 1), i.e., 2-methyl-3-butyn-2-ol (C-β$_1$, Cβ$_2$ and C-β$_3$) > 3-butyn-2-ol (C-β$_1$ and C-β$_2$) > 3-butyn-1-ol (C-β$_1$). Alkyne hydrogenation has been proposed to proceed via an electrophilic mechanism [12,33]. Reactive hydrogen is provided by dissociative chemisorption of H$_2$ on Pd [34]. The hydroxyl function can serve to deactivate the triple bond for electrophilic attack through inductive effects by decreasing the overall electron density due to

-C≡C- → -OH electron transfer [35,36]. The presence of (electron donating [37]) methyl substituent(s) bonded to the C-α serves to decrease the "electron-release" from the triple bond (see charge transfer in Figure 1) which favours the electrophilic attack. Our results are in line with the work of Karavanov and Gryaznov [12] who studying the liquid phase hydrogenation of functionalised tertiary alkynols over a Pd-Ru alloy membrane catalyst reported a (40%) enhanced activity as the electron donating character of the substituent increased (i.e., -CH$_2$OH < -H < -CH$_3$).

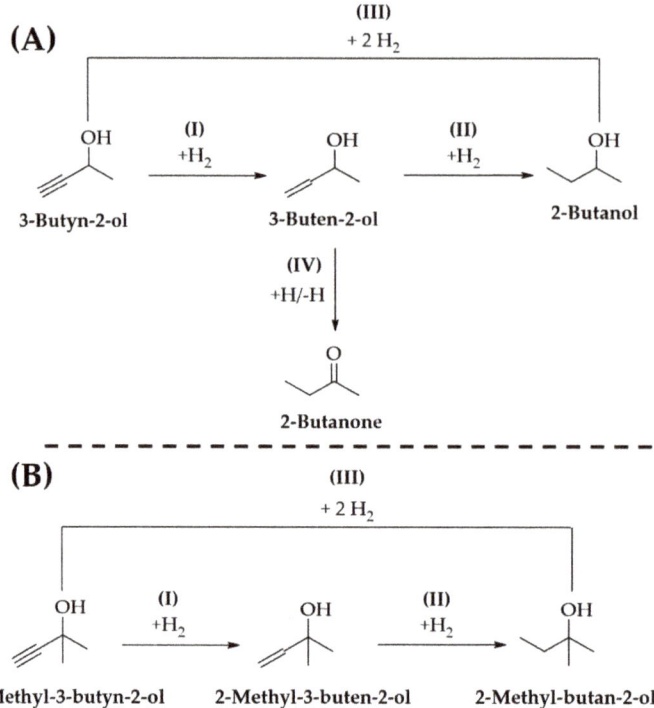

Figure 5. Reaction schemes for the hydrogenation of (**A**) secondary (3-butyn-2-ol) and (**B**) tertiary (2-methyl-3-butyn-2-ol) alkynols.

The results of product selectivity as a function of alkynol conversion for the three C$_4$ alkynols under consideration are presented in Figure 6B,C. We observe 100% selectivity in terms of -C≡C- → -C=C- bond reduction in the transformation of 3-butyn-1-ol and 3-butyn-2-ol at low conversions where the alkenol selectivity vs. conversion profiles (Figure 6B) for secondary and tertiary alkynols follow a linear decrease of the intermediate concentration as conversion increases, suggesting that they follow the same consecutive hydrogenation route as the primary (steps (**I-II**) in Figures 3 and 5A,B). In each case, regardless of the degree of conversion, greater alkenol selectivity was recorded in the transformation of the primary ~ secondary > tertiary. The lower alkenol selectivity recorded for the tertiary alkynol can be ascribed to direct formation of 2-methyl-butan-2-ol ($S_{\text{2-Methyl-butan-2-ol}}$ = 14% at X ~5%) following step (**III**) in Figure 5B. Semagina et al. [38] using monodispersed Pd nanoparticles in the liquid phase hydrogenation of 2-methyl-3-butyn-2-ol reached a similar conclusion and suggested direct hydrogenation to 2-methyl-butan-2-ol based on the $S_{\text{2-Methyl-3-buten-2-ol}} < 99\%$ at low X. Alkynol dissociative adsorption on (low coordination) Pd sites [39] can lead to direct -C≡C- → -C-C- bond hydrogenation [7] following hydrogen attack of the surface (multi-coordinated) alkilidyne intermediate [3]. This intermediate is generated by H abstraction at the "external" carbon in the -C≡C- bond (e.g., C-δ in 3-butyn-1-ol, Figure 1). Alternatively, double bond migration [8] with aldehyde/ketone formation (steps **(IV–V)** in Figure 3 and **(IV)** in Figure 5A) can occur as a result of hydrogen addition to the surface π-allyl intermediate generated by hydrogen removal from the carbon bonded to the triple bond functionality (e.g., C-$β_1$ in 3-butyn-1-ol) [40]. The lower activation energy for the formation of the π-allyl (*vs.* alkylidyne) intermediate [41,42] can account for the absence of direct -C≡C- → -C-C- bond hydrogenation in the conversion of 3-butyn-1-ol and 3-butyn-2-ol. In contrast, hydrogen abstraction in 2-methyl-3-butyn-2-ol is only possible at the (external -C≡C- carbon) C-γ (i.e., no C-α hydrogen) to generate 2-methyl-butan-2-ol. Alkenol double bond migration (*via* hydrogen addition to the external carbon, i.e., C-δ in 3-butyn-1-ol and C-γ in 3-butyn-2-ol, Figure 1, of the π-allyl intermediate [5]) was promoted to a lesser extent in the transformation of 3-butyn-1-ol vs. 3-butyn-2-ol, i.e., higher selectivity to 2-butanone relative to crotyl alcohol + butyraldehyde at all conversions (Figure 6C). Likewise, Bianchini et al. [43] reported a lower isomerisation yield in the liquid phase hydrogenation of 3-buten-1-ol (relative to 3-buten-2-ol) over a Rh complex catalyst. We examined crotyl alcohol and butyraldehyde reactivity in order to assess 1-butanol formation via hydrogenation (steps **(VI)** and **(VII)**, respectively, in Figure 3) and probe selectivity responses. Under similar reaction conditions, we recorded no conversion of butyraldehyde, a response that is consistent with the low capacity of -C=O group (e.g., methyl vinyl ketone and benzalacetone [44]) hydrogenation by Pd [45]. Conversion of crotyl alcohol generated 1-butanol as the sole product but at an appreciable higher (by a factor of 2) rate when compared with that recorded for the 3-butyn-1-ol reaction. The lower double bond migration in terms of crotyl alcohol + butyraldehyde (vs. 2-butanone) generation must result from a more facile transformation of the crotyl alcohol intermediate. Indeed, lack of activity was observed for the conversion of 2-butanone over Pd/Al_2O_3. We acknowledge that catalytic response may not governed by inductive effect alone and the dynamics of surface interactions by the hydrogen reactant can have a major bearing. Future work will be carried out to evaluate the effect of H_2 content in the feed (i.e., Alkynol: H_2 molar ratio) on catalytic performance.

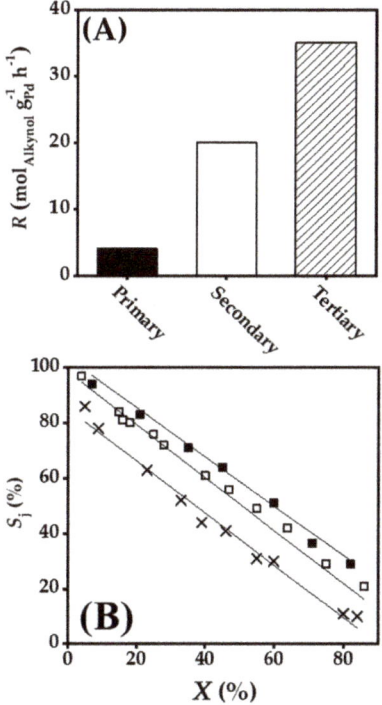

Figure 6. *Cont.*

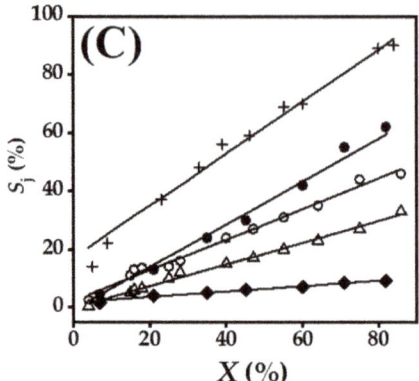

Figure 6. (**A**) Reaction rate (R, mol$_{Alkynol}$ g$_{Pd}^{-1}$ h^{-1}) and variation of selectivity (S_j, %) as a function of conversion (X, %) for products from (**B**) –C≡C– → –C=C– bond partial reduction and (**C**) hydrogen bond migration/reduction in the hydrogenation of primary (solid bar and solid symbols), secondary (open bar and open symbols) and tertiary (hatched bar and crossed symbols) C$_4$ alkynols over Pd/Al$_2$O$_3$; 3-buten-1-ol (■),1-butanol (●), crotyl alcohol + butyraldehyde (◆), 3-buten-2-ol (□), 2-butanol (○), 2-butanone (△), 2-methyl-3-buten-2-ol (×) and 2-methyl-butan-2-ol (+). *Note:* solid lines provide a guide to aid visual assessment. *Reaction conditions:* T = 373 K, p = 1 atm.

3. Materials and Methods

3.1. Catalyst Characterisation

A commercial 1.2% wt. Pd/Al$_2$O$_3$ (Sigma-Aldrich, Saint Louis, MO, USA) served as model catalyst. Before use, the catalysts were sieved (ATM fine test sieves) to mean particle diameter = 75 μm and thermally treated in 60 cm^3 min^{-1} H$_2$ (BOC, Beijing, China, ≥99.99%) at 2 K min^{-1} to 573 K to ensure reduction to Pd^0 [46]. *Post*-activation, the sample was cooled to ambient temperature and passivated in 1% *v/v* O$_2$/He (30 cm^3 min^{-1}) for 1 h for ex situ characterisation. Metal particle size and shape was examined by scanning transmission electron microscopy (STEM) using a JEOL 2200FS operated at an accelerating voltage of 200 kV and employing Gatan Digital Micrograph 1.82 for data acquisition/manipulation. The sample was deposited on a holey Cu grid (300 mesh) after dispersion in acetone. The number weighted mean Pd diameter (d) was determined as described elsewhere [47] from a count of 800 particles. X-ray photoelectron spectroscopy (XPS) analyses were conducted on an Axis Ultra instrument (Kratos Analytical, Manchester, UK) under ultra-high vacuum conditions (<10^{-8} Torr) employing a monochromatic Al Kα X-ray source (1486.6 eV). The emitted photoelectrons (source power = 150 W) were sampled from a 750 × 350 μm^2 area at a take-off angle = 90°. The survey (0–1000 eV) and high-resolution spectra (Pd $3d_{5/2}$ and $3d_{3/2}$) were collected with analyser pass energies of 80 and 40 eV, respectively. Charging effects were compensated using the adventitious carbon 1s peak calibrated at 284.5 eV as an internal standard. Curve-fitting served to identify/quantify Pd species with modified electronic properties using CasaXPS software in which the Pd $3d$ spectra were fitted with abstraction of the Shirley background using the Gaussian–Lorentzian function with a fixed full width at half maximum (FWHM) of 2.4 and Pd $3d_{5/2}$ intensity of +1.5-fold with respect to Pd $3d_{3/2}$ peak [48]. The goodness of data fitting was based on residual standard deviation; acceptable value ≤ 0.71 [49].

3.2. Catalytic Procedure

All reactions were carried out at T = 373 K at P = 1 atm in situ after activation (in H$_2$) in a continuous flow fixed bed vertical tubular glass reactor (15 mm i.d.). The operating conditions and catalytic reactor were selected to ensure negligible heat/mass transport limitations. A layer of borosilicate glass beads (1 mm diameter) served as a *pre*-heating zone. A butanolic solution of the alkynol (3-butyn-1-ol+2-butanol; 3-butyn-2-ol+1-butanol; 2-methyl-3-butyn-2-ol+1-butanol) was vaporised and reached 373 K before contacting the catalyst. In order to maintain isothermal conditions (±1 K) the catalyst bed was diluted with ground glass (75 µm diameter). Reaction temperature was monitored continuously using a thermocouple inserted in a thermowell within the catalyst bed. The reactant was delivered at a fixed calibrated flow rate via a glass/teflon air-tight syringe and Teflon line using a microprocessor-controlled infusion pump (Model 100 kd Scientific). A co-current flow of H_2/N_2 ($P_{H2} \sim 7 \times 10^{-2}$ atm) and alkynol was maintained at $GHSV = 1 \times 10^4 \, h^{-1}$. The flow rate was continuously monitored with a Humonics (Model 520) digital flowmeter. Molar metal Pd (n) to inlet alkynol molar feed rate (n/F) spanned the range $3 \times 10^{-5} - 368 \times 10^{-4}$ h. In blank tests, reactions in the absence of catalyst or over the Al_2O_3 support alone did not result in any measurable conversion. The reactor effluent was condensed in an ice-bath trap for analysis on a Perkin-Elmer Auto System XL gas chromatograph equipped with a programmed split/splitless injector and a flame ionisation detector using a Stabilwax (fused silica) 30 m × 0.32 mm i.d., 0.25 µm film thickness capillary column (RESTEK, Bellefonte, PA, USA). Data acquisition and manipulation was performed using the TotalChrom Workstation Version 6.1.2 (for Windows) chromatography data system. The solvents (2-butanol (Alpha Aesar, Haverhill, MA, USA, 99%) and 1-butanol (Fisher, The Hamptons, NH, USA, 99.4%)), reactants (3-butyn-1-ol (Aldrich, Beijing, China, 97%), 3-butyn-2-ol (Aldrich, 97%) and 2-methyl-3-butyn-2-ol (Aldrich, 98%)) and products (3-buten-1-ol (Aldrich, 96%), 1-butanol (Aldrich, 99%), crotyl alcohol (Aldrich, 96%), butyraldehyde (Aldrich, 96%), 3-buten-2-ol (Aldrich, 97%), 2-butanol (Aldrich, 99.5%), 2-butanone (Aldrich, 99%), 2-methyl-3-buten-2-ol (Aldrich, 98%), 2-methyl-butan-2-ol (Aldrich, 99%)) were used without further purification. Reactant and product molar fractions (x_i) were obtained using detailed calibration plots (not shown). Catalytic performance is considered in terms of conversion (X) at steady state after 3 h on-stream:

$$X(\%) = \frac{[\text{Alkynol}]_{in} - [\text{Alkynol}]_{out}}{[\text{Alkynol}]_{in}} \times 100 \quad (1)$$

while selectivity to product j (S_j) was obtained from:

$$S_j(\%) = \frac{[\text{Product}_j]_{out}}{[\text{Alkynol}]_{in} - [\text{Alkynol}]_{out}} \times 100 \quad (2)$$

where the subscripts "in" and "out" represent the inlet and outlet streams. Catalytic activity is also quantified in terms of alkynol consumption rate (R, $mol_{Alkynol} \, g_{Pd}^{-1} \, h^{-1}$) according to the procedure described elsewhere [50]. Reactions were repeated with the same batch of catalyst delivering a carbon mass balance and raw data reproducibility within ±6%.

3.3. Thermodynamic Analysis

The thermodynamic analysis of catalytic processes provides critical information in terms of highest conversion/selectivity possible under specific operating conditions. All the reactant and products involved in the hydrogenation of 3-butyn-1-ol (as representative) were considered (3-butyn-1-ol, 3-buten-1-ol, 1-butanol, crotyl alcohol, butyraldehyde and H_2). The inlet 3-butyn-1-ol was set at 1 mol and product distribution determined at equilibrium where $T = 373$ K, $P = 1$ atm and H_2: Alkynol molar ratio = 2 to mimic catalytic reaction conditions. Aspen Plus was used to make the equilibrium calculations [51] in order to extract product distribution in a system with minimised Gibbs free energy using the method of group contribution [52].

4. Conclusions

We have examined the effect of -OH group position on catalytic gas phase hydrogenation of C_4 alkynols over Pd/Al_2O_3 ($Pd^{\delta-}$ nanoparticles with mean (number weighted) size = 3 nm). A correlation between the number of electron-donating (-CH$_3$) groups and catalytic activity has been established consistent with the following decreasing activity sequence: tertiary (2-methyl-3-butyn-2-ol) > secondary (3-butyn-2-ol) > primary (3-butyn-1-ol). The conversion of primary and secondary alkynols follows a stepwise (alkynol → alkenol → alkanol) reaction mechanism while direct alkynol → alkanol transformation was a feature of 2-methyl-3-butyn-2-ol hydrogenation. Double bond migration is promoted to a greater extent in the transformation of 3-butyn-2-ol relative to 3-butyn-1-ol consistent with crotyl alcohol hydrogenation. The results in this work establish the role of -C≡C- polarity in determining the activity/selectivity pattern for the synthesis of valuable alkenols.

Author Contributions: F.C.-L. conceived the idea and was in charge of overall direction and planning of the project; A.G.-F. carried out the experiments; A.G.-F. and C.P. wrote the manuscript with input from F.C.-L.

Funding: This research was funded by the Engineering and Physical Sciences Research Council EPRSC grant number EP/L016419/1 [studentship to Alberto González-Fernández and Chiara Pischetola, CRITICAT program].

Acknowledgments: Open Access Funding by Heriot-Watt University.

Conflicts of Interest: The authors declare no conflict of interest.

References

1. Delgado, J.A.; Benkirane, O.; Claver, C.; Curulla-Ferré, D.; Godard, C. Advances in the Preparation of Highly Selective Nanocatalysts for the Semi-Hydrogenation of Alkynes Using Colloidal Approaches. *Dalton Trans.* **2017**, *46*, 12381–12403. [CrossRef]
2. Rajaram, J.; Narula, A.P.S.; Chawla, H.P.S.; Dev, S. Semihydrogenation of Acetylenes. *Tetrahedron* **1983**, *39*, 2315–2322. [CrossRef]
3. Molnár, Á.; Sárkány, A.; Varga, M. Hydrogenation of Carbon-Carbon Multiple Bonds: Chemo-, Regio- and Stereo-Selectivity. *J. Mol. Catal. A Chem.* **2001**, *173*, 185–221. [CrossRef]
4. Mei, D.; Sheth, P.; Neurock, M.; Smith, C. First-Principles-Based Kinetic Monte Carlo Simulation of the Selective Hydrogenation of Acetylene over Pd(111). *J. Catal.* **2006**, *242*, 1–15. [CrossRef]
5. Karpiński, Z. Catalysis by Supported, Unsupported, and Electron-Deficient Palladium. In *Advances in Catalysis*; Academic Press: Warsaw, Poland, 1990; Volume 37, pp. 45–100.
6. Semagina, N.; Kiwi-Minsker, L. Recent Advances in the Liquid-Phase Synthesis of Metal Nanostructures with Controlled Shape and Size for Catalysis. *Catal. Rev.* **2009**, *51*, 147–217. [CrossRef]
7. Sárkány, A.; Weiss, A.H.; Guczi, L. Structure Sensitivity of Acetylene-Ethylene Hydrogenation over Pd Catalysts. *J. Catal.* **1986**, *98*, 550–553. [CrossRef]
8. Hub, S.; Touroude, R. Mechanism of Catalytic Hydrogenation of But-1-yne on Palladium. *J. Catal.* **1988**, *114*, 411–421. [CrossRef]
9. Semagina, N.; Renken, A.; Kiwi-Minsker, L. Palladium Nanoparticle Size Effect in 1-Hexyne Selective Hydrogenation. *J. Phys. Chem. C* **2007**, *111*, 13933–13937. [CrossRef]
10. Boitiaux, J.P.; Cosyns, J.; Vasudevan, S. Hydrogenation of Highly Unsaturated Hydrocarbons over Highly Dispersed Pd Catalyst. Part II: Ligand Effect of Piperidine. *Appl. Catal.* **1985**, *15*, 317–326. [CrossRef]
11. Terasawa, M.; Yamamoto, H.; Kaneda, K.; Imanaka, T.; Teranishi, S. Selective Hydrogenation of Acetylenes to Olefins Catalyzed by Polymer-Bound Palladium(II) Complexes. *J. Catal.* **1979**, *57*, 315–325. [CrossRef]
12. Karavanov, A.N.; Gryaznov, V.M. Effect of the Structure of Substituted Propargyl and Allyl Alcohols on the Rate of their Liquid Phase Hydrogenation on a Pd-Ru Alloy Membrane Catalyst. *Bull. Acad. Sci. USSR Division Chem. Sci.* **1989**, *38*, 1593–1596. [CrossRef]
13. Bonrath, W.; Netscher, T. Catalytic Processes in Vitamins Synthesis and Production. *Appl. Catal. A Gen.* **2005**, *280*, 55–73. [CrossRef]
14. Bonrath, W.; Medlock, J.; Schütz, J.; Wüstenberg, B.; Netscher, T. Hydrogenation in the Vitamins and Fine Chemicals Industry—An Overview. In *Hydrogenation*; InTech: Rijeka, Croatia, 2012; pp. 69–90.
15. Bonrath, W.; Eggersdorfer, M.; Netscher, T. Catalysis in the Industrial Preparation of Vitamins and Nutraceuticals. *Catal. Today* **2007**, *121*, 45–57. [CrossRef]
16. Crespo-Quesada, M.; Cárdenas-Lizana, F.; Dessimoz, A.L.; Kiwi-Minsker, L. Modern Trends in Catalyst and Process Design for Alkyne Hydrogenations. *ACS Catal.* **2012**, *2*, 1773–1786. [CrossRef]

17. Izumi, Y.; Tanaka, Y.; Urabe, K. Selective Catalytic Hydrogenation of Propargyl Alcohol with Heteropoly Acid-Modified Palladium. *Chem. Lett.* **1982**, *11*, 679–682. [CrossRef]
18. Uberman, P.M.; Costa, N.J.S.; Philippot, K.; Carmona, R.; dos Santos, A.A.; Rossi, L.M. A Recoverable Pd Nanocatalyst for Selective Semi-Hydrogenation of Alkynes: Hydrogenation of Benzyl-Propargylamines as a Challenging Model. *Green Chem.* **2014**, *16*, 4566–4574. [CrossRef]
19. Navarro-Fuentes, F.; Keane, M.; Ni, X.-W. A Comparative Evaluation of Hydrogenation of 3-Butyn-2-ol over Pd/Al$_2$O$_3$ in an Oscillatory Baffled Reactor and a Commercial Parr Reactor. *Org. Process Res. Dev.* **2019**, *23*, 38–44. [CrossRef]
20. Hou, R.; Wang, T.; Lan, X. Enhanced Selectivity in the Hydrogenation of Acetylene due to the Addition of a Liquid Phase as a Selective Solvent. *Ind. Eng. Chem. Res.* **2013**, *52*, 13305–13312. [CrossRef]
21. Pérez, D.; Olivera-Fuentes, C.; Curbelo, S.; Rodríguez, M.J.; Zeppieri, S. Study of the Selective Hydrogenation of 1,3-Butadiene in Three Types of Industrial Reactors. *Fuel* **2015**, *149*, 34–45. [CrossRef]
22. Prestianni, A.; Crespo-Quesada, M.; Cortese, R.; Ferrante, F.; Kiwi-Minsker, L.; Duca, D. Structure Sensitivity of 2-Methyl-3-Butyn-2-ol Hydrogenation on Pd: Computational and Experimental Modeling. *J. Phys. Chem. C* **2014**, *118*, 3119–3128. [CrossRef]
23. Yarulin, A.; Yuranov, I.; Cárdenas-Lizana, F.; Abdulkin, P.; Kiwi-Minsker, L. Size-Effect of Pd-(Poly(*N*-vinyl-2-pyrrolidone)) Nanocatalysts on Selective Hydrogenation of Alkynols with Different Alkyl Chains. *J. Phys. Chem. C* **2013**, *117*, 13424–13434. [CrossRef]
24. Smirnov, M.Y.; Klembovskii, I.O.; Kalinkin, A.V.; Bukhtiyarov, V.I. An XPS Study of the Interaction of a Palladium Foil with NO$_2$. *Kinet. Catal.* **2018**, *59*, 786–791. [CrossRef]
25. Juan-Juan, J.; Román-Martínez, M.C.; Illán-Gómez, M.J. Catalytic Activity and Characterization of Ni/Al$_2$O$_3$ and NiK/Al$_2$O$_3$ Catalysts for CO$_2$ Methane Reforming. *Appl. Catal. A Gen.* **2004**, *264*, 169–174. [CrossRef]
26. Nag, N.K. A Study on the Dispersion and Catalytic Activity of Gamma Alumina-Supported Palladium Catalysts. *Catal. Lett.* **1994**, *24*, 37–46. [CrossRef]
27. McCarty, J.G. Kinetics of PdO Combustion Catalysis. *Catal. Today* **1995**, *26*, 283–293. [CrossRef]
28. Weissman, D.L.; Shek, M.L.; Spicer, W.E. Photoemission Spectra and Thermal Desorption Characteristics of Two States of Oxygen on Pd. *Surf. Sci.* **1980**, *92*, L59–L66. [CrossRef]
29. Papp, A.; Molnár, Á.; Mastalir, Á. Catalytic Investigation of Pd Particles Supported on MCM-41 for the Selective Hydrogenations of Terminal and Internal Alkynes. *Appl. Catal. A Gen.* **2005**, *289*, 256–266. [CrossRef]
30. Da Silva, F.P.; Rossi, L.M. Palladium on Magnetite: Magnetically Recoverable Catalyst for Selective Hydrogenation of Acetylenic to Olefinic Compounds. *Tetrahedron* **2014**, *70*, 3314–3318. [CrossRef]
31. Derrien, M.L. Selective Hydrogenation Applied to the Refining of Petrochemical Raw Materials Produced by Steams Cracking. *Stud. Surf. Sci. Catal.* **1986**, *27*, 613–666.
32. Nikolaev, S.A.; Zanaveskin, L.N.; Smirnov, V.V.; Averyanov, V.A.; Zanaveskin, K.L. Catalytic Hydrogenation of Alkyne and Alkadiene Impurities from Alkenes. Practical and Theoretical Aspects. *Russ. Chem. Rev.* **2009**, *78*, 231–247. [CrossRef]
33. Morrill, C.; Grubbs, R.H. Highly Selective 1,3-Isomerization of Allylic Alcohols via Rhenium Oxo Catalysis. *J. Am. Chem. Soc.* **2005**, *127*, 2842–2843. [CrossRef]
34. Jewell, L.; Davis, B. Review of Absorption and Adsorption in the Hydrogen–Palladium System. *Appl. Catal. A Gen.* **2006**, *310*, 1–15. [CrossRef]
35. Nikoshvili, L.; Shimanskaya, E.; Bykov, A.; Yuranov, I.; Kiwi-Minsker, L.; Sulman, E. Selective Hydrogenation of 2-Methyl-3-Butyn-2-ol over Pd-Nanoparticles Stabilized in Hypercrosslinked Polystyrene: Solvent Effect. *Catal. Today* **2014**, *241*, 179–188. [CrossRef]
36. Sulman, E.M. Selective Hydrogenation of Unsaturated Ketones and Acetylene Alcohols. *Russ. Chem. Rev.* **1994**, *63*, 923–936. [CrossRef]
37. Hansch, C.; Leo, A.; Taft, R.W. A Survey of Hammett Substituent Constants and Resonance and Field Parameters. *Chem. Rev.* **1991**, *91*, 165–195. [CrossRef]
38. Semagina, N.; Renken, A.; Laub, D.; Kiwi-Minsker, L. Synthesis of Monodispersed Palladium Nanoparticles to Study Structure Sensitivity of Solvent-Free Selective Hydrogenation of 2-Methyl-3-butyn-2-ol. *J. Catal.* **2007**, *246*, 308–314. [CrossRef]

39. Boitiaux, J.P.; Cosyns, J.; Robert, E. Liquid Phase Hydrogenation of Unsaturated Hydrocarbons on Palladium, Platinum and Rhodium Catalysts. Part II: Kinetic Study of 1-Butene, 1,3-Butadiene and 1-Butyne Hydrogenation on Rhodium; Comparison with Platinum and Palladium Part II. *Appl. Catal.* **1987**, *32*, 169–183. [CrossRef]
40. Morrill, T.C.; D'Souza, C.A. Efficient Hydride-Assisted Isomerization of Alkenes via Rhodium Catalysis. *Organometallics* **2003**, *22*, 1626–1629. [CrossRef]
41. Karlsson, E.A.; Bäckvall, J.-E. Mechanism of the Palladium-Catalyzed Carbohydroxylation of Allene-Substituted Conjugated Dienes: Rationalization of the Recently Observed Nucleophilic Attack by Water on a (π-Allyl)palladium Intermediate. *Chem. A Eur. J.* **2008**, *14*, 9175–9180. [CrossRef]
42. Behm, R.J.; Penka, V.; Cattania, M.-G.; Christmann, K.; Ertl, G. Evidence for '"Subsurface"' Hydrogen on Pd(110): An Intermediate Between Chemisorbed and Dissolved Species. *J. Chem. Phys.* **1983**, *78*, 7486–7490. [CrossRef]
43. Bianchini, C.; Meli, A.; Oberhauser, W. Isomerization of Allylic Alcohols to Carbonyl Compounds by Aqueous-Biphase Rhodium Catalysis. *New J. Chem.* **2001**, *25*, 11–12. [CrossRef]
44. Ide, M.S.; Hao, B.; Neurock, M.; Davis, R.J. Mechanistic Insights on the Hydrogenation of α,β-Unsaturated Ketones and Aldehydes to Unsaturated Alcohols over Metal Catalysts. *ACS Catal.* **2012**, *2*, 671–683. [CrossRef]
45. Ponec, V. On the Role of Promoters in Hydrogenations on Metals; α,β-Unsaturated Aldehydes and Ketones. *Appl. Catal. A Gen.* **1997**, *149*, 27–48. [CrossRef]
46. Amorim, C.; Keane, M.A. Palladium Supported on Structured and Nonstructured Carbon: A Consideration of Pd Particle Size and the Nature of Reactive Hydrogen. *J. Colloid Interface Sci.* **2008**, *322*, 196–208. [CrossRef] [PubMed]
47. Cárdenas-Lizana, F.; Wang, X.; Lamey, D.; Li, M.; Keane, M.A.; Kiwi-Minsker, L. An Examination of Catalyst Deactivation in *p*-Chloronitrobenzene Hydrogenation over Supported Gold. *Chem. Eng. J.* **2014**, *255*, 695–704. [CrossRef]
48. Venezia, A.M.; Liotta, L.F.; Deganello, G.; Schay, Z.; Guczi, L. Characterization of Pumice-Supported Ag–Pd and Cu–Pd Bimetallic Catalysts by X-Ray Photoelectron Spectroscopy and X-Ray Diffraction. *J. Catal.* **1999**, *182*, 449–455. [CrossRef]
49. Fu, Z.; Hu, J.; Hu, W.; Yang, S.; Luo, Y. Quantitative analysis of Ni^{2+}/Ni^{3+} in $Li[Ni_xMn_yCo_z]O_2$ Cathode Materials: Non-linear Least-squares Fitting of XPS Spectra. *Appl. Surf. Sci.* **2018**, *441*, 1048–1056. [CrossRef]
50. Cárdenas-Lizana, F.; Lamey, D.; Perret, N.; Gómez-Quero, S.; Kiwi-Minsker, L.; Keane, M.A. Au/Mo_2N as a New Catalyst Formulation for the Hydrogenation of *p*-Chloronitrobenzene in Both Liquid and Gas Phases. *Catal. Commun.* **2012**, *21*, 46–51. [CrossRef]
51. Ye, G.; Xie, D.; Qiao, W.; Grace, J.R.; Lim, C.J. Modelling of Fluidized Bed Membrane Reactors for Hydrogen Production from Steam Methane Reforming with Aspen Plus. *Int. J. Hydrogen Energy* **2009**, *34*, 4755–4762. [CrossRef]
52. Joback, K.G.; Reid, R.C. Estimation of Pure-Component Properties from Group-Contributions. *Chem. Eng. Commun.* **1987**, *57*, 233–243. [CrossRef]

© 2019 by the authors. Licensee MDPI, Basel, Switzerland. This article is an open access article distributed under the terms and conditions of the Creative Commons Attribution (CC BY) license (http://creativecommons.org/licenses/by/4.0/).

Article

Synthesis, Structure, and Catalytic Reactivity of Pd(II) Complexes of Proline and Proline Homologs

David B. Hobart Jr., Joseph S. Merola *, Hannah M. Rogers, Sonia Sahgal, James Mitchell, Jacqueline Florio and Jeffrey W. Merola

Department of Chemistry, Virginia Tech, Blacksburg, VA 24061, USA; dhobart@vt.edu (D.B.H.J.); hannahm@vt.edu (H.M.R.); sonia.sahgal5@gmail.com (S.S.); jhhmvt@gmail.com (J.M.); jflo5392@vt.edu (J.F.); senrath@gmail.com (J.W.M.)
* Correspondence: jmerola@vt.edu

Received: 7 May 2019; Accepted: 6 June 2019; Published: 10 June 2019

Abstract: Palladium(II) acetate reacts with proline and proline homologs in acetone/water to yield square planar bis-chelated palladium amino acid complexes. These compounds are all catalytically active with respect to oxidative coupling of olefins and phenylboronic acids. Some enantioselectivity is observed and formation of products not reported in other Pd(II) oxidative couplings is seen. The crystal structures of nine catalyst complexes were obtained. Extended lattice structures arise from N-H••O or O••(HOH)••O hydrogen bonding. NMR, HRMS, and single-crystal XRD data obtained on all are evaluated.

Keywords: palladium; chelate; amino acid; proline; N-methylproline; azetidine; pipecolinic acid; 4-fluoroproline; 4-hydroxyproline; 2-α-benzylproline; hydrogen bonding; oxidative coupling; X-ray crystallography

1. Introduction

Oxidative coupling reactions are some of the most utilized reactions in modern synthetic chemistry, and transition metal catalyzed oxidations are well known [1]. Palladium(II) oxidative coupling catalysis is a huge field [1–6], and this introduction cannot give even a cursory overview of it [7–13]. Some recent reviews may be the best way of relaying important background information. Focusing on the general, reactions such as the Heck, Suzuki, and Sonogashira couplings are known to proceed via a Pd^0 species, with oxidative addition/reductive elimination yielding the desired products. Oxidative palladium(II) catalysis differs from these in that it utilizes molecular oxygen to regenerate the active catalyst in palladium(II) catalyzed coupling reactions. There are two proposed mechanisms for the catalytic cycle [14–20].

There are many different types of coupling reactions noted to proceed via palladium(II) oxidative catalysis. There are hundreds of examples in the current literature of carbon-carbon [2,5,21–39], carbon-oxygen [40–47], carbon-nitrogen [47–55], carbon-sulfur [42,56,57], and carbon-phosphorous [58] couplings that are catalyzed by palladium(II) oxidative catalysis. These coupling reactions are used in the manufacture of many pharmaceuticals, natural products, fine chemicals, and polymers. In addition, palladium catalysts are known for their functional group tolerance, mild reaction conditions, and low sensitivity to air and water. Pairing these advantages with an abundant and easily accessible oxidant source shows the great utility and economic benefit that these systems can provide.

Also important for this discussion is a significant history of using amino acids as chelating ligands, especially for Pd(II) and Pt(II). Wolfgang Beck has had a five-decade career publishing the series "Metal Complexes of Biologically Important Ligands," consisting of over 175 articles [59]. Many of those papers deal with amino acid complexes of metals including palladium. Recently, proline has come to the fore as a "co-catalyst" used along with palladium in a number of organic reactions [60].

More specifically related to this paper are recent publications that describe a bis-proline complex of palladium as catalyst for various organic transformations. Blum et al. [61] show that proline complexes of Pd(II) catalyze various coupling reactions, while Chatterjee et al. [62] showed that bis-proline pd(II) complexes were useful in the Suzuki-Miyaura cross-coupling reaction in water. The formation of biaryl products from palladium-catalyzed cross-couplings is well known, and there are examples from the literature [21,50,63–67] that demonstrate that these biaryls can be formed as desired reaction products or as undesired side products. For reactions where the cross-coupled product is desired, the formation of biaryls is an unwanted side reaction and catalyst systems of this type where the biaryl formation is minimized or eliminated are preferred.

This paper describes the synthesis and catalytic activity of palladium(II)-amino acid chelates where proline, N-methylproline, 4-fluoroproline, 4-hydroxyproline, 2-benzylproline, azetidine-2-carboxylic acid, and pipecolinic acid were used as the chelating ligands. In earlier work, we showed that the amino acid complexes of rhodium and iridium piano stools were useful for asymmetric hydrogenation [68]. We have previously reported on the simpler glycine complexes [69], and these proline and proline homologs represent another unique subset of amino acid ligands where the R-group of the amino acid is a cyclic ring moiety. Subsequent papers will discuss our work with beta-amino acid complexes and amino acids where the R-group is a linear substituent.

2. Results and Discussion

2.1. Characterization and Hydrogen Bonding Interactions

In the following discussions, compounds **1–9** (Figure 1, above) were synthesized as shown in the reaction scheme in Figure 2:

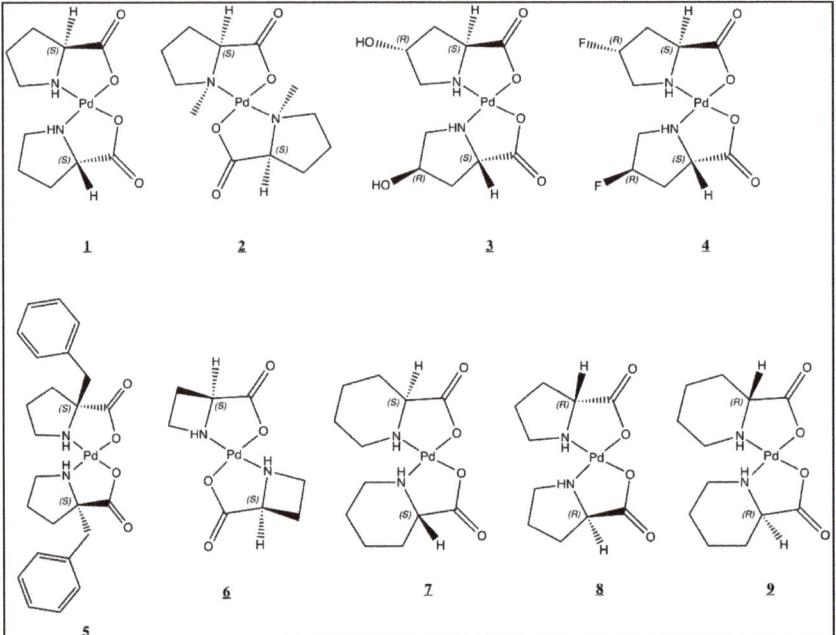

Figure 1. Compound structures and numbering scheme for proline and proline homolog complexes. Stereochemistry is shown at all chiral centers.

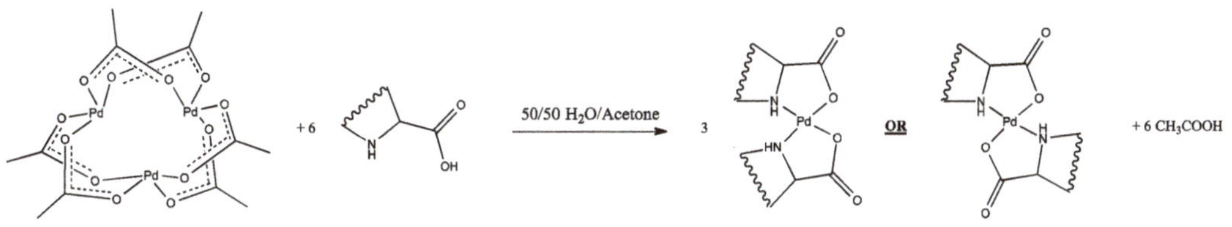

Figure 2. General reaction scheme for the synthesis of *cis* and *trans* palladium(II) proline/cyclic complexes.

The most common of the cyclic amino acids is L-proline, one of 20 naturally occurring α-amino acids. Compound **1** was prepared as the *cis* isomer and confirmed by X-ray crystal structure analysis (Figure 3).

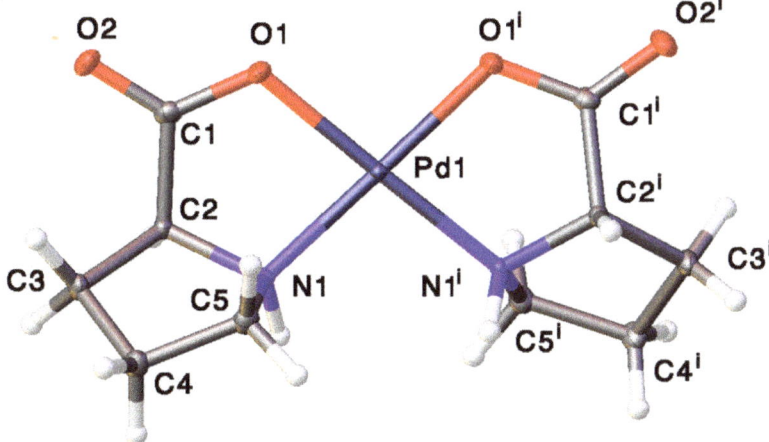

Figure 3. Thermal ellipsoid plot of the molecular structure of crystalline *cis*-bis(L-prolinato)palladium(II), **1**. Atoms labelled "i" are generated by a C_2 rotation. Thermal ellipsoids are shown at the 50% probability level. CCDC:1913626.

The complex crystallizes in the $C222_1$ space group. Pd-N and Pd-O bond lengths are 2.0105 Å and 2.0193 Å, respectively. N-Pd-O bond angles are 82.65° for each chelate ring and 96.27° between the chelate rings (O-Pd-O). There were no unusual bond lengths and angles in complex **1** (see the Supplementary Materials for the full listing). Intermolecular hydrogen bonding is common for palladium amino acid complexes; the exact nature is dependent on the amino acid and any substitution on the amino acid backbone. For complex **1**, intermolecular hydrogen bonding is observed in the crystal lattice between the amine protons and the non-coordinated carboxyl oxygen atoms (Figure 4). In this case, the palladium complex molecules arrange themselves and can approach closely enough for this purely complex to complex H-bonding. This compound was reported previously by Ito et al., but at room temperature in a non-standard space group [70]. The bond lengths and angles of the compounds reported here may be compared with those reported previously in the literature [71–74].

The ^{1}H NMR spectrum in D_2O shows three multiplets at δ 4.08–3.63, 3.37–2.73, and 2.28–1.52, with integrated ratios of 1:2:4, respectively. Palladium's isotope distribution pattern was observed in the HRMS spectrum (see Supplementary Materials).

D-proline was used to prepare *cis*-bis(D-prolinato)palladium(II), compound **8**. Characterization data for **8** was the same as that seen for **1**, with the stereochemistry of the chiral carbon reversed (see Supplementary Materials), as is expected for enantiomeric species.

In *N*-methylproline the amine proton in proline is replaced with a methyl group. The resultant complex formed with this ligand is *trans* bis-(*N*-methylprolinato) palladium(II) (Figure 5), confirmed by X-ray crystallographic analysis.

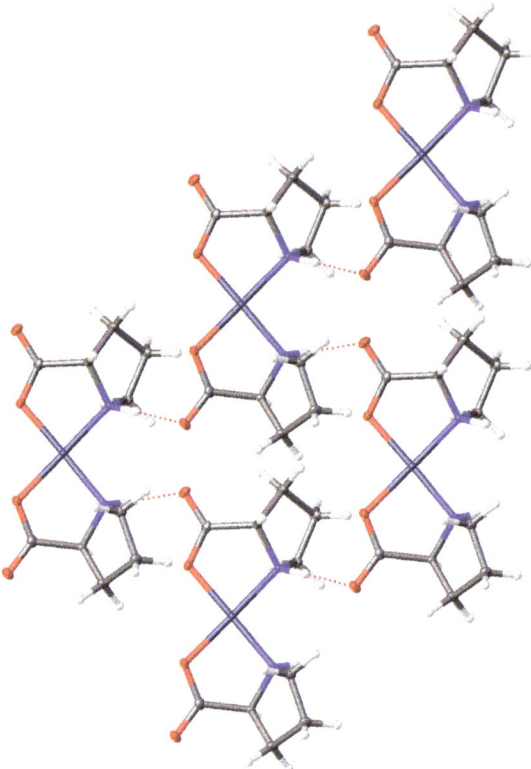

Figure 4. Crystal packing diagram for complex **1** viewed along [001] showing the intermolecular hydrogen bonding motif.

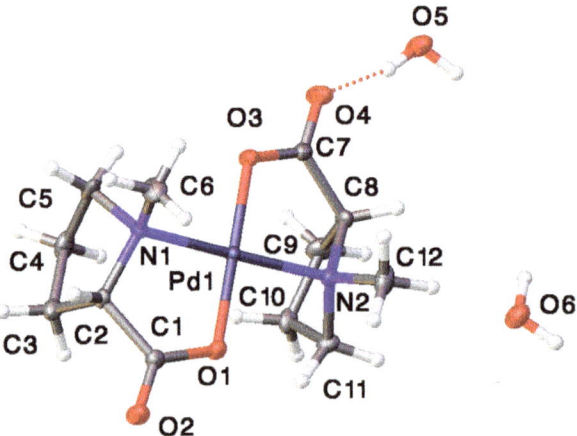

Figure 5. Thermal ellipsoid plot of the molecular structure of crystalline *trans*-bis-(*N*-methylprolinato) palladium(II) dihydrate, **2**. Thermal ellipsoids are shown at the 50% probability level. CCDC:1913622.

As with the glycine complexes, replacement of the amine hydrogen atom with a methyl group results in a degree of steric crowding that disfavors formation of the *cis* isomer [69]. All attempts to synthesize the *cis* isomer from $PdCl_2$ were unsuccessful. Complex **2** crystallizes in the $P2_12_12_1$ space group with Pd-N and Pd-O bond lengths of 2.051 Å and 1.9900 Å, respectively. N-Pd-O bond angles are 83.90° in the chelate ring. N-Pd-O bond angles between the chelate rings are 95.42°. There were no unusual bond lengths and angles in complex **1** (see Supplementary Materials for full listing). Because there are no H-bonding donors in the molecule due to the N-methyl substitution, there is no intermolecular hydrogen bonding between complex molecules in the lattice. However, water is now incorporated in the lattice and there is hydrogen bonding between water molecules and complex molecules (Figure 6). Intermolecular hydrogen-bonded water molecules are observed to bridge between the coordinated carboxylate oxygen atom of one complex molecule and the carbonyl oxygen of an adjacent complex molecule. It is also interesting to note that both of the pyrrolidine rings are turned down such that they are orientated towards the same face of the chelate plane.

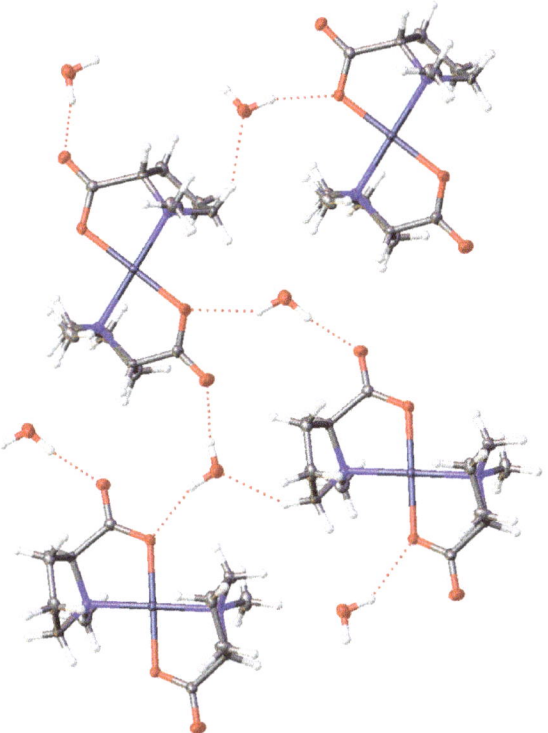

Figure 6. Crystal packing diagram of (**2**) as viewed along [010] showing the intermolecular hydrogen bonding motif.

The ^{1}H NMR spectrum in D_2O shows the expected singlet for the methyl protons at δ 2.71 ppm. The remaining proton resonances are present in the expected ratios; however, the splitting patterns are complex. Palladium's isotope distribution pattern is observed in the HRMS spectrum with peaks at 361.0526, 362.0542, 363.0532, 365.0531, and 367.0541 amu.

Hydroxyproline and fluoroproline have more electron-withdrawing substituents on their backbones than their strictly alkyl homologs, and this influence was probed by synthesizing their respective complexes **3** and **4**. *Cis*-bis-(*trans*-4-hydroxyprolinato)palladium(II) was synthesized (Compound **3**, Figure 7) using *trans*-4-hydroxyproline as the ligand.

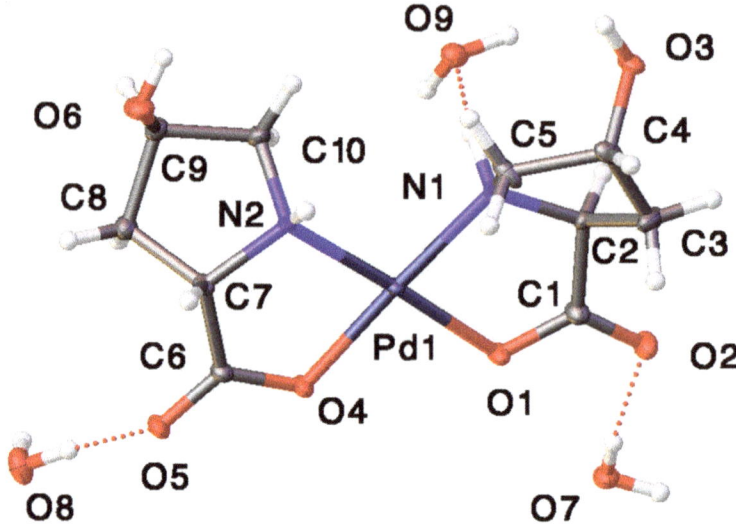

Figure 7. Thermal ellipsoid plot of the molecular structure of crystalline *cis*-bis-(*trans*-4-hydroxyprolinato)palladium(II) trihydrate, (**3**). Thermal ellipsoids are shown at the 50% probability level. CCDC:1913624.

Complex **3** crystallizes in the $P2_1$ space group with 3 hydrogen bonded water molecules per complex molecule in the lattice. The addition of another H-bond donor and acceptor complicates the H-bonding picture in the crystal lattice. There is intermolecular hydrogen bonding between one of the 4-hydroxyl group hydrogen atoms and the carbonyl oxygen of an adjacent molecule. The hydroxyl oxygen atom is hydrogen bonded to a lattice water molecule that in turn hydrogen bonds to a coordinated carboxylate oxygen of the adjacent molecule. The other 4-hydroxyl group is hydrogen bonded to two lattice water molecules that also hydrogen bind to carbonyl oxygen atoms on adjacent complex molecules in the lattice. The amine hydrogens are hydrogen bonded to lattice waters.

Pd-N and Pd-O bond lengths are 2.0153 Å and 2.0006 Å, respectively. N-Pd-O bond angles are 83.92° in the chelate ring with N-Pd-N bond angles between the chelate rings are 97.84°. All bond lengths and angles are within the ranges reported for similar d^8 metal chelates (see Supplementary Materials). The ^{1}H NMR spectrum in D_2O shows a singlet at 4.41 ppm, indicating that the hydroxyl proton does not exchange, or exchanges very slowly. All other resonances are as expected. Palladium's isotope distribution pattern is observed in the HRMS.

Cis-bis-(*trans*-4-fluoroprolinato)palladium(II) (Figure 8) crystallizes in the C2 space group. As is the case with the parent complex **1**, pure complex-to-complex H-bonding occurs and there are no water molecules in the lattice; the hydrogen bonding arrangement is quite different from that seen with the hydroxyproline complex. For the fluoroproline complex, there is hydrogen bonding from each amine hydrogen atom to a carbonyl oxygen atom on separate, adjacent complex molecules in the lattice (See Supplementary Material). The fluorine atoms do not participate in a hydrogen bonding interaction.

The Pd-N and Pd-O bond lengths in compound **4** are 2.006 Å and 2.017 Å, respectively. The N-Pd-O bond angle in the chelate ring is 82.092°, with the N-Pd-N bond angle between the chelate rings at 98.544° (see Supplementary Materials). As with the previous complexes, these values are in good agreement with other square planar palladium *N*,*O* chelates.

The proton NMR spectrum of complex **4** shows a complicated set of multiplets due to ^{1}H-^{1}H and ^{1}H-^{19}F coupling; however, integration does show the expected ratios of protons. The ^{13}C NMR spectrum is somewhat easier to interpret, showing the expected five carbon resonances with ^{19}F coupling constants observed on the order of 130–170 Hz. The ^{19}F NMR shows a singlet at −179.33 ppm with ^{13}C-^{19}F coupling of 141 Hz. HRMS shows the expected palladium isotopic pattern.

2-Benzylproline adds additional steric demands to the proline ligand. Compound **5**, *trans*-bis-(2-benzylprolinato)palladium(II) (Figure 9), was prepared using 2-benzylproline hydrochloride as the ligand.

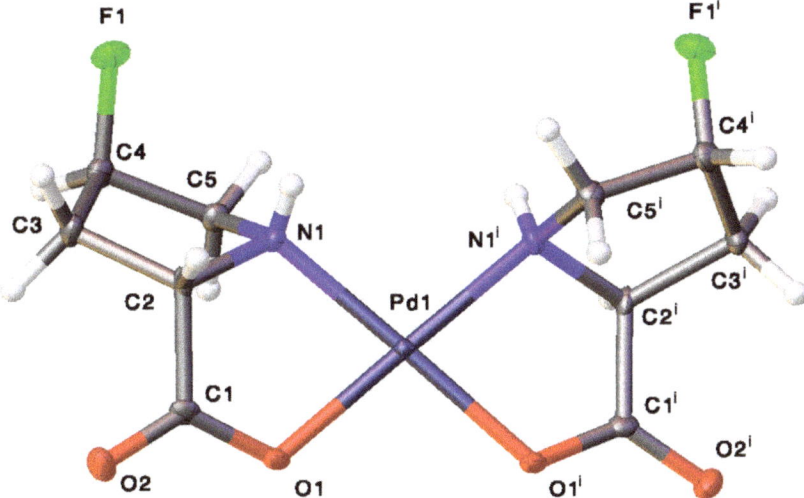

Figure 8. *Cis*-bis-(*trans*-4-fluoroprolinato)palladium(II), (**4**). Thermal ellipsoids are drawn at the 50% probability level. Atoms labeled with superscript "I" are generated by a C$_2$ rotation. CCDC:1913621.

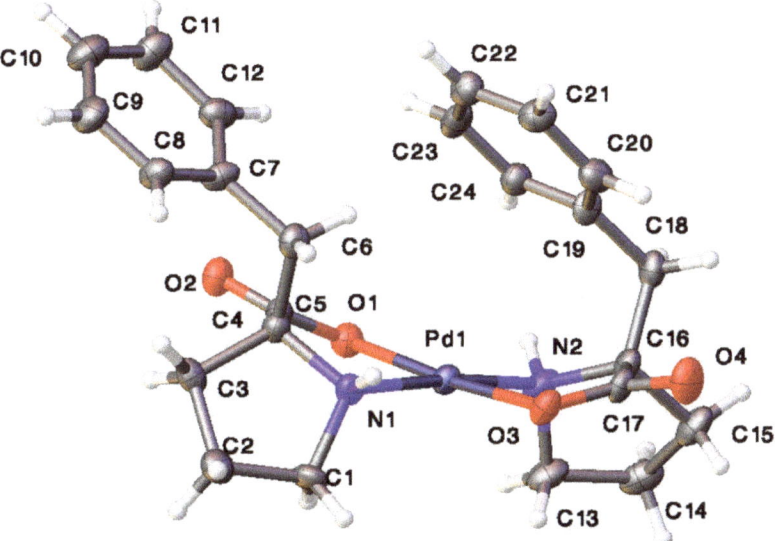

Figure 9. Thermal ellipsoid plot of the molecular structure of crystalline *trans*-bis-(2-benzylprolinato)palladium(II), **5**. Thermal ellipsoids are shown at the 50% probability level. CCDC:1913619.

Crystallizing in the P2$_1$2$_1$2$_1$ space group, *trans*-bis-(2-benzylprolinato)palladium(II) has Pd-N bond lengths of 2.024 Å and 2.037 Å. Pd-O bond lengths are 2.006 Å and 2.004 Å (see Supplementary Materials). The chelate rings are slightly twisted out of the square plane. The N-Pd-O angles between the chelate rings are 98.3 and 97.2°. The N-Pd-O angles in the chelate rings are 82.6°. The benzyl groups on the ligands are oriented up and away from the proline ring, with one of the benzyl groups laying over the square plane. This is the same arrangement reported by Sabat [75] for the palladium(II)-tyrosine complex; however, in the case of **5** the second benzyl group does not lie over an adjacent metal center, but rather in the lattice space between complex molecules. This arrangement does suggest that there is a π-d interaction occurring between the metal and the aromatic ring of the ligand. Two of the carbon atoms in the benzyl ring lie closer to the metal center than their calculated Van Der Waals radii. The Pd-C(19) contact distance is 3.452 Å and the Pd-C(24) contact distance is 3.472 Å. The calculated Van Der Waals radius [76] for a Pd-C bond is 3.91 Å, or approximately 0.45 Å more than what is observed in the crystal structure. This reduction in the Pd-C contact distances suggests an energetically favorable interaction between the π electron cloud of the benzyl ring and the empty d$_z^2$ orbital on the metal center. The other Pd-C contact distances within the benzyl ring are in the range of 4.009–4.565 Å. Hydrogen bonding occurs between the amine hydrogen atoms and the coordinated carboxylate oxygen atom of the adjacent molecule (see Supplementary Materials). There are no water molecules in the lattice, which is not surprising given the hydrophobicity of the benzyl groups.

The ^{1}H NMR spectrum of complex **5** is somewhat complicated. The aromatic benzyl protons show a multiplet at 7.25 ppm with the benzyl methylene protons resonating as a pair of doublets at 3.42 and 3.00 ppm. The integrated ratio of the benzyl protons is the expected 5:2. The pyrrolidine ring protons show multiplets at 3.29, 2.43, 2.02, and 1.87 ppm in a ratio of 2:1:2:1. The expected mass and isotopic splitting pattern is once again observed in the HRMS for complex **5** with the [M+H]$^+$ peak at 515.1175 amu.

The proline ring is a five-membered moiety, and both four- and six-membered ring homologs are known. The four-membered ring homolog, L-azetidine-2-carboxylic acid, was used to prepare *trans*-bis-(L-azetidine-2-carboxylato)palladium(II) (Compound **6**, Figure 10). The crystal structure of this compound shows some unique phase-change characteristics and will be the subject of a separate crystallographic paper. The ^{1}H NMR data show the expected ratios of integrated resonances, and the ^{13}C NMR spectrum shows possible evidence of aquo complex formation. As seen with the glycine complexes [69] discussed in a prior paper, the carbon NMR data for **6** shows two peaks for each carbon. The HRMS is as expected for a palladium complex.

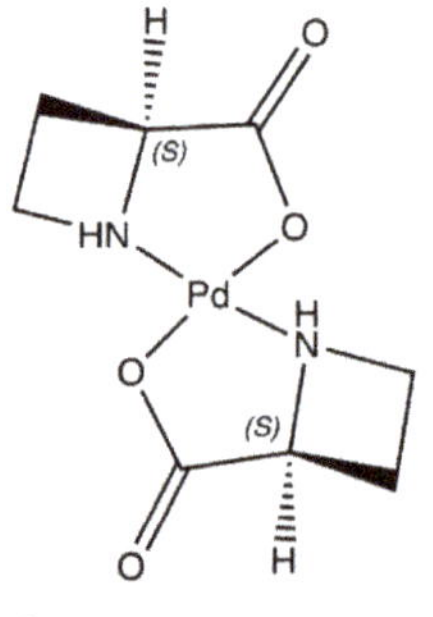

Figure 10. Molecular structure of *trans*-bis-(L-azetidine-2-carboxylato)palladium(II), (**6**).

The six-membered ring homolog, L-pipecolinic acid, was used to prepare *cis*-bis-(L-pipecolinato)palladium(II) (Compound **7**, Figure 11). This complex crystallizes in the C2 space group. Pd-N and Pd-O bond lengths are approximately equivalent at 2.01–2.03Å, comparable to the other complexes discussed within. The piperidine ring adopts the classic "chair" formation seen in cyclohexyl ring systems.

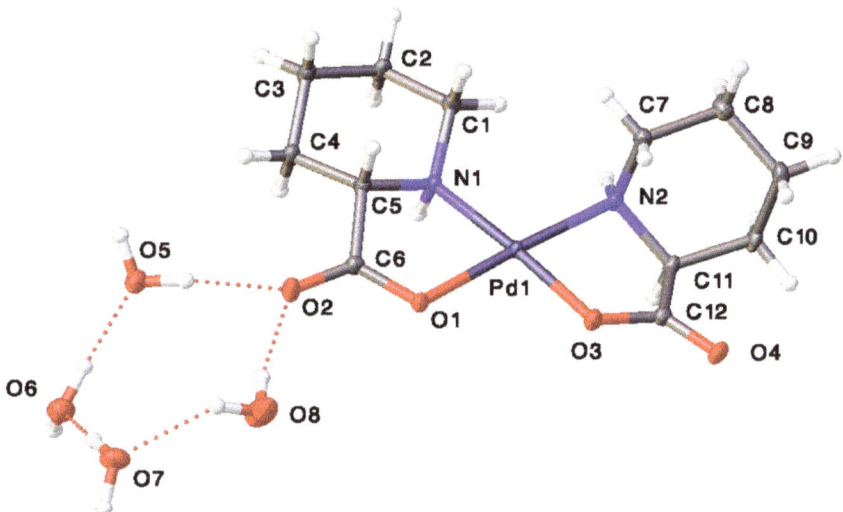

Figure 11. ORTEP plot of *cis*-bis-(L-pipecolinato)palladium(II), (**7**). Thermal ellipsoids are shown at the 50% probability level. CCDC:1913623.

There are four hydrogen bonded water molecules per complex unit in the lattice that form a pentagonal ring structure with a carbonyl oxygen of the complex. One amine hydrogen atom is hydrogen bonded to the opposite carboxyl oxygen of the adjacent molecule in the lattice. The other amine hydrogen is hydrogen bonded to one of the water molecules within the pentagonal water structure. While the specific features of these hydrogen-bonding motifs in the solid state say little about solution-state structures, they do indicate that H-bonding is most likely taking place in any solvent that contains either an H-bond donor or an H-bond acceptor or both.

D-pipecolinic acid was used to prepare cis-bis(D-pipicolinato)palladium(II), Compound **9**. Characterization data for **9**, as expected, is the same as that seen for **7**, but with the stereochemistry of the chiral carbon reversed (Figure 12).

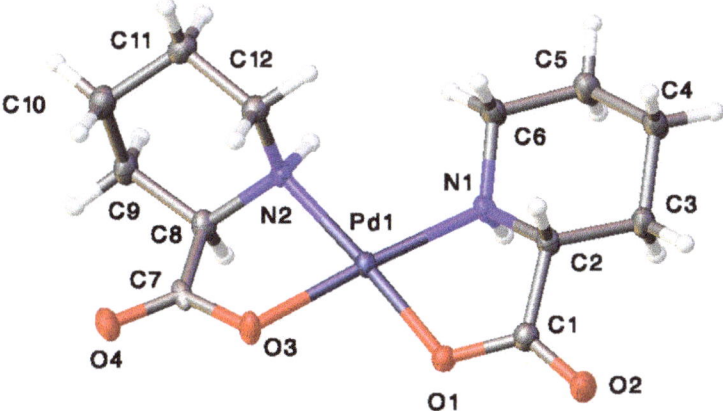

Figure 12. ORTEP plot of cis-bis(D-pipicolinato)palladium(II), (**9**). Lattice water molecules have been removed for clarity. Thermal ellipsoids are shown at the 50% probability level. CCDC:1913625.

2.2. Catalytic Activity

Asymmetric carbon-carbon bond formation is one of the most useful transformations in synthetic chemistry [77]. A palladium(II) catalyzed coupling reaction between phenylboronic acid and methyl tiglate was chosen as a model to evaluate the catalytic reactivity of these new palladium(II)-amino acid complexes and whether or not any asymmetric induction was possible (Figure 13).

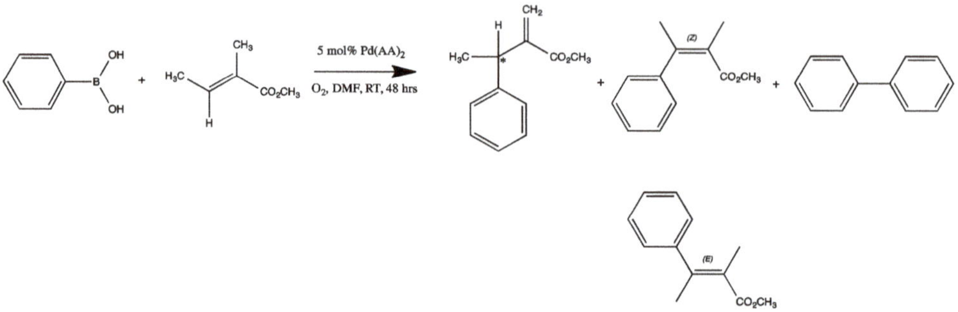

Figure 13. Bis(amino acid)Pd(II) catalyzed cross-coupling of phenylboronic acid and methyl tiglate.

2.3. Oxidative Coupling of Phenylboronic Acids and Alkenes

The standard coupling reaction that was used to evaluate the catalytic potential for each of the catalyst complexes was the aforementioned methyl tiglate and phenylboronic acid coupling [29,30]. This substrate was chosen because of the literature references already available to allow for comparison. All of the complexes described in this paper, except the N-methylproline complex, catalyzed this reaction and those data are summarized in Table 1 below. We have previously postulated that only the cis complexes are catalytically active, based on our observations with the glycine complexes described in a previous paper [67]. We see here, however, that the azetidine complex catalyzes the reaction even though it exists as the trans isomer. This suggests that N-alkylation, and not cis/trans geometry, may be the limiting factor in the catalytic ability of these complexes.

Table 1. Coupling reaction product distributions for catalysts **1, 2, 3, 4, 5, 6, 7**.

Complex	R/S Yield, %	% ee	E/Z Yield, %	Biaryl, %
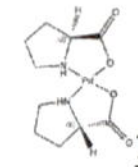 **1**	42	24	28	30
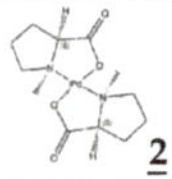 **2**	\multicolumn{4}{c}{No Reaction Observed}			
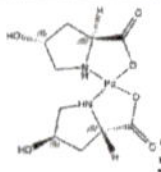 **3**	91	14	9	0
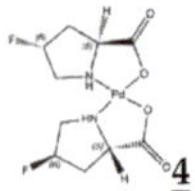 **4**	94	11	6	0
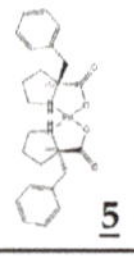 **5**	33	2	67	0

Table 1. Cont.

Complex	R/S Yield, %	% ee	E/Z Yield, %	Biaryl, %
6	66	11	30	4
7	97	1	2	1

Some general observations regarding product distributions and catalyst structure can be made based on our results. The presence of an electronegative group, –F or –OH, on the proline ring leads to a decrease in the formation of the E/Z products. The presence of purely alkyl functionality on the proline ring leads to an increase of the E/Z yield with corresponding loss of R/S product. The exception here is with the pipecolinic acid complex. This complex generates almost all R/S product, albeit with no enantioselectivity, and very little E/Z or homocoupled product. These general observations notwithstanding, there is still a great deal of variability in the product distributions that does not seem to follow any general trend. This suggests that the particular steric environment about the metal center during the catalytic cycle likely plays an important role in determining which products will form. As is often the case in examining various ligands for catalysis, it is difficult to separate the interplay between steric and electronic effects.

2.4. Proposed Mechanism of Pd-AA$_2$ Oxidative Coupling

The following mechanism is proposed for the palladium(II)-amino acid complex catalyzed oxidative coupling of phenylboronic acids to olefins (Figure 14, below). Step 1 involves the transmetallation of phenylboronic acid onto the palladium center. This is accomplished by an associative mechanism whereby the carboxylate group of one of the ligands de-coordinates to maintain a four-coordinate intermediate. The now-free carboxylate acts as a base towards the free boronic acid group, thus no addition of a base is required as is seen in a typical Suzuki coupling. The lack of catalytic activity of complex **2**, the N-methylated version of L-proline, shows the importance of the N-H bond for activity and the proposed mechanism suggests that H-bonding to a substrate is needed in this cycle.

DFT calculations show that the transmetallated intermediate has a geometry such that the metal center is completely occluded with the exception of a lobe of the empty d_{z^2} orbital that lies above the palladium atom (Figure 15).

The dissociated carboxylate end of the aminoacidato ligand wraps under the metal and covers the other d_{z^2} lobe. The remaining empty d_{z^2} lobe is then free to coordinate a neutral olefin, maintaining charge neutrality. Insertion of the phenyl group into the olefin double bond, followed by β-hydride elimination, yields the observed products. There are two possible pathways for beta-hydride elimination. Hydride elimination from the methyl carbon yields the R/S product, while hydride elimination from the methine carbon yields the E/Z product. To regenerate the catalyst and begin the cycle again, molecular oxygen abstracts the hydride, generating a peroxide. Qualitative peroxide test strips do indicate the presence of minute quantities of peroxide in the 0–25 ppm range.

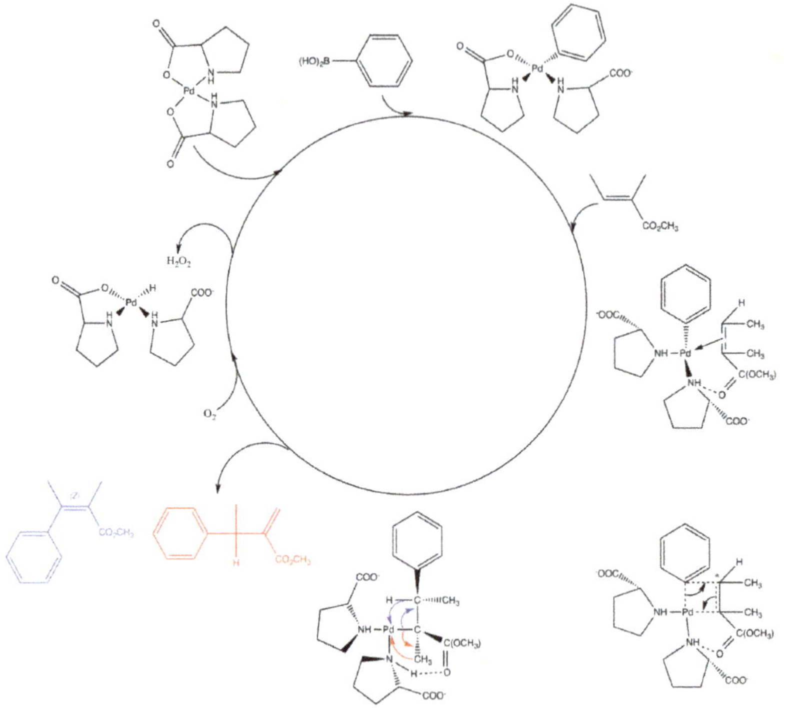

Figure 14. Proposed mechanism of the palladium(II) -amino acid complex catalyzed oxidative coupling of phenylboronic acids to olefins.

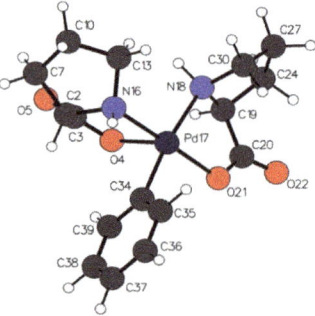

Figure 15. DFT-optimized geometry of the transmetallated intermediate. The non-occluded lobe of the d_{z^2} orbital projects out of the page towards the reader.

2.5. Biaryl Formation

Biaryl formation results from the coupling of two phenyl boronic acid substrates. Biaryl formation was noted to occur for every catalyst; however, the degree of biaryl formation varied greatly. Steric considerations about the metal center must therefore allow for both of these groups to orient themselves cis to each other. The mechanism proposed above can be slightly modified to allow for this possibility. If we consider a second transmetallation step to occur rather than olefin coordination, the two phenyl groups are oriented cis to each other. Elimination of the biaryl yields a Pd^0 center, which is then oxidized by molecular oxygen back to a Pd^{II} center. It is not clear which factors may dampen biaryl formation, but all of the substituted L-proline complexes as well as the pipicolinate and azetidine showed little to no biaryl formation, a useful feature for an atom-economic process.

2.6. Multiple Insertions

A unique aspect of this coupling/catalyst system is the ability for the products to undergo additional coupling cycles. The initial alkene products of the coupling reaction can in turn enter the catalytic cycle again and undergo an additional phenylboronic acid addition. This second product can also re-enter the cycle for a third phenylboronic acid addition. We have observed one, two, and three phenylboronic acid addition products for these catalyst systems; however, a fourth addition product has not been observed for any catalyst. This is likely due to steric concerns whereby the third coupling product is simply too bulky to coordinate to the metal center. Given that the initial reaction conditions begin with a 3:1 excess of alkene to phenylboronic acid, noting products from multiple additions is particularly fascinating and suggests that the product of the first addition is activated toward further additions, a finding that will be the focus of a future study. In order to maximize additional couplings, the ratio was reversed to be 3:1 excess of phenylboronic acid to methyl tiglate. In the GC-MS of these coupling reactions, we observe three peaks of mass 190.2, six peaks of mass 266.3, and two peaks of mass 342.4; the fourth coupling product would have a mass of 418.5 if formed (Figure 16). For the actual chromatograms, see the Supplementary Materials.

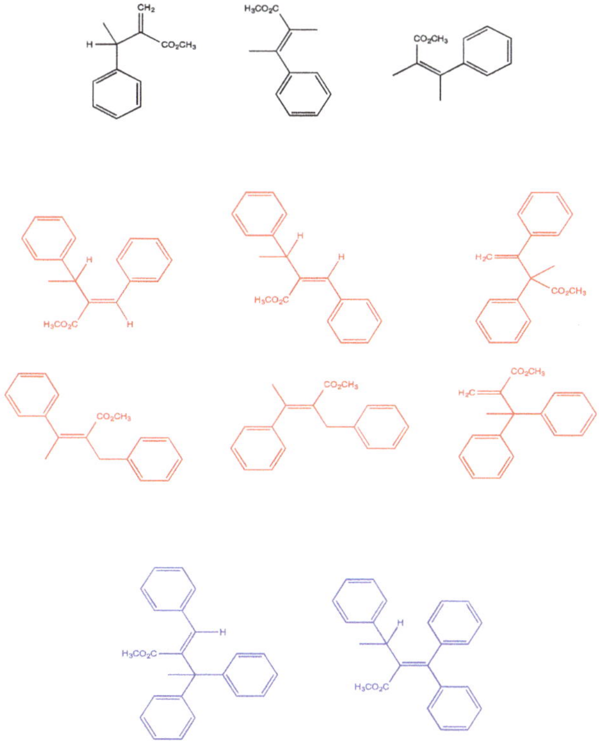

Figure 16. Postulated structures of multiple phenylboronic acid additions to the products of the bis(amino acid)Pd(II) catalyzed coupling of phenylboronic acid and methyl tiglate. First addition = black, second addition = red, third addition = blue.

2.7. Temperature Effects

Temperature has a significant effect on the enantioselectivity of the coupling reaction. The coupling reaction was carried out with the standard set of reaction conditions using the bis-proline complex as the catalyst at temperatures of 0, 25, and 65 °C. Enantioselectivities were noted to increase significantly with decreasing temperature as shown in Table 2, below. The equipment available for this study did not allow the reaction to be carried out below 0 °C and this could be an interesting study for the future with the proper equipment.

Table 2. Enantioselectivity versus temperature for the bis(amino acid)palladium(II) catalyzed oxidative coupling of phenylboronic acid to methyl tiglate.

Reaction Temperature, °C	%ee
65	~1
25	20
0	41

2.8. Solvent Effects

The standard coupling reaction was carried out in N,N-dimethylformamide, toluene, dichloromethane, and water solvents using the bis-proline complex as the catalyst. By far, DMF proved to be the superior solvent for this system. As a polar aprotic solvent, DMF has a hydrogen bond acceptor that greatly facilitates dissolution of the catalyst, which has unusually poor solubility in most common solvents. Subsequent trials were made with DMSO and acetonitrile as the solvents, but neither of these solvents gave appreciable product formation. As coordinating solvents, it is highly likely that solvent coordination to the complex blocks the active sites on the metal center required for reactivity. DMF, as a poorly-coordinating solvent, does not suffer this effect. No reaction was noted for either the dichloromethane (DCM) or toluene systems. DCM is a slightly polar aprotic solvent but lacks a hydrogen bond acceptor/donor, and toluene is a non-polar solvent. Neither of these solvents were observed to dissolve the catalyst, therefore the lack of any observed reactivity is not surprising. Water proved to be an interesting solvent choice. The catalyst is soluble in water, as is the phenylboronic acid substrate, but biphenyl formation was noted as the only reaction product. Methyl tiglate is extremely water-insoluble and the lack of PBA-MT cross-coupling products can be attributed to the lack of alkene solubility in water. This suggests that water may indeed be a "green" solvent choice for these systems so long as appropriate water-soluble substrates can be identified. Water as a solvent was used successfully for biaryl formation with a Pd(II) proline complex [78].

2.9. Pd(II)-Amino Acid Complexes as Polymerization Catalysts

Given the observation that these catalysts facilitate multiple substrate additions, it was hoped that they might also serve as novel polymerization catalysts. A suitable monomer containing both alkene and phenylboronic acid moieties, 4-(trans-3-methoxy-3-oxo-1-propen-1-yl)benzene boronic acid, was identified (Figure 17) and obtained for study.

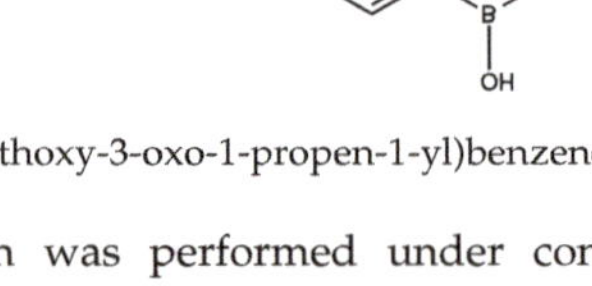

Figure 17. 4-(trans-3-methoxy-3-oxo-1-propen-1-yl)benzene boronic acid monomer.

The polymerization reaction was performed under conditions identical to the normal phenylboronic acid-methyl tiglate coupling using the *cis*-bis-(L-pipecolinato)palladium(II) complex as the catalyst. This complex was chosen due to the fact that it exhibited the least amount of homocoupling. While high molecular weight polymer was not isolated from the reaction, high-resolution time-of-flight mass spectrometric analysis of the reaction provides evidence of oligomer formation. Mass spectral peaks corresponding to oligomer masses where n = 2, 3, 4, 5, and 6 were observed (n = number of monomeric repeat units) (Figure 18). The *cis*-bis-(L-pipecolinato)palladium(II) catalyst once again showed no formation of homocoupled monomer.

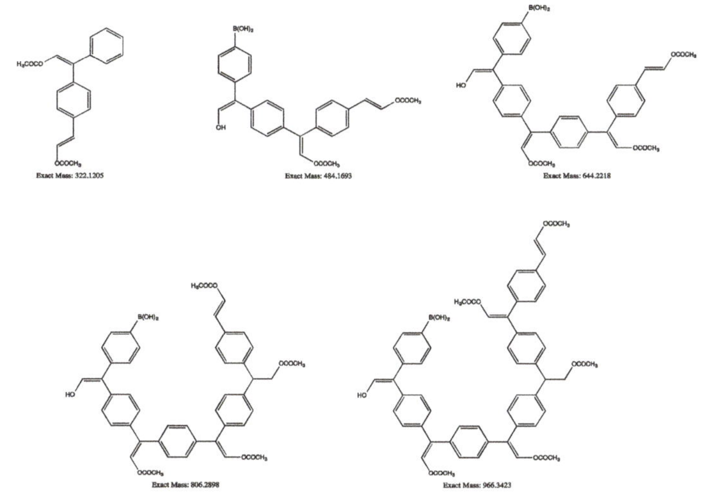

Figure 18. Proposed oligomer structures corresponding to HR-TOF MS data. From top left, n = 2, 3, 4, 5, and 6 where n = number of monomer units in the oligomer.

2.10. Other Coupling Substrates

There are hundreds, if not thousands, of possible boronic acid/olefin combinations that could be studied with our palladium(II)-amino acid catalytic systems. In an effort to probe some of the other possibilities of these systems, several substituted phenyl boronic acids and olefins were also examined as substrates for the coupling reaction.

An electron-withdrawing group on the phenyl boronic acid was introduced in the form of the trifluoromethyl group in 4-(trifluoromethyl)phenylboronic acid. The coupling reaction between this boronic acid and methyl tiglate was carried out as before with the bis-(L-prolinato)palladium(II) catalyst. The reaction proceeded smoothly with complete consumption of the phenylboronic acid substrate within the 48-hour reaction time. Product distributions were as follows: 71% R/S product with an enantiomeric excess of 11%, 24% homocoupled biaryl, 1% of the Z-alkene, and 4% of the secondary addition product (Figure 19).

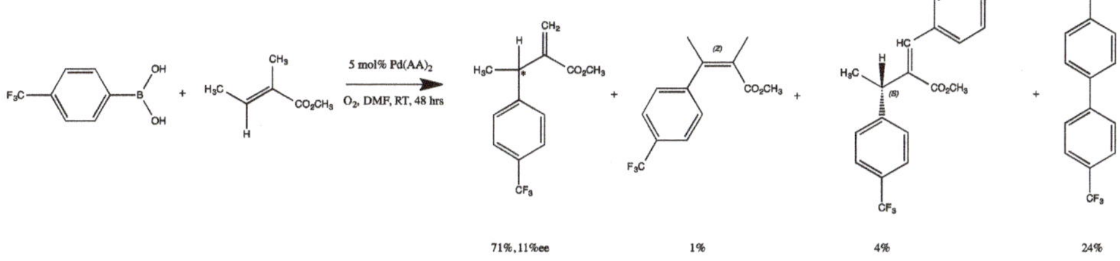

Figure 19. Reaction scheme and product distributions of the coupling reaction between 4-(trifluoromethyl)phenylboronic acid and methyl tiglate.

The same coupling reaction was also carried out with a phenylboronic acid with an electron donating group in the para position. In this case 4-methoxyphenylboronic acid was used (Figure 20).

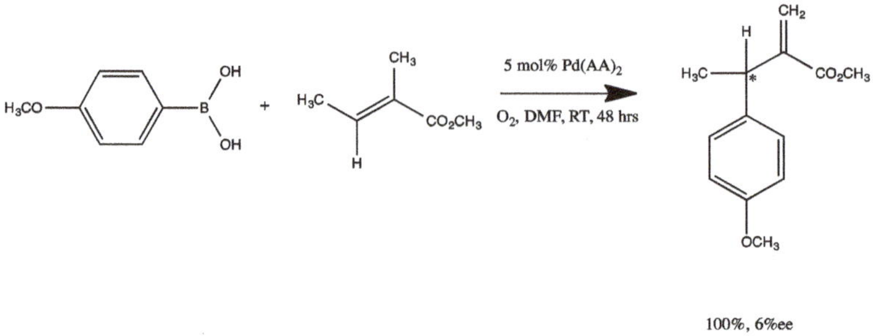

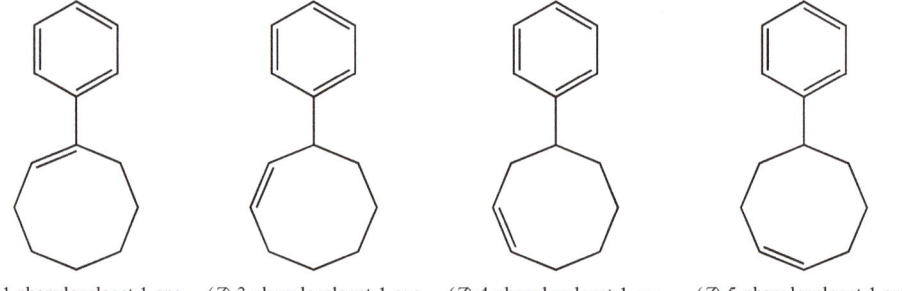

Figure 20. Reaction scheme and product distributions of the coupling reaction between 4-methoxyphenylboronic acid and methyl tiglate.

In this case, as before, complete consumption of the phenylboronic acid was observed. Interestingly, there was no evidence of homocoupling, alkene formation, or multiple phenylboronic acid additions noted for this reaction. The only product detected was the R/S product with an enantiomeric excess of 6%.

Methyl tiglate is considered to be an activated alkene, and it was hoped that our catalysts would also be useful for coupling non-activated alkenes. To this end *cis*-cyclooctene, 1,5-cyclooctadiene, and 1,5-hexadiene were evaluated with phenyl boronic acid in the standard coupling reaction. To our delight, all three alkenes coupled with phenylboronic acid when the reaction was catalyzed by the *cis*-bis-(L-prolinato)palladium(II) catalyst. The *cis*-cyclooctene coupling can generate four possible products (Figure 21), and four product peaks of the correct mass are observed in the GC-MS analysis of the reaction. The 1,5-cyclooctadiene coupling has two possible products (Figure 22), and here again we see two peaks of appropriate mass in the GC-MS trace. Finally, the 1,5-hexadiene coupling also has two possible products (Figure 23) and two peaks of correct mass are observed by GC-MS.

(*E*)-1-phenylcyclooct-1-ene   (*Z*)-3-phenylcyclooct-1-ene   (*Z*)-4-phenylcyclooct-1-ene   (*Z*)-5-phenylcyclooct-1-ene

Figure 21. Possible product structures for the coupling reaction between phenylboronic acid and *cis*-cyclooctene.

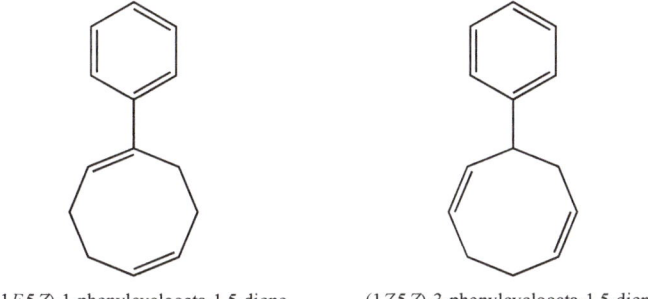

(1*E*,5*Z*)-1-phenylcycloocta-1,5-diene   (1*Z*,5*Z*)-3-phenylcycloocta-1,5-diene

Figure 22. Possible product structures for the coupling reaction between phenylboronic acid and 1,5-cyclooctadiene.

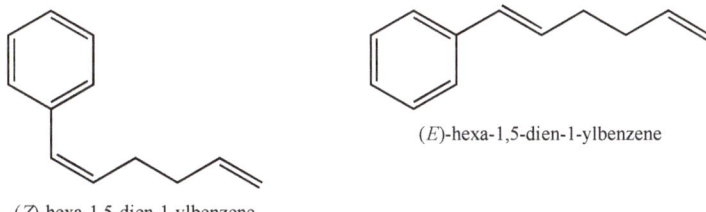

(*E*)-hexa-1,5-dien-1-ylbenzene

(*Z*)-hexa-1,5-dien-1-ylbenzene

Figure 23. Possible product structures for the coupling reaction between phenylboronic acid and 1,5-hexadiene.

3. Materials and Methods

All reagents were purchased from commercial suppliers and used as received. Palladium(II) acetate was obtained from Pressure Chemical, Pittsburgh, PA, USA. Proline, N-methylproline, azetidine, and pipecolinic acid were purchased from Sigma-Aldrich, St. Louis, MO, USA. 4-fluoroproline, 4-hydroxyproline, and 2-α-benzylproline were purchased from Chem-Impex International, Inc., Wood Dale, IL, USA. Reagent grade solvents (ether, acetone, ethyl acetate, DMF) were purchased from Sigma-Aldrich. Deuterated solvents for NMR spectroscopy were obtained from Cambridge Isotope Laboratories, Tewksbury, MA, USA.

^{1}H and ^{13}C NMR spectra were collected on either a Varian MR-400 or a Bruker Avance III 600 MHz NMR spectrometer. High-Resolution Mass Spectra (HRMS) were collected on an Agilent 6220 (Santa Clara, CA, USA). Accurate Mass TOF LC-MS. X-ray crystallographic data were collected at 100 K on an Oxford Diffraction Gemini diffractometer with an EOS CCD detector and Mo Kα radiation. Data collection and data reduction were performed using Agilent's CrysAlisPro software (Yarnton, Oxfordshire, UK) [79]. Structure solution and refinement were performed with ShelX [80,81], and Olex2 was used for graphical representation of the data [82].

All molecular modeling calculations were performed using Gaussian 09[83] using the WebMO interface. Full geometry optimizations and single-point energy calculations of all structures in water were performed via density functional theory (DFT) with the Becke three-parameter exchange functional [84] and the Lee–Yang–Parr correlation functional [85,86]. Because palladium is not covered in the cc-pVDZ basis set used, computations involving Pd employed Stuttgart/Dresden quasi-relativistic pseudopotentials [87].

3.1. General Procedure for the Synthesis of Palladium(II) Amino Acid Complexes

All reactions proceeded in very much identical ways and the following is the general procedure for all synthesis: An appropriately sized vial was fitted with a magnetic stir bar and charged with palladium(II) acetate and an appropriate volume of 50/50 (v/v) acetone/water. The mixture was stirred until all solids had dissolved. To this we added the amino acid and stirred the mixture overnight. The reaction solutions turned from a clear red-orange to a clear pale-yellow supernatant with a pale-yellow precipitate. The supernatant was transferred via pipette to a clean vial and allowed to evaporate to give clear yellow needles. The pale-yellow precipitate was washed with water and dried under a vacuum. The combined yield of single crystals and precipitate was measured and the resulting solid was characterized by ^{1}H, ^{13}C, HRMS, C,H analysis, and single-crystal X-ray diffractometry where possible.

3.2. Synthesis of cis-bis-(L-prolinato)palladium(II) (1)

Following the general procedure, the following amounts were used: 55.7 mg palladium(II) acetate (0.2481 mmol), 3.0 mL of 50/50 (v/v) acetone/water and 57.1 mg L-proline (0.4960 mmol). Yield: 79.3 mg of product (0.2369 mmol, 96% yield). *Cis*-Pd(C$_5$H$_8$NO$_2$)$_2$ (**1**) was identified on the basis of the following data: ^{1}H NMR (400 MHz, D$_2$O) δ 4.08–3.63 (m, 1H), 3.37–2.73 (m, 2H), 2.28–1.52 (m, 4H). ^{13}C NMR (101 MHz, D$_2$O) δ 186.49, 64.89, 52.58, 29.31, 24.68. HRMS/ESI+ (m/z): [M+H]+ calcd for Pd(C$_5$H$_8$NO$_2$)$_2$, 335.0218; found, 335.0224. Anal. Calcd. for Pd(C$_5$H$_8$NO$_2$)$_2$: C, 35.89%; H, 4.82%; N, 8.37%. Found: C, 35.98%; H, 4.83%; N, 8.35%. X-ray crystallographic data -CCDC: 1913626.

3.3. Synthesis of trans-bis-(N-methyl-L-prolinato)palladium(II) (2)

Following the general procedure, the following amounts were used: 35.2 mg palladium(II) acetate (0.1568 mmol), 3.0 mL of 50/50 (v/v) acetone/water and N-methyl-L-proline (42.7 mg, 0.3306 mmol). Yield: 53.6 mg of product (0.1477 mmol, 94% yield). *Trans*-Pd(C$_6$H$_{10}$NO$_2$)$_2$ (**2**) was identified on the basis of the following data: ^{1}H NMR (400 MHz, D$_2$O) δ 3.28 (dd, J = 10.4, 7.0 Hz, 1H), 3.12 (ddd, J = 10.9, 7.0, 2.8 Hz, 1H), 2.71 (s, 3H), 2.61–2.47 (m, 1H), 2.42–2.13 (m, 3H), 1.99 (dtt, J = 12.9, 6.7, 3.4 Hz, 1H). HRMS/ESI+ (m/z): [M+H]+ calcd for Pd(C$_6$H$_{10}$NO$_2$)$_2$, 363.0531; found, 363.0532. Anal. Calcd. for Pd(C$_6$H$_{10}$NO$_2$)$_2$·2H$_2$O: C, 36.15%; H, 6.07%; N, 7.03%. Found: C, 37.61%; H, 5.85%; N, 7.29%. X-ray crystallographic data -CCDC: 1913622.

3.4. Synthesis of cis-bis-(trans-4-hydrox-L-yprolinato)palladium(II) (3)

Following the general procedure, the following amounts were used: 59.4 mg palladium(II) acetate (0.2646 mmol), 2.0 mL of 50/50 (v/v) acetone/water and 77.1 mg 4-hydroxy-L-proline (0.5880 mmol). Yield: 94.3 mg of product (0.2572 mmol, 97% yield). *Cis*-Pd(C$_5$H$_8$NO$_3$)$_2$ (**3**) was identified on the basis of the following data: ^{1}H NMR (400 MHz, D$_2$O) δ 4.41 (s, 1H), 4.10 (t, J = 9.1 Hz, 1H), 3.35–3.28 (m, 1H), 3.27–3.15 (m, 2H), 3.10 (d, J = 12.7 Hz, 1H), 2.22–2.06 (m, 2H). HRMS/ESI+ (m/z): [M+H]+ calcd for Pd(C$_5$H$_8$NO$_3$)$_2$, 367.0116; found, 367.0130. Anal. Calcd. for Pd(C$_5$H$_8$NO$_3$)$_2$: C, 32.76%; H, 4.40%. Found: C, 32.88%; H, 4.42%. X-ray crystallographic data -CCDC: 1913624.

3.5. Synthesis of cis-bis-(trans-4-fluoro-L-prolinato)palladium(II) (4)

Following the general procedure, the following amounts were used: 49.6 mg palladium(II) acetate (0.2209 mmol), 3.0 mL of acetone and 64.6 mg *trans*-4-fluoro-L-proline (0.4853 mmol). Yield: 78.1 mg of product (0.2107 mmol, 95% yield). *Cis*-Pd(C$_5$H$_7$FNO$_2$)$_2$ (**4**) was identified on the basis of the following data: ^{1}H NMR (400 MHz, D$_2$O) δ 5.26–5.09 (m, 1H), 4.16–4.00 (m, 1H), 3.45–3.14 (m, 2H), 2.54–1.89 (m, 3H). ^{13}C NMR (101 MHz, D$_2$O) δ 186.09 (d, J = 140.5 Hz), 92.48 (d, J = 174.5 Hz), 62.48 (d, J = 150.3 Hz), 56.87 (dd, J = 131.2, 21.7 Hz), 36.10 (dd, J = 28.7, 21.6 Hz). ^{19}F NMR (471 MHz, D$_2$O) δ −179.33 (d, J = 140.9 Hz). HRMS/ESI+ (m/z): [M+H]+ calcd for Pd(C$_5$H$_7$FNO$_2$)$_2$, 371.0029; found, 371.0036. Anal. Calcd. for Pd(C$_5$H$_7$FNO$_2$)$_2$: C, 32.41%; H, 3.81%; N, 7.56%. Found: C, 32.99%; H, 3.92%; N, 7.53%. X-ray crystallographic data -CCDC: 1913621.

3.6. Synthesis of trans-bis-(2-benzylprolinato)palladium(II) (5)

Following the general procedure, the following amounts were used: 21.2 mg palladium(II) acetate (0.0944 mmol), 3.0 mL of 50/50 (v/v) acetone/water and 50.2 mg 2-benzylproline hydrochloride (0.2077 mmol). Yield: 44.7 mg of product (0.0868 mmol, 92% yield). Trans-Pd($C_{12}H_{14}NO_2$)$_2$ (**5**) was identified on the basis of the following data: ^{1}H NMR (400 MHz, D_2O) δ 7.32–7.15 (m, 5H), 3.42 (d, J = 14.6 Hz, 1H), 3.36–3.23 (m, 2H), 3.00 (d, J = 14.6 Hz, 1H), 2.48–2.37 (m, 1H), 2.09–1.96 (m, 2H), 1.87 (pd, J = 9.7, 8.8, 3.6 Hz, 1H). HRMS/ESI+ (m/z): [M+H]+ calcd for Pd($C_{12}H_{14}NO_2$)$_2$, 515.1157; found, 515.1175. Anal. Calcd. for Pd($C_{12}H_{14}NO_2$)$_2$: C, 55.98%; H, 5.48%; N, 5.44%. Found: C, 55.95%; H, 5.52%; N, 5.37%. X-ray crystallographic data -CCDC: 1913619.

3.7. Synthesis of trans-bis-(L-azetidine-2-carboxylato)palladium(II) (6)

Following the general procedure, the following amounts were used: 49.9 mg palladium(II) acetate (0.2223 mmol), 2.0 mL of 50/50 (v/v) acetone/water and 51.2 mg L-azetidine-2-carboxylic acid (0.5064 mmol). Yield: 66.7 mg of product (0.2175 mmol, 98% yield). Trans-Pd($C_4H_6NO_2$)$_2$ (**6**) was identified on the basis of the following data: ^{1}H NMR (400 MHz, D_2O) δ 4.44 (dt, J = 17.3, 8.8 Hz, 1H), 3.75–3.64 (m, 2H), 2.81–2.68 (m, 1H), 2.67–2.55 (m, 2H). ^{13}C NMR (101 MHz, D_2O) δ 187.86, 186.59, 63.36, 61.42, 50.32, 48.79, 24.61, 24.58. HRMS/ESI+ (m/z): [M+H]+ calcd for Pd($C_4H_6NO_2$)$_2$, 515.1157; found, 515.1175. Anal. Calcd. for Pd($C_4H_6NO_2$)$_2$: C, 55.98%; H, 5.48%; N, 5.44%. Found: C, 55.95%; H, 5.52%; N, 5.37%.

3.8. Synthesis of cis-bis-(L-pipecolinato)palladium(II) (7)

Following the general procedure, the following amounts were used: 107.5 mg palladium(II) acetate (0.4788 mmol), 3.0 mL of 50/50 (v/v) acetone/water, and 126.6 mg L-pipecolinic acid (0.9802 mmol). Yield: 152.4 mg of product (0.4202 mmol, 88% yield). Cis-Pd($C_6H_{10}NO_2$)$_2$ (**7**) was identified on the basis of the following data: ^{1}H NMR (400 MHz, D_2O) δ 3.78–3.57 (m, 1H), 2.96–2.63 (m, 2H), 1.94–1.08 (m, 6H). HRMS/ESI+ (m/z): [M+H]+ calcd for Pd($C_6H_{10}NO_2$)$_2$, 363.0531; found, 363.0543. Anal. Calcd. for Pd($C_6H_{10}NO_2$)$_2$·4H_2O: C, 33.15%; H, 6.49%; N, 6.44%. Found: C, 33.30%; H, 6.50%; N, 6.45%. X-ray crystallographic data -CCDC: 1913623.

3.9. Synthesis of cis-bis-(D-prolinato)palladium(II) (8)

Following the general procedure, the following amounts were used: 50.2 mg palladium(II) acetate (0.2236 mmol), 2.0 mL of 50/50 (v/v) acetone/water, and 58.7 mg D-proline (0.5099 mmol). Yield: 71.1 mg of product (0.2124 mmol, 95% yield). Cis-Pd($C_5H_8NO_2$)$_2$ (**8**) was identified on the basis of the following data: ^{1}H NMR (400 MHz, D_2O) δ 3.79 (dd, J = 9.1, 7.6 Hz, 1H), 3.10–2.96 (m, 2H), 2.17–2.06 (m, 1H), 1.99–1.78 (m, 2H), 1.67–1.54 (m, 1H). HRMS/ESI+ (m/z): [M+H]+ calcd for Pd($C_5H_8NO_2$)$_2$, 335.0218; found, 335.0222. Anal. Calcd. for Pd($C_5H_8NO_2$)$_2$: C, 35.89%; H, 4.82%; N, 8.37%. Found: C, 36.10%; H, 4.72%; N, 8.45%. X-ray crystallographic data -CCDC: 1913620.

3.10. Synthesis of cis-bis-(D-pipecolinato)palladium(II) (9)

Following the general procedure, the following amounts were used: 112.8 mg palladium(II) acetate (0.5024 mmol), 2.0 mL of 50/50 (v/v) acetone/water and 132.2 mg D-pipecolinic acid (1.0235 mmol). Yield: 151.9 mg of product (0.4188 mmol, 83% yield). Cis-Pd($C_6H_{10}NO_2$)$_2$ (**9**) was identified on the basis of the following data: ^{1}H NMR (400 MHz, D_2O) δ 3.75–3.56 (m, 1H), 2.97–2.66 (m, 2H), 2.08–1.08 (m, 6H). HRMS/ESI+ (m/z): [M+H]+ calcd for Pd($C_6H_{10}NO_2$)$_2$, 363.0531; found, 363.0520. Anal. Calcd. for Pd($C_6H_{10}NO_2$)$_2$·4H_2O: C, 33.15%; H, 6.49%; N, 6.44%. Found: C, 35.09%; H, 6.02%; N, 6.85%. X-ray crystallographic data -CCDC: 1913625.

3.11. General Procedure for Catalytic Reactions

The couplings were carried out in DMF solvent under an O_2 atmosphere with a 3:1 alkene:boronic acid ratio and 5 mol % catalyst loading, based on the boronic acid. Coupling reactions were stirred under O_2 for 48 h. The reaction work-up consisted of dilution with water followed by extraction with ethyl acetate and drying over anhydrous magnesium sulfate. Analysis included chiral GC using an Agilent CP-ChiralSil-Dex CB column (Agilent Technologies, Santa Clara, CA, USA).

In order to better analyze the multiple coupling products, the methyl tiglate to boronic acid ratio was changed to make the alkene:boronic acid ratio 1:3.

4. Conclusions

Nine palladium(II) bis-amino acid chelates with aliphatic ring structures for their R-group have been synthesized, characterized, and tested for catalytic activity for the oxidative coupling of phenylboronic acid with olefins. The amino acids employed include L-proline, D-proline, N-methylproline, azetidine, L-pipecolinic acid, D-pipecolinic acid, 2-α-benzylproline, 4-hydroxyproline, and 4-fluoroproline. The N-methylproline, 2-α-benzylproline, and azetidine complexes exist as the *trans* isomer, with all other complexes being *cis*. All of these complexes are square planar, C_2 symmetric molecules that exhibit varying degrees of intermolecular hydrogen bonding. All complexes are catalytically active with respect to the oxidative coupling of phenylboronic acids to olefins, with the exception of the N-methylproline complex. Enantioselectivities are modest with the best example, *cis*-bis(prolinato)palladium(II), yielding an enantiomeric excess of 24% with enantioselectivity increasing with decreasing temperature. These complexes couple a wide variety of both electron-rich and electron-deficient phenylboronic acids and activated and non-activated olefins. The finding of multiple cross-couplings on a single substrate is a fascinating finding that will be the subject of future studies.

Supplementary Materials: The following are available online at http://www.mdpi.com/2073-4344/9/6/515/s1, Figure S1: Example of mass spectrum showing Pd isotope pattern, Figure S2. Packing diagram for Complex 3 showing hydrogen-bonding motif; Figure S3. Packing diagram for Complex 4 showing hydrogen-bonding motif; Figure S4. Packing diagram for Complex 5 showing hydrogen-bonding motif. Figure S5. Packing diagram for Complex 7 showing hydrogen-bonding motif; Figure S6. Typical GC-MS trace of the first oxidative coupling of phenylboronic acid with methyl tiglate; Figure S7. Typical GC-MS trace of the second oxidative coupling products of phenylboronic acid with methyl tiglate; Figure S8. Typical GC-MS trace of the Third oxidative coupling products of phenylboronic acid with methyl tiglate; Report I. Complete crystallographic experimental parameters and tables of bond lengths and angles for complex 1. Report II. Complete crystallographic experimental parameters and tables of bond lengths and angles for complex 2. Report III. Complete crystallographic experimental parameters and tables of bond lengths and angles for complex 3. Report IV. Complete crystallographic experimental parameters and tables of bond lengths and angles for complex 4. Report V. Complete crystallographic experimental parameters and tables of bond lengths and angles for complex 5. Report VI. Complete crystallographic experimental parameters and tables of bond lengths and angles for complex 6. Report VII. Complete crystallographic experimental parameters and tables of bond lengths and angles for complex 8. Report VIII. Complete crystallographic experimental parameters and tables of bond lengths and angles for complex 9. Example of HRMS showing Pd isotope pattern, figures showing crystal lattice hydrogen-bonding motifs for select complexes, GC-MS traces showing the multiple cross-coupling analysis and full experimental data and complete listing of bond lengths and angles for compounds 1-9. In addition, CCDC numbers 1913619-1913626 contain the full supplementary .cif files for this paper. These data can be obtained free of charge from the Cambridge Crystallographic Data Centre via www.ccdc.cam.ac.uk/structures.

Author Contributions: Conceptualization, J.S.M. and D.B.H.J. Synthesis and characterization of compounds, D.B.H.J., H.M.R., S.S., J.M., and J.W.M.

Funding: This research received funding in the form of a $1000 grant for syringes from the Hamilton Syringe Company. The authors gratefully acknowledge the Virginia Tech Open Access Subvention Fund for funding the open access fee for this article.

Conflicts of Interest: The authors declare no conflict of interest.

References

1. Jin, L.Q.; Lei, A.W. Mechanistic aspects of oxidation of palladium with O_2. *Sci. China Chem.* **2012**, *55*, 2027–2035. [CrossRef]
2. Beccalli, E.M.; Broggini, G.; Martinelli, M.; Sottocornola, S. C-C, C-O, C-N Bond Formation on sp2 Carbon by Palladium(II)-Catalyzed Reactions Involving Oxidant Agents. *Chem. Rev.* **2007**, *107*, 5318–5365. [CrossRef]
3. Obora, Y.; Ishii, Y. Pd(II)/HPMoV-catalyzed direct oxidative coupling reaction of benzene derivatives with olefins. *Molecules* **2010**, *15*, 1487–1500. [CrossRef] [PubMed]
4. Stahl, S.S. Palladium oxidase catalysis. Selective oxidation of organic chemicals by direct dioxygen-coupled turnover. *Angew. Chem. Int. Ed.* **2004**, *43*, 3400–3420. [CrossRef] [PubMed]
5. Wu, W.; Jiang, H. Palladium-Catalyzed Oxidation of Unsaturated Hydrocarbons Using Molecular Oxygen. *Acc. Chem. Res.* **2012**, *45*, 1736–1748. [CrossRef] [PubMed]
6. Zeni, G.; Larock, R.C. Synthesis of heterocycles via palladium-catalyzed oxidative addition. *Chem. Rev.* **2006**, *106*, 4644–4680. [CrossRef] [PubMed]
7. Lu, Y.; Goldstein, E.L.; Stoltz, B.M. Palladium-Catalyzed Enantioselective Csp^3-Csp^3 Cross-Coupling for the Synthesis of (Poly)fluorinated Chiral Building Blocks. *Org. Lett.* **2018**, *20*, 5657–5660. [CrossRef] [PubMed]
8. Khan, F.; Dlugosch, M.; Liu, X.; Banwell, M.G. The Palladium-Catalyzed Ullmann Cross-Coupling Reaction: A Modern Variant on a Time-Honored Process. *Acc. Chem. Res.* **2018**, *51*, 1784–1795. [CrossRef] [PubMed]
9. Christoffel, F.; Ward, T.R. Palladium-Catalyzed Heck Cross-Coupling Reactions in Water: A Comprehensive Review. *Catal. Lett.* **2018**, *148*, 489–511. [CrossRef]
10. Roy, D.; Uozumi, Y. Recent Advances in Palladium-Catalyzed Cross-Coupling Reactions at ppm to ppb Molar Catalyst Loadings. *Adv. Synth. Catal.* **2018**, *360*, 602–625. [CrossRef]
11. Biffis, A.; Centomo, P.; Del Zotto, A.; Zecca, M. Pd Metal Catalysts for Cross-Couplings and Related Reactions in the 21st Century: A Critical Review. *Chem. Rev.* **2018**, *118*, 2249–2295. [CrossRef] [PubMed]
12. Devendar, P.; Qu, R.Y.; Kang, W.M.; He, B.; Yang, G.F. Palladium-Catalyzed Cross-Coupling Reactions: A Powerful Tool for the Synthesis of Agrochemicals. *J. Agric. Food Chem.* **2018**, *66*, 8914–8934. [CrossRef] [PubMed]
13. Sherwood, J.; Clark, J.H.; Fairlamb, I.J.S.; Slattery, J.M. Solvent effects in palladium catalysed cross-coupling reactions. *Green Chem.* **2019**, *21*, 2164–2213. [CrossRef]
14. Gligorich, K.M.; Cummings, S.A.; Sigman, M.S. Palladium-catalyzed reductive coupling of styrenes and organostannanes under aerobic conditions. *J. Am. Chem. Soc.* **2007**, *129*, 14193–14195. [CrossRef] [PubMed]
15. Adamo, C.; Amatore, C.; Ciofini, I.; Jutand, A.; Lakmini, H. Mechanism of the Palladium-Catalyzed Homocoupling of Arylboronic Acids: Key Involvement of a Palladium Peroxo Complex. *J. Am. Chem. Soc.* **2006**, *128*, 6829–6836. [CrossRef]
16. Canovese, L.; Visentin, F.; Chessa, G.; Santo, C.; Levi, C.; Uguagliati, P. Oxidative coupling of activated alkynes with palladium(0) olefin complexes: Side production of the highly symmetric hexamethyl mellitate species under mild conditions at low alkyne/complex molar ratios. *Inorg. Chem. Commun.* **2006**, *9*, 388–390. [CrossRef]
17. Hull, K.L.; Lanni, E.L.; Sanford, M.S. Highly Regioselective Catalytic Oxidative Coupling Reactions: Synthetic and Mechanistic Investigations. *J. Am. Chem. Soc.* **2006**, *128*, 14047–14049. [CrossRef]
18. Hull, K.L.; Sanford, M.S. Determining the mechanism of palladium-catalyzed oxidative coupling reactions. In Proceedings of the 239th ACS National Meeting & Exposition, San Francisco, CA, USA, 21–25 March 2010.
19. Lu, Y.; Wang, D.-H.; Engle, K.M.; Yu, J.-Q. Pd(II)-Catalyzed Hydroxyl-Directed C-H Olefination Enabled by Monoprotected Amino Acid Ligands. *J. Am. Chem. Soc.* **2010**, *132*, 5916–5921. [CrossRef]
20. Muzart, J. Molecular oxygen to regenerate Pd(II) active species. *Chem. Asian J.* **2006**, *1*, 508–515. [CrossRef]
21. Lei, A.; Zhang, X. A novel palladium-catalyzed homocoupling reaction initiated by transmetalation of palladium enolates. *Tetrahedron Lett.* **2002**, *43*, 2525–2528. [CrossRef]
22. Liegault, B.; Fagnou, K. Palladium-Catalyzed Intramolecular Coupling of Arenes and Unactivated Alkanes in Air. *Organometallics* **2008**, *27*, 4841–4843. [CrossRef]
23. Liu, C.; Jin, L.; Lei, A. Transition-metal-catalyzed oxidative cross-coupling reactions. *Synlett* **2010**, *2010*, 2527–2536.

24. Martinez, C.; Alvarez, R.; Aurrecoechea, J.M. Palladium-Catalyzed Sequential Oxidative Cyclization/Coupling of 2-Alkynylphenols and Alkenes: A Direct Entry into 3-Alkenylbenzofurans. *Org. Lett.* **2009**, *11*, 1083–1086. [CrossRef] [PubMed]
25. Prateeptongkum, S.; Driller, K.M.; Jackstell, R.; Spannenberg, A.; Beller, M. Efficient Synthesis of Biologically Interesting 3,4-Diaryl-Substituted Succinimides and Maleimides: Application of Iron-Catalyzed Carbonylations. *Chem. Eur. J.* **2010**, *16*, 9606–9615. [CrossRef] [PubMed]
26. Van Aeken, S.; Verbeeck, S.; Deblander, J.; Maes, B.U.W.; Tehrani, K.A. Synthesis of 3-substituted benzo[g]isoquinoline-5,10-diones: A convenient one-pot Sonogashira coupling/iminoannulation procedure. *Tetrahedron* **2011**, *67*, 2269–2278. [CrossRef]
27. Venkatraman, S.; Huang, T.; Li, C.-J. Carbon-carbon bond formation via palladium-catalyzed reductive coupling of aryl halides in air and water. *Adv. Synth. Catal.* **2002**, *344*, 399–405. [CrossRef]
28. Wakioka, M.; Mutoh, Y.; Takita, R.; Ozawa, F. A highly selective catalytic system for the cross-coupling of (E)-styryl bromide with benzene boronic acid: Application to the synthesis of all-trans poly(arylenevinylene)s. *Bull. Chem. Soc. Jpn.* **2009**, *82*, 1292–1298. [CrossRef]
29. Yoo, K.S.; O'Neill, J.; Sakaguchi, S.; Giles, R.; Lee, J.H.; Jung, K.W. Asymmetric Intermolecular Boron Heck-Type Reactions via Oxidative Palladium(II) Catalysis with Chiral Tridentate NHC-Amidate-Alkoxide Ligands. *J. Org. Chem.* **2010**, *75*, 95–101. [CrossRef]
30. Yoo, K.S.; Park, C.P.; Yoon, C.H.; Sakaguchi, S.; O'Neill, J.; Jung, K.W. Asymmetric Intermolecular Heck-Type Reaction of Acyclic Alkenes via Oxidative Palladium(II) Catalysis. *Org. Lett.* **2007**, *9*, 3933–3935. [CrossRef]
31. Chen, Q.; Li, C. Activation of the Vinylic C-Cl Bond by Complexation of Fe(CO)$_3$: Palladium-Catalyzed Coupling Reactions of (η^4-Chlorodiene)tricarbonyliron Complexes. *Organometallics* **2007**, *26*, 223–229. [CrossRef]
32. Yoo, K.S.; Yoon, C.H.; Jung, K.W. Oxidative Palladium(II) Catalysis: A Highly Efficient and Chemoselective Cross-Coupling Method for Carbon-Carbon Bond Formation under Base-Free and Nitrogenous-Ligand Conditions. *J. Am. Chem. Soc.* **2006**, *128*, 16384–16393. [CrossRef] [PubMed]
33. Heck, R.F.; Nolley, J., Jr. P. Palladium-catalyzed vinylic hydrogen substitution reactions with aryl, benzyl, and styryl halides. *J. Org. Chem.* **1972**, *37*, 2320–2322. [CrossRef]
34. Herr, R.J.; Dowling, M.S.; Scampini, A.C.; Smith, T.M. Iridium- and Palladium-Catalyzed Syntheses of (S)(+) and (R)(-) Coniine from Enantiopure Allylic Alcohols. In Proceedings of the 35th Northeast Regional Meeting of the American Chemical Society, Burlington, VT, USA, 29 June–2 July 2008.
35. Horiguchi, H.; Tsurugi, H.; Satoh, T.; Miura, M. Palladium/phosphite or phosphate catalyzed oxidative coupling of arylboronic acids with alkynes to produce 1,4-diaryl-1,3-butadienes. *Adv. Synth. Catal.* **2008**, *350*, 509–514. [CrossRef]
36. Jin, L.; Zhao, Y.; Wang, H.; Lei, A. Palladium-catalyzed R(sp^3)-Zn/R(sp)-SnBu$_3$ oxidative cross-coupling. *Synthesis* **2008**, *2008*, 649–654.
37. Johnson, T.; Lautens, M. Palladium(II)-Catalyzed Enantioselective Synthesis of α-(Trifluoromethyl)arylmethylamines. *Org. Lett.* **2013**, *15*, 4043–4045. [CrossRef]
38. Jordan-Hore, J.A.; Sanderson, J.N.; Lee, A.-L. Mild and Ligand-Free Pd(II)-Catalyzed Conjugate Additions to Hindered γ-Substituted Cyclohexenones. *Org. Lett.* **2012**, *14*, 2508–2511. [CrossRef] [PubMed]
39. Khabibulin, V.R.; Kulik, A.V.; Oshanina, I.V.; Bruk, L.G.; Temkin, O.N.; Nosova, V.M.; Ustynyuk, Y.A.; Bel'skii, V.K.; Stash, A.I.; Lysenko, K.A.; et al. Mechanism of the Oxidative Carbonylation of Terminal Alkynes at the C-H Bond in Solutions of Palladium Complexes. *Kinet. Catal.* **2007**, *48*, 228–244. [CrossRef]
40. Alvarez, R.; Martinez, C.; Madich, Y.; Denis, J.G.; Aurrecoechea Jose, M.; de Lera Angel, R. A general synthesis of alkenyl-substituted benzofurans, indoles, and isoquinolones by cascade palladium-catalyzed heterocyclization/oxidative Heck coupling. *Chemistry* **2010**, *16*, 12746–12753. [CrossRef]
41. Aouf, C.; Thiery, E.; Le Bras, J.; Muzart, J. Palladium-Catalyzed Dehydrogenative Coupling of Furans with Styrenes. *Org. Lett.* **2009**, *11*, 4096–4099. [CrossRef]
42. Beccalli, E.M.; Borsini, E.; Broggini, G.; Rigamonti, M.; Sottocornola, S. Intramolecular palladium-catalyzed oxidative coupling on thiophene and furan rings. Determinant role of the electronic availability of the heterocycle. *Synlett* **2008**, *2008*, 1053–1057.
43. Maehara, A.; Satoh, T.; Miura, M. Palladium-catalyzed direct oxidative vinylation of thiophenes and furans under weakly basic conditions. *Tetrahedron* **2008**, *64*, 5982–5986. [CrossRef]

44. Thiery, E.; Harakat, D.; Le Bras, J.; Muzart, J. Palladium-Catalyzed Oxidative Coupling of 2-Alkylfurans with Olefins through C-H Activation: Synthesis of Difurylalkanes. *Organometallics* **2008**, *27*, 3996–4004. [CrossRef]
45. Xi, P.; Yang, F.; Qin, S.; Zhao, D.; Lan, J.; Gao, G.; Hu, C.; You, J. Palladium(II)-Catalyzed Oxidative C-H/C-H Cross-Coupling of Heteroarenes. *J. Am. Chem. Soc.* **2010**, *132*, 1822–1824. [CrossRef] [PubMed]
46. Yamashita, M.; Hirano, K.; Satoh, T.; Miura, M. Synthesis of Condensed Heteroaromatic Compounds by Palladium-Catalyzed Oxidative Coupling of Heteroarene Carboxylic Acids with Alkynes. *Org. Lett.* **2009**, *11*, 2337–2340. [CrossRef] [PubMed]
47. Yang, S.-D.; Sun, C.-L.; Fang, Z.; Li, B.-J.; Li, Y.-Z.; Shi, Z.-J. Palladium-catalyzed direct arylation of (hetero)arenes with aryl boronic acids. *Angew. Chem. Int. Ed.* **2008**, *47*, 1473–1476. [CrossRef] [PubMed]
48. Bardhan, S.; Wacharasindhu, S.; Wan, Z.-K.; Mansour, T.S. Heteroaryl ethers by oxidative palladium catalysis of pyridotriazol-1-yloxy pyrimidines with arylboronic acids. *Org. Lett.* **2009**, *11*, 2511–2514. [CrossRef]
49. Belitsky, J.M. Palladium catalyzed homocoupling of indole and aryl boronic acids. In Proceedings of the 236th ACS National Meeting, Philadelphia, PA, USA, 17–21 August 2008.
50. Belitsky, J.M. Palladium Catalyzed Homocoupling of Indole and Aryl Boronic Acids. In Proceedings of the Abstract Central Cent. Regional Meeting of the American Chemical Society, Cleveland, OH, USA, 20–23 May 2009.
51. Clawson, R.W.; Deavers, R.E.; Akhmedov, N.G.; Soederberg, B.C.G. Palladium-catalyzed synthesis of 3-alkoxysubstituted indoles. *Tetrahedron* **2006**, *62*, 10829–10834. [CrossRef]
52. Djakovitch, L.; Rouge, P. New homogeneously and heterogeneously [Pd/Cu]-catalysed C3-alkenylation of free NH-indoles. *J. Mol. Catal. A Chem.* **2007**, *273*, 230–239. [CrossRef]
53. Gong, X.; Song, G.; Zhang, H.; Li, X. Palladium-Catalyzed Oxidative Cross-Coupling between Pyridine N-Oxides and Indoles. *Org. Lett.* **2011**, *13*, 1766–1769. [CrossRef]
54. He, C.-Y.; Fan, S.; Zhang, X. Pd-catalyzed oxidative cross-coupling of perfluoroarenes with aromatic heterocycles. *J. Am. Chem. Soc.* **2010**, *132*, 12850–12852. [CrossRef]
55. Wang, Z.; Li, K.; Zhao, D.; Lan, J.; You, J. Palladium-Catalyzed Oxidative C-H/C-H Cross-Coupling of Indoles and Pyrroles with Heteroarenes. *Angew. Chem. Int. Ed.* **2011**, *50*, 5365–5369. [CrossRef] [PubMed]
56. Henke, A.; Srogl, J. Pd^{2+} and Cu^{2+} catalyzed oxidative cross-coupling of mercaptoacetylenes and arylboronic acids. *Chem. Commun.* **2011**, *47*, 4282–4284. [CrossRef] [PubMed]
57. Kirchberg, S.; Tani, S.; Ueda, K.; Yamaguchi, J.; Studer, A.; Itami, K. Oxidative biaryl coupling of thiophenes and thiazoles with arylboronic acids through palladium catalysis: Otherwise difficult C4-selective C-H arylation enabled by boronic acids. *Angew. Chem. Int. Ed.* **2011**, *50*, 2387–2391. [CrossRef] [PubMed]
58. Schwan, A.L. Palladium catalyzed cross-coupling reactions for phosphorus-carbon bond formation. *Chem. Soc. Rev.* **2004**, *33*, 218–224. [CrossRef] [PubMed]
59. Wagner-Schuh, B.; Beck, W. Metal Complexes of Biologically Important Ligands, CLXXVII. Dichlorido Platinum(II) and Palladium(II) Complexes with Long Chain Amino Acids and Amino Acid Amides. *Z. Anorg. Allg. Chem.* **2017**, *643*, 632–635. [CrossRef]
60. Liu, R.R.; Li, B.L.; Lu, J.; Shen, C.; Gao, J.R.; Jia, Y.X. Palladium/l-Proline-Catalyzed Enantioselective α-Arylative Desymmetrization of Cyclohexanones. *J. Am. Chem. Soc.* **2016**, *138*, 5198–5201. [CrossRef]
61. Tsvelikhovsky, D.; Popov, I.; Gutkin, V.; Rozin, A.; Shvartsman, A.; Blum, J. On the involvement of palladium nanoparticles in the Heck and Suzuki reactions. *European J. Org. Chem.* **2009**, 98–102. [CrossRef]
62. Chatterjee, A.; Ward, T.R. Recent Advances in the Palladium Catalyzed Suzuki-Miyaura Cross-Coupling Reaction in Water. *Catal. Letters* **2016**, *146*, 820–840. [CrossRef]
63. Klaerner, C.; Greiner, A. Synthesis of polybenzyls by Suzuki Pd-catalyzed crosscoupling of boronic acids and benzyl bromides. Model reactions and polyreactions. *Macromol. Rapid Commun.* **1998**, *19*, 605–608. [CrossRef]
64. Wu, N.; Li, X.; Xu, X.; Wang, Y.; Xu, Y.; Chen, X. Homocoupling reaction of aryl boronic acids catalyzed by $Pd(OAc)_2/K_2CO_3$ in water under air atmosphere. *Lett. Org. Chem.* **2010**, *7*, 11–14. [CrossRef]
65. Xu, Z.; Mao, J.; Zhang, Y. $Pd(OAc)_2$-catalyzed room temperature homocoupling reaction of arylboronic acids under air without ligand. *Catal. Commun.* **2007**, *9*, 97–100. [CrossRef]
66. Yamamoto, Y. Homocoupling of arylboronic acids with a catalyst system consisting of a palladium(II) N-heterocyclic carbene complex and p-benzoquinone. *Synlett* **2007**, 1913–1916. [CrossRef]
67. Zhou, L.; Xu, Q.X.; Jiang, H.F. Palladium-catalyzed homo-coupling of boronic acids with supported reagents in supercritical carbon dioxide. *Chin. Chem. Lett.* **2007**, *18*, 1043–1046. [CrossRef]

68. Morris, D.M.D.M.; McGeagh, M.; De Peña, D.; Merola, J.S. Extending the range of pentasubstituted cyclopentadienyl compounds: The synthesis of a series of tetramethyl(alkyl or aryl)cyclopentadienes (Cp*R), their iridium complexes and their catalytic activity for asymmetric transfer hydrogenation. *Polyhedron* **2014**, *84*, 120–135. [CrossRef]
69. Hobart, D.B.; Berg, M.A.G.G.; Merola, J.S. Bis-glycinato complexes of palladium(II): Synthesis, structural determination, and hydrogen bonding interactions. *Inorg. Chim. Acta* **2014**, *423*, 21–30. [CrossRef]
70. Ito, T.; Marumo, F.; Saito, Y. The crystal structure of bis-(L-prolinato)palladium(II). *Acta Crystallogr. Sect. B Struct. Crystallogr. Cryst. Chem.* **2002**, *27*, 1062–1066. [CrossRef]
71. Chernova, N.N.; Strukov, V.V.; Avetikyan, G.B.; Chernonozhkin, V.N. Synthesis and structure of complex palladium(II) bis(histidinates). *Zh. Neorg. Khim.* **1980**, *25*, 1569–1574.
72. Jarzab, T.C.; Hare, C.R.; Langs, D.A. cis-Bis(L-tyrosinato)palladium(II) hemihydrate, $C_{36}H_{42}N_4O_{13}Pd_2$. *Cryst. Struct. Commun.* **1973**, *2*, 399–403.
73. Jarzab, T.C.; Hare, C.R.; Langs, D.A. cis-Bis(L-valinato)palladium(II) monohydrate, $C_{10}H_{22}N_2O_5Pd$. *Cryst. Struct. Commun.* **1973**, *2*, 395–398.
74. Komorita, T.; Hidaka, J.; Shimura, Y. Metal complexes with amino acid amides. III. Geometrical structures and electronic spectra of bis(α-amino acid-amidato)palladium(II), -nickel(II), and -copper(II). *Bull. Chem. Soc. Jpn.* **1971**, *44*, 3353–3363. [CrossRef]
75. Sabat, M.; Jezowska, M.; Kozlowski, H. X-Ray Evidence of the Metal-Ion Tyrosine Aromatic Ring Interaction in Bis(L-Tyrosinato)Palladium(Ii). *Inorg. Chim. Acta* **1979**, *37*, L511–L512. [CrossRef]
76. Batsanov, S.S. Van der Waals radii of elements. *Inorg. Mater.* **2001**, *37*, 871–885. [CrossRef]
77. Bhowmick, S.; Bhowmick, K.C. Catalytic asymmetric carbon-carbon bond-forming reactions in aqueous media. *Tetrahedron Asymmetry* **2011**, *22*, 1945–1979. [CrossRef]
78. Zhang, G.; Luan, Y.; Han, X.; Wang, Y.; Wen, X.; Ding, C. Pd(L-proline)$_2$ complex: An efficient catalyst for Suzuki-Miyaura coupling reaction in neat water. *Appl. Organomet. Chem.* **2014**, *28*, 332–336. [CrossRef]
79. Rigaku Oxford Diffraction CrysAlisPro Software System. 2018. Available online: https://www.rigakuxrayforum.com/forumdisplay.php?fid=57 (accessed on 2 June 2019).
80. Sheldrick, G.M. SHELXT—Integrated space-group and crystal-structure determination. *Acta Crystallogr. Sect. A Found. Adv.* **2015**, *71*, 3–8. [CrossRef] [PubMed]
81. Sheldrick, G.M. A short history of SHELX A short history of SHELX. *Acta Crystallogr. Sect. A* **2008**, *64*, 112–122. [CrossRef] [PubMed]
82. Dolomanov, O.V.; Bourhis, L.J.; Gildea, R.J.; Howard, J.A.K.; Puschmann, H. OLEX2: A complete structure solution, refinement and analysis program. *J. Appl. Crystallogr.* **2009**, *42*, 339–341. [CrossRef]
83. Frisch, M.J.; Trucks, G.W.; Schlegel, H.B.; Scuseria, G.E.; Robb, M.A.; Cheeseman, J.R.; Scalmani, G.; Barone, V.; Petersson, G.A.; Nakatsuji, H.; et al. Gaussian 9 Citation. Available online: https://gaussian.com/g03citation/ (accessed on 4 May 2016).
84. Becke, A.D. Density-functional thermochemistry. III. The role of exact exchange. *J. Chem. Phys.* **1993**, *98*, 5648–5652. [CrossRef]
85. Lee, C.; Yang, W.; Parr, R.G. Development of the Colle-Salvetti correlation-energy formula into a functional of the electron density. *Phys. Rev. B* **1988**, *37*, 785–789. [CrossRef]
86. Stephens, P.J.; Devlin, F.J.; Chabalowski, C.F.; Frisch, M.J. Ab Initio calculation of vibrational absorption and circular dichroism spectra using density functional force fields. *J. Phys. Chem.* **1994**, *98*, 11623–11627. [CrossRef]
87. Andrae, D.; Häußermann, U.; Dolg, M.; Stoll, H.; Preuss, H. Energy-Adjusted Abinitio Pseudopotentials for the 2nd and 3rd Row Transition-Elements. *Theor. Chim. Acta* **1990**, *77*, 123–141. [CrossRef]

© 2019 by the authors. Licensee MDPI, Basel, Switzerland. This article is an open access article distributed under the terms and conditions of the Creative Commons Attribution (CC BY) license (http://creativecommons.org/licenses/by/4.0/).

Article

Facile Synthesis of P25@Pd Core-Shell Catalyst with Ultrathin Pd Shell and Improved Catalytic Performance in Heterogeneous Enantioselective Hydrogenation of Acetophenone

Xiuyun Gao, Lulu He, Juntong Xu, Xueying Chen * and Heyong He

Department of Chemistry and Shanghai Key Laboratory of Molecular Catalysis and Innovative Materials, Fudan University, Shanghai 200433, China; 16210220034@fudan.edu.cn (X.G.); 14110220027@fudan.edu.cn (L.H.); 15307110311@fudan.edu.cn (J.X.); heyonghe@fudan.edu.cn (H.H.)
* Correspondence: xueyingchen@fudan.edu.cn; Tel.: +86-21-3124-2978

Received: 16 May 2019; Accepted: 5 June 2019; Published: 9 June 2019

Abstract: Heterogeneous enantioselective hydrogenation is an ideal method for synthesizing important chiral compounds in pesticides and pharmaceuticals. Up to the present, supported noble-metal catalysts are most widely studied in heterogeneous enantioselective hydrogenations. However, it is found that the weak interactions existing on the surface of support may have negative effects on the enantioselectivity. Herein, a new category of TiO_2 (Aeroxide® P25) supported Pd catalyst with ultrathin Pd shell was successfully prepared via a simple strategy based on the reduction of Pd^I carbonyl complex. Characterization results show that a well-dispersed ultrathin Pd shell with an average thickness of ~1.0 nm and a Pd loading of 36 wt.% was formed over the surface of P25 support. By excluding the negative weak interactions from the support, the P25@Pd core-shell catalyst with unique electronic properties of Pd exhibits higher activity and enantioselectivity than that of Pd/P25 catalyst prepared by the impregnation method and unsupported Pd black catalyst in the enantioselective hydrogenation of acetophenone.

Keywords: P25@Pd; core-shell; heterogeneous enantioselective hydrogenation; acetophenone

1. Introduction

With the wide application of optically pure chiral compounds in pesticides, pharmaceuticals and fragrances, it is particularly important to develop effective preparation methods for synthesizing single-enantiomer compounds [1–3]. Among the many established methods, enantioselective hydrogenation over heterogeneous catalysts is one of the most ideal strategies owing to its inherent operational and economic advantages, e.g., atom economy, easy separation and recovery of catalysts. Therefore, it has shown great potential in industrial research and caused extensive concern in academic circles [1–4].

The enantioselective hydrogenation of acetophenone is a probe reaction widely chosen for heterogeneous asymmetric hydrogenation studies, since it is a good example of competitive reaction and its target product (chiral 1-phenylethanol) is related to the production of pharmaceutical intermediates [5–9]. However, the enantioselectivity reported so far over traditional supported noble-metal (e.g., Pd, Pt) catalysts is generally low and further improvements are still needed.

Nowadays, chirally modified supported noble-metal catalysts have been most widely used in heterogeneous enantioselective hydrogenations [2–7,9–14]. It is well known that the catalytic performance of the supported metal catalysts strongly depends on the metal particle size. Thus, to make the metal nanoparticles disperse well for achieving high activity, the metal loading of the catalysts studied in the literature is generally less than 5 wt.%. However, it is found that the various weak interactions (such as hydrogen bond, physical adsorption and van der Waals force) existing on the surface of the support are in the same energy range with the energy difference between the two transition states of R- and S-products in chiral reactions (<15 kJ/mol), which may affect the chiral recognition process and thus have a negative effect on the enantioselectivity [15]. Baiker et al. found that the acidity and basicity of the support could significantly affect both the chemoselectivity and the enantioselectivity in the enantioselective hydrogenation of activated ketones [16]. Therefore, the exposure of support surface adds the complexity in elucidating the nature of the chiral recognition on metal catalysts in heterogeneous enantioselective hydrogenation reactions. Thus, to reduce the difficulty in heterogeneous asymmetric hydrogenation studies and achieve higher enantioselectivity, it is highly desirable to exclude the negative weak interactions existing on the surface of the support.

It is well known that supported metal catalysts with core-shell structure have attracted much attention because of their unique structure and better performance in some catalytic reactions [17,18]. Therefore, if we could form an ultrathin metal layer on the surface of the support, it may not only eliminate the adverse effect of weak interactions from the support on heterogeneous asymmetric hydrogenations, but also affect the electronic properties of noble-metal catalysts, which is of great significance in both fundamental and practical aspects.

In this work, a new category of TiO_2 (Aeroxide® P25) supported Pd catalyst with unique core-shell structure (P25 core and Pd ultrathin shell, denoted as P25@Pd) was tactically prepared by a facile strategy under mild conditions based on the reduction of Pd^I carbonyl complex [19] over the surface of the P25 support. The as-prepared P25@Pd catalyst exhibits a well-dispersed ultrathin Pd shell with an average thickness of ~1.0 nm and a Pd loading of 36 wt.%. The formation of the ultrathin Pd shell over the surface of the P25 support excludes the negative weak interactions from support and the resulting P25@Pd core-shell catalyst exhibits improved activity and enantioselectivity in the enantioselective hydrogenation of acetophenone. The unique electronic properties of Pd influenced by the interaction between the ultrathin Pd shell and the P25 support also probably have a positive effect on the enantioselectivity. The new category of supported noble-metal catalyst with ultrathin metal shell could be promising for heterogeneous asymmetric catalytic studies.

2. Results and Discussion

2.1. Morphology of the P25@Pd Catalyst

Figure 1 shows the representative transmission electron microscopy (TEM) images of the P25 support and the as-prepared P25@Pd core-shell catalyst. The P25 support displays a nanoparticle morphology with an average particle size of ca. 21 nm (Figure 1a). As compared to the P25 support, the much darker outlines of the nanoparticles in P25@Pd catalyst (Figure 1b) indicate that Pd metal is deposited over the surface of the P25 support. To further visualize the core-shell structure of the P25@Pd catalyst, a high angle annular dark field scanning TEM (HAADF-STEM) image is presented in Figure 1c, which clearly exhibits the coverage of the Pd shells (brighter areas in contrast). The average thickness of the Pd shell calculated based on the statistic calculation (Figure 1d) is ca. 1.0 nm. Figure 1e provides a typical high resolution TEM (HRTEM) image of a P25@Pd nanoparticle. The interplanar spacing of 0.23 nm can be observed in the Pd shell on the P25@Pd nanoparticle, ascribable to (111) lattice spacing of face-centered cubic (fcc) Pd [20]. A d-spacing of 0.35 nm observed in the P25 core corresponds to the (101) planes of anatase TiO_2 [21]. Figure 1f presents the typical elemental mapping images of the P25@Pd catalyst. The results clearly demonstrate the successful coverage of Pd shell over the surface of the P25 support, unambiguously confirming the P25 core/Pd shell structure. As detected by inductively coupled plasma atomic emission spectroscopy (ICP-AES), the weight percentage of Pd in the P25@Pd catalyst is 36%, which is in good accordance with the energy dispersive X-ray emission (EDX) analysis (Pd/Ti molar ratio of 0.43).

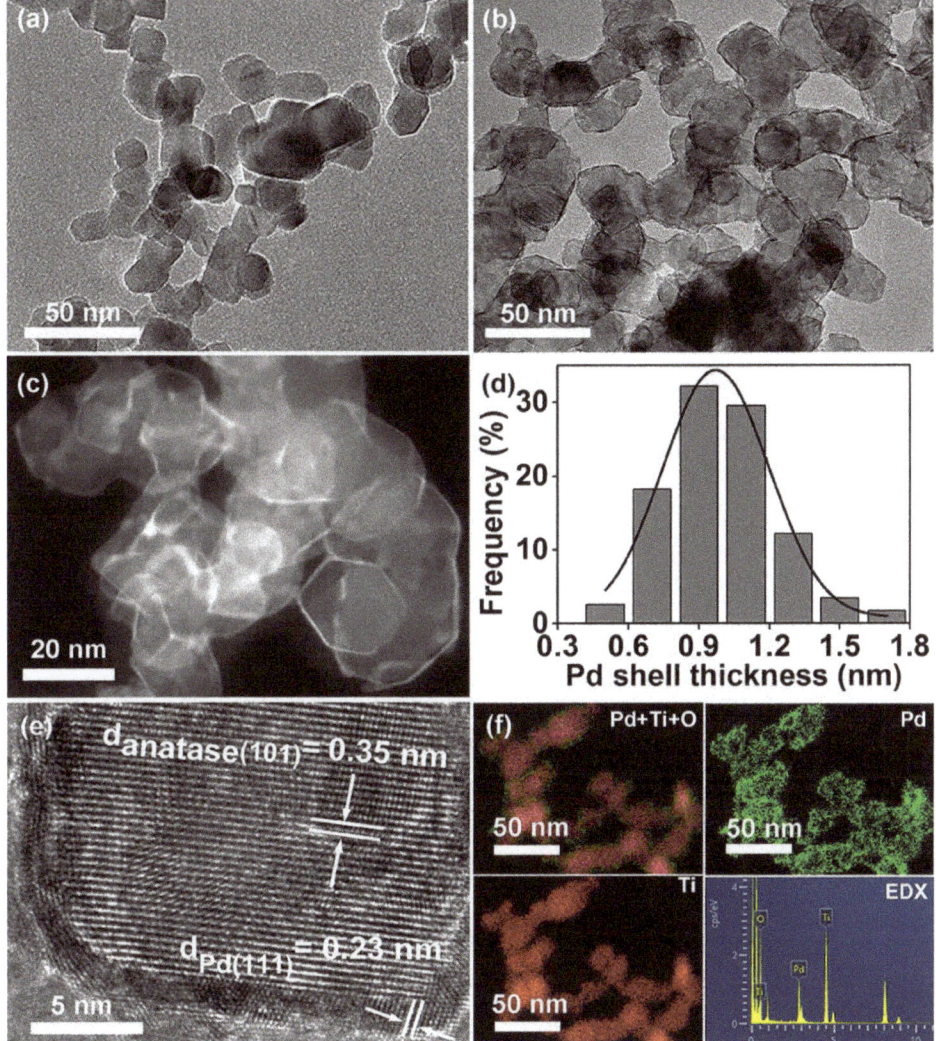

Figure 1. Transmission electron microscopy (TEM) images of (**a**) P25 support and (**b**) P25@Pd core-shell catalyst; (**c**) high angle annular dark field scanning (HAADF-STEM) image, (**d**) Pd shell thickness distribution, (**e**) high resolution TEM (HRTEM) image, and (**f**) elemental mapping results of P25@Pd core-shell catalyst.

2.2. Structural Properties of the P25@Pd Catalyst

The Brunauer–Emmett–Teller (BET) specific surface area of the P25@Pd catalyst is 53.4 m^2 g^{-1}, which is slightly larger than that of the parent P25 support (49.7 m^2 g^{-1}). The X-ray diffraction (XRD) measurements were carried out to identify the phase and lattice structures of the parent P25 support and P25@Pd core-shell catalyst. As shown in Figure 2a, there are several diffraction peaks in the parent P25 support, which can be fully indexed to anatase and rutile TiO$_2$ (JCPDS card no. 89–4921, 89–4920) [21]. In addition to the diffraction peaks of the parent P25 support [21], two additional broad diffraction peaks at 2θ of 40.0, and 46.5° can be clearly observed, corresponding to (111) and (200) reflections of fcc Pd (JCPDS card no. 89–4897). The lattice constant of Pd in P25@Pd core-shell catalyst calculated by peak fitting (Figure 2b) is ca. 0.389 nm, which is similar to that of the Pd black sample. According to the fitting results, the intensity ratio of the Pd(111) and Pd(200) peaks is 4.14, which is 1.8 times that of the Pd black sample (2.27), indicating that the ultrathin Pd shell in the P25@Pd catalyst has (111) preferred orientation.

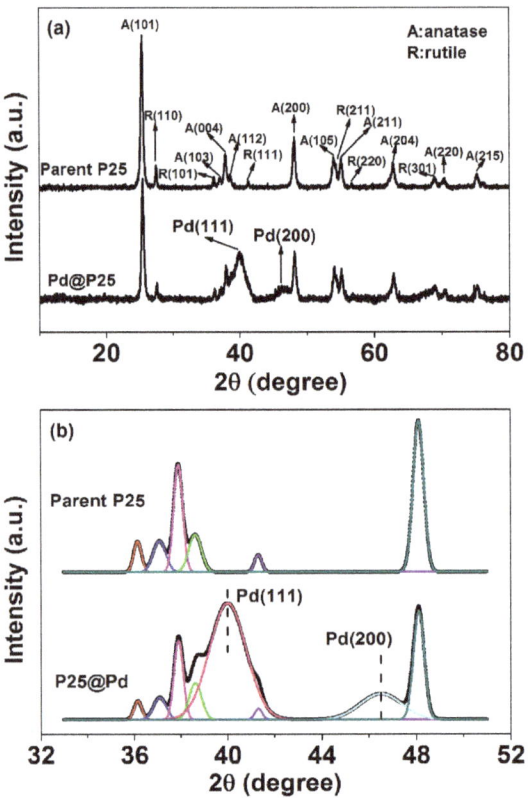

Figure 2. X-ray diffraction (XRD) patterns of parent P25 support and P25@Pd core-shell catalyst: (**a**) Experimental data, (**b**) peak fitting results.

2.3. Study on the Formation of P25@Pd Core-Shell Structure

In order to investigate the formation of P25@Pd core-shell structure, a set of comparative experiments were carefully carried out. Without the addition of P25 support, only ultrathin Pd nanosheets are formed by the reduction of Pd^I carbonyl complex [19] (Figure 3a). In the presence of P25 support, the reduction of Pd^I carbonyl complex [19] completely occurs on the surface of P25 support to form the P25@Pd core-shell catalyst and no isolated Pd nanosheets could be observed. However, when we replace the P25 support with mesoporous carbon, γ-Al$_2$O$_3$ or SiO$_2$ support, instead of forming core-shell structure, isolated ultrathin Pd nanosheets aggregated together or small Pd nanoparticles located on the surface of the support are observed (Figure 3b–d). Thus, the preferred deposition of Pd over the P25 support to form the core-shell structure could be possibly attributed to the unique interaction between Pd and the P25 support [22–24]. Due to the interaction between Pd and the P25 support, the Pd^I carbonyl complex [19] is prone to be reduced over the surface of each P25 nanoparticle, facilitating the uniform coating of Pd ultrathin shell over the P25 nanoparticle surface to form the core-shell structure. The detailed formation mechanism of the core-shell structure still requires further study. For comparison, we also synthesized Pd/P25 catalyst with the same Pd loading by impregnation method with $N_2H_4 \cdot H_2O$ as the reductant. As shown in Figure 3e, Pd nanoparticles on P25 support are varied in size and tend to aggregate together in the Pd/P25 catalyst, which is totally different from that of the P25@Pd core-shell catalyst. The molar ratio of Pd/Ti is similar to that of the P25@Pd catalyst as verified by EDX analysis (Figure 3f).

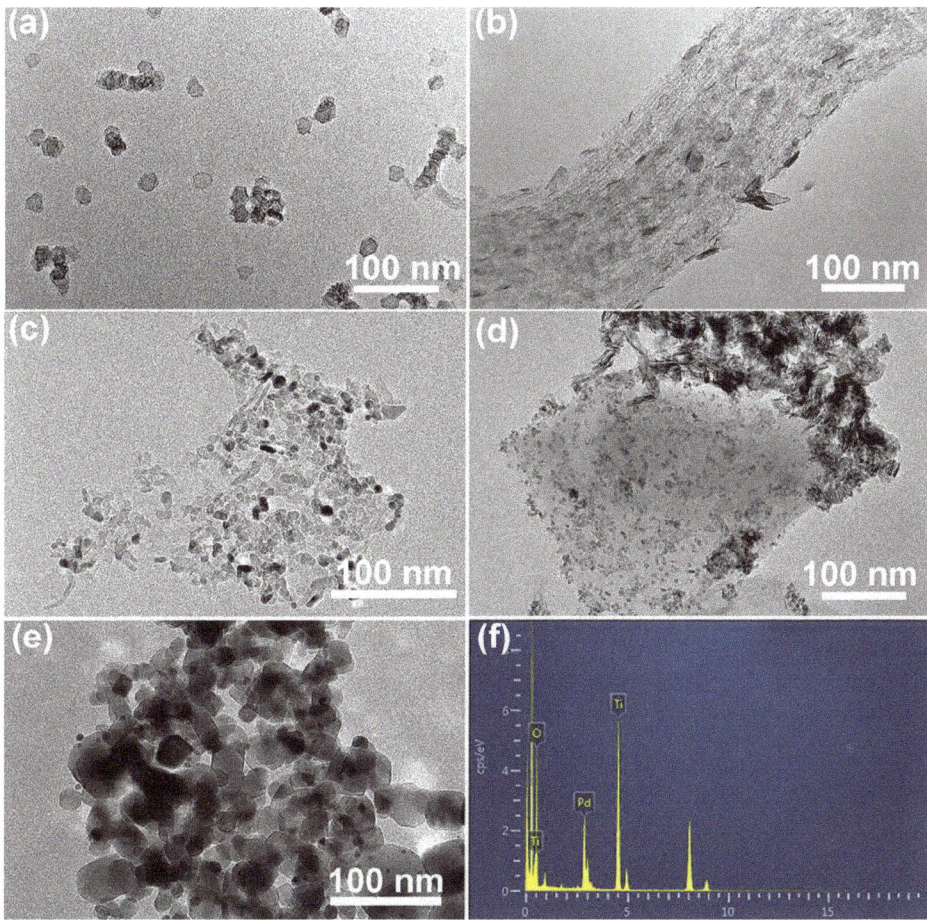

Figure 3. TEM images of (**a**) Pd nanosheets obtained in the absence of P25 support, (**b**, **c**, **d**) Pd catalyst prepared by the same reduction procedure except using mesoporous carbon, commercial γ-Al_2O_3, and SiO_2 as support, respectively; (**e**) TEM image and (**f**) energy dispersive X-ray emission (EDX) results of Pd/P25 catalyst prepared by traditional impregnation method.

2.4. Electronic Properties of the P25@Pd Catalyst

X-ray photoelectron spectroscopy (XPS) was utilized to detect the electronic state of elements in P25@Pd catalyst. Figure 4a presents the Pd 3d core level spectrum of the P25@Pd core-shell catalyst. The Pd 3d core-level line could be fitted with two main doublets with the binding energy (BE) of the Pd$3d_{5/2}$ peaks at 334.9 and 336.1 eV, corresponding to metallic Pd and Pd$^{\delta+}$ species, respectively [23,25,26]. The Ti 2p spectrum of P25@Pd core-shell catalyst (Figure 4b) is dominated by species in the Ti^{4+} oxidation state at the BE of 458.8 and 464.5 eV, which is consistent with the typical Ti $2p_{3/2}$ and Ti $2p_{1/2}$ values for

P25 support [27–29]. The surface Pd/Ti atomic ratio of the P25@Pd catalyst derived from XPS data is evaluated to be 0.99, which is more than twice that of the bulk composition measured by ICP and EDX, reflecting the surface enrichment of the Pd shell.

To investigate whether the presence of $Pd^{\delta+}$ species is mainly due to the electron transfer from Pd to P25 support [23,25,30] or just Pd oxidation, cyclic voltammeter (CV) studies of P25@Pd catalyst were performed. Cyclic voltammograms (Figure 5) show that there is no Pd reduction peak in the first cycle while obvious Pd reduction peak at 0.47 V is clearly observed in the second cycle. The results indicate that the presence of $Pd^{\delta+}$ species is unlikely due to Pd oxidation but rather to electron transfer from Pd to P25 support, which is similar to previous literature reports [23,25,30].

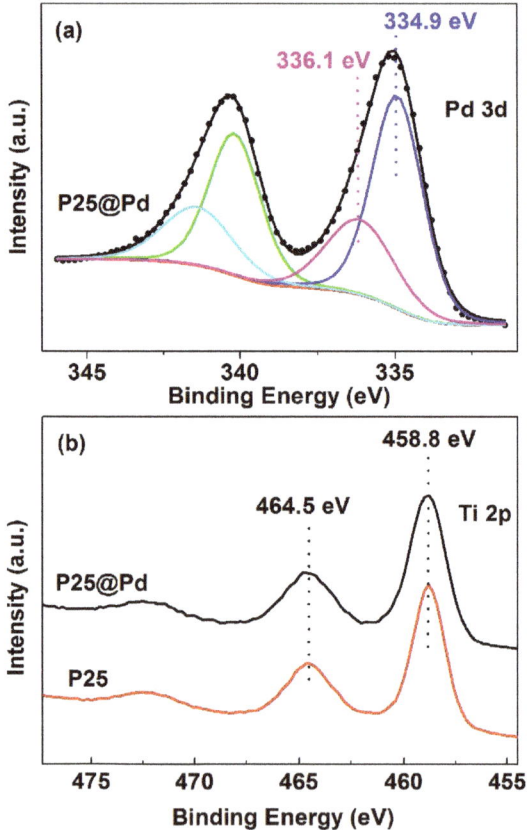

Figure 4. (a) X-ray photoelectron spectroscopy (XPS) Pd 3d spectra of P25@Pd core-shell catalyst. Circle symbols are experimental data. Pd^0 peaks at 334.9 eV and 340.2 eV, and $Pd^{\delta+}$ peaks at 336.1 eV and 341.4 eV are fitting results. (b) XPS Ti 2p spectra of P25@Pd catalyst and P25 support.

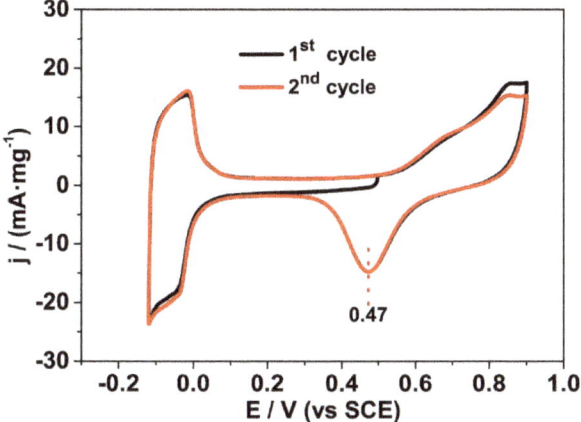

Figure 5. Cyclic voltammograms at a scanning rate of 50 mV s^{-1} measured on a glass carbon electrode modified with P25@Pd core-shell catalyst in a 0.5 M H_2SO_4 solution.

2.5. Enantioselective Hydrogenation of Acetophenone

The catalytic performance of P25@Pd catalyst was evaluated by utilizing the enantioselective hydrogenation of acetophenone as a probe reaction. As an α,β-unsaturated ketone, both the aromatic ring and the carbonyl group may undergo hydrogenation. As shown in Scheme 1, the selective hydrogenation of acetophenone involves several competitive and consecutive reactions, which may produce side products such as acetylcyclohexane, ethylbenzene and 1-cyclohexylethanol in addition to the target product, 1-phenylethanol [31–33]. Therefore, it is a great challenge to develop a catalyst with both high chemoselectivity and enantioselectivity to chiral 1-phenylethanol.

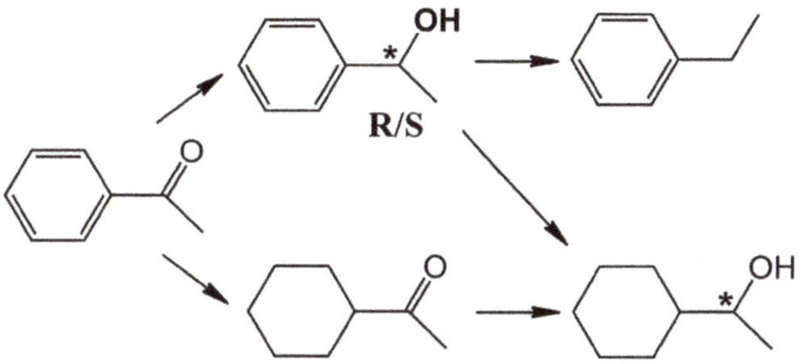

Scheme 1. Reasonable products in selective hydrogenation of acetophenone.

Figure 6 shows the catalytic performance of the P25@Pd catalyst in the enantioselective hydrogenation of acetophenone. For comparison, the catalytic behaviours of Pd/P25 with the same metal loading and unsupported Pd black catalysts were also tested under the same reaction conditions and the catalytic results are shown in Figure 6. Over P25@Pd catalyst, the yield of 1-phenylethanol increases steeply up to ~100% in a reaction time of 165 min and then keeps unchanged at a prolonged reaction time, demonstrating the excellent selectivity of the P25@Pd catalyst to the C=O group. Our previous study showed that, under the present reaction conditions, the liquid phase hydrogenation of acetophenone is zero-order for acetophenone [34], which is consistent with other literature reports [35,36]. Therefore, we calculated the reaction rate from the slope of the acetophenone conversion versus time plots based on our previous report [34]. The reaction rate is 36.5 mmol h^{-1} g$_{Pd}^{-1}$ on P25@Pd core-shell catalyst (Table 1, Entry 1), which is 36.5 and 30.2 times that of impregnated

Pd/P25 (1.00 mmol h^{-1} g$_{Pd}^{-1}$, Table 1, Entry 2) and Pd black catalysts (1.21 mmol h^{-1} g$_{Pd}^{-1}$, Table 1, Entry 3), respectively, demonstrating the superior activity of the P25@Pd catalyst.

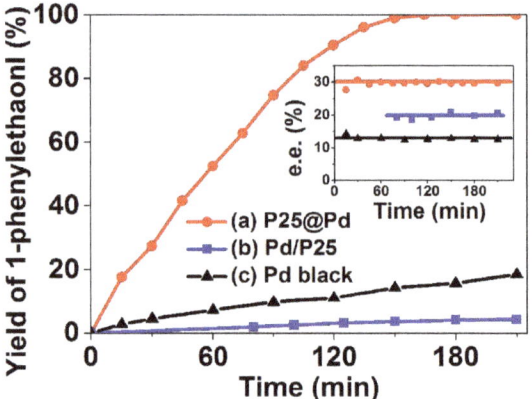

Figure 6. Catalytic results of acetophenone enantioselective hydrogenation over (**a**) P25@Pd, (**b**) Pd/P25, and (**c**) Pd black catalysts. Reaction conditions: 273 K, 7 mg of catalyst, 20 µL of acetophenone, 400 mg of S-proline, 17 mL of methanol, and H$_2$ (60 mL·min^{-1}).

Table 1. Catalytic performance of P25@Pd, Pd/P25 and Pd black catalysts in the enantioselective hydrogenation of acetophenone.

Entry	Catalyst	Reaction Rate (mmol h^{-1} g$_{Pd}^{-1}$)	e.e. (%)	Dominant Enantiomer
1	P25@Pd	36.5	30	R
2	Pd/P25	1.00	20	R
3	Pd black	1.21	13	R

Reaction conditions: 273 K, 7 mg of catalyst, 20 µL of acetophenone, 400 mg of S-proline, 17 mL of methanol, and H$_2$ (60 mL·min^{-1}).

As shown in the inset of Figure 6, the impregnated Pd/P25 catalyst gives an average enantioselectivity of ~20% (R-enantiomer dominant) which is similar to the value of supported Pd or Pt catalysts reported in literature [5–7]. Under the same reaction conditions, the P25@Pd core-shell catalyst exhibits an average enantioselectivity of 30% (R-enantiomer dominant), which is 50% higher than that of the impregnated Pd/P25 catalyst. The improved enantioselectivity of P25@Pd core-shell catalyst as compared to the impregnated Pd/P25 catalyst in the heterogeneous enantioselective hydrogenation of acetophenone could probably be attributed to the exclusion of negative weak interactions existing on the surface of support. As for the unsupported Pd black catalyst, the enantioselectivity is only 13% (R-enantiomer dominant), which is 43% of the value on P25@Pd core-shell catalyst. This demonstrates that the electronic properties of Pd in the P25@Pd core-shell catalyst affected by the interaction between the ultrathin Pd shell and the P25 support probably also play a positive role in the improvement of enantioselectivity. If we could further replace commercial P25 by pure anatase or rutile TiO$_2$ support with specific facets, the electronic state of Pd may be tuned by changing the interaction between Pd and the specific TiO$_2$ facets, which may be promising in heterogeneous enantioselective hydrogenations.

3. Materials and Methods

3.1. Chemicals and Materials

Commercial TiO_2 (Aeroxide® P25) was purchased from ACROS (Geel, Belgium). γ-Al_2O_3 was purchased from Alfa Aesar (Tewksbury, MA, USA). Mesoporous carbon was synthesized based on the method reported by Ryoo et al. [37]. Palladium chloride ($PdCl_2$) was purchased from J&K Scientific Ltd. (Beijing, China). Hydrochloric acid, SiO_2, $N_2H_4 \cdot H_2O$ and dimethyl formamide (DMF) were purchased from Sinopharm Chemical Regent Co. Ltd. (Shanghai, China). The aqueous solution of H_2PdCl_4 (1.0 M) was perpared by dissovling $PdCl_2$ in a concentrated hydrochloric acid solution.

3.2. Catalyst Synthesis

P25@Pd catalyst: The synthesis of the P25@Pd core-shell catalyst was based on the reduction of Pd^I carbonyl complex [19]. In a typical synthesis, 40 mg of P25 was added in 8.00 mL of distilled water and then ultrasonicated for 60 min to obtain well-dispersed P25 aqueous suspension. Next, 240 µL of H_2PdCl_4 aqueous solution (1.0 M) was added to 80 mL of anhydrous DMF in a three-necked glass flask at ambient temperature under stirring, and a brownish red solution was obtained. Then, carbon monoxide at atmospheric pressure with a flow rate of 100 mL min^{-1} was purged through the solution for 20 min and the brownish red solution turned bright yellow due to the reduction of H_2PdCl_4 to Pd^I carbonyl complex [19]. Then, 8.00 mL of ultrasonicated well-dispersed aqueous suspension of P25 was added to the bright yellow solution and dark blue suspensions were formed due to the reduction of Pd^I carbonyl complex to metallic Pd [19]. After 20 min, the dark blue suspensions were centrifuged. The final catalyst was washed with methanol three times, and dried in a vacuum for activity test.

Pd/P25 catalyst: The Pd/P25 catalyst is synthesized by a traditional impregnation method. First, 26.4 µL of H_2PdCl_4 (1.0 M) was pipetted and diluted with 1.0 mL of distilled water. Then, 5 mg of commercial P25 support was added and the mixture was ultrasonicated at ambient temperature for 1 h. The mixture was evaporated and dried at 393 K for 2 h. The resulting brown powders were reduced by 10 mL of aqueous solution of $N_2H_4 \cdot H_2O$ (30%) with stirring at ambient conditions. The products were washed with distilled water three times and then with methanol three times. The catalyst was kept in methanol for activity test.

Pd black catalyst: 468 µL of H_2PdCl_4 (1.0 M) was pipetted and diluted with 20 mL of distilled water. Then, 200 mL of aqueous solution of $N_2H_4 \cdot H_2O$ (30%) was added dropwise under stirring at ambient conditions. The black suspensions were centrifuged and the obtained Pd black catalyst was washed with distilled water three times and then with methanol three times. The catalyst was kept in methanol for activity test.

3.3. Characterization

The transmission electron microscopy (TEM), the high-resolution TEM (HRTEM) images, the scanning transmission electron microscopy (STEM), and the energy dispersive X-ray emission (EDX) mapping results were obtained on a field-emission transmission electron microscope (FEI Tecnai G^2 F20 S-Twin, Hillsboro, OR, USA, 200 kV). The element content of catalysts was determined by inductively coupled plasma atomic emission spectroscopy (ICP-AES, Thermo Elemental IRIS Intrepid, Waltham, MA, USA). The Brunauer–Emmett–Teller (BET) specific surface areas were analyzed by N_2 adsorption on a Tristar II 3020 apparatus (Norcross, GA, USA). The sample was degassed under N_2 flow at 383 K for 2 h before the measurement. The X-ray diffraction (XRD) patterns were acquired on a Bruker AXS D8 Advance X-ray diffractometer (Karlsruhe, Germany, Cu-Kα radiation, λ = 0.15418 nm, 40 kV, 40 mA) with a scanning rate of 1° min^{-1} at 2θ ranging from 20° to 90°. X-ray photoelectron spectroscopy (XPS) was utilized to detect the electronic state of the samples on a Perkin–Elmer PHI 5000C instrument (Waltham, MA, USA, 14 kV, 250 W) with Mg Kα (hν = 1253.6 eV) as the excitation source. Prior to the measurement, the samples were degassed at 298 K for 12 h in a vacuum chamber. The binding energy (BE) values were calibrated by referring to the C 1s peak (284.6 eV) of contaminant carbon.

Cyclic voltammeter (CV) with a scanning rate of 50 mV s^{-1} was performed in a N$_2$-saturated 0.5 M H$_2$SO$_4$ solution at ambient temperature by using a standard three-electrode electrochemical cell on a Bio-Logic science instrument (SP-300, Grenoble, France). Then, 10 μl of suspension containing P25@Pd core-shell catalyst (Pd 0.55 mg mL^{-1}) was pipetted onto a polished glass carbon electrode (d = 5 mm) and dried in air at ambient temperature. Then, 5 μl of Nafion solution (1 wt.%) was pipetted onto it and the Pd working electrode was obtained. A conventional three-electrode cell was used, including a saturated calomel electrode (SCE) as the reference electrode, a graphite rod as the counter electrode and the as-prepared Pd electrode as the working electrode. The Pd loading was 28 μg cm^{-2}.

3.4. Activity Test

Similar to our previous report [34], the liquid phase enantioselective hydrogenation of acetophenone was carried out at 273 K in a three-necked glass flask under atmospheric H$_2$ pressure (60 mL·min^{-1}). First, 400 mg of S-proline and 7 mg of catalyst were dispersed in 17 mL of methanol. Then, 20 μL of reactant acetophenone was pipetted into the mixture. The reaction was stirred with a rate of 1000 rpm for the elimination of diffusion effects. In the reaction process, the supernatant was sampled at intervals and detected with a gas chromatography (Agilent 7820A, Waltham, MA, USA) equipped with a chiral capillary column (CP-CHIRASIL-Dex CB, 25 m × 0.25 μm × 0.32 mm) and a flame ionization detector (FID). The retention times of products were identified with the standard chemicals. The enantiomeric excess was expressed as e.e. % = |(R − S)|/(R + S) × 100.

4. Conclusions

P25@Pd core-shell catalyst with an ultrathin Pd shell of ca. 1 nm and a high Pd loading of ~36 wt.% was successfully prepared by a facile strategy under mild conditions based on the reduction of Pd^I carbonyl complex on the surface of the P25 support. By excluding the negative weak interactions existing on the surface of the support, the P25@Pd core-shell catalyst exhibits higher activity and enantioselectivity than that of the impregnated Pd/P25 catalyst in the enantioselective hydrogenation of acetophenone. The much higher enantioselectivity over the P25@Pd core-shell catalyst than that on unsupported Pd black catalyst indicates that the presence of the P25 core is necessary to affect the electronic properties of Pd, which also probably have a positive effect on the enantioselectivity. The work provides understanding for the designed synthesis of effective enantioselective supported metal catalyst for heterogeneous asymmetric catalysis, which also opens up a new strategy for the synthesis of supported noble-metal catalyst with both high metal loading and well-dispersed metal.

Author Contributions: Conceptualization, X.C and X.G.; investigation, X.G., L.H. and J.X.; writing—original draft preparation, X.G. and X.C.; writing—review and editing, X.C. and H.H.; supervision, X.C. and H.H.; project administration, X.C. and H.H.; funding acquisition, X.C.

Funding: This research was funded by the National Natural Science Foundation of China, grant number 21573045, 21773034.

Conflicts of Interest: The authors declare no conflict of interest.

References

1. Blaser, H.U.; Pugin, B. Scope and Limitations of the Application of Heterogeneous Enantioselective Catalysis. In *Chiral Reactions in Heterogeneous Catalysis*; Jannes, G., Dubois, V., Eds.; Plenum Press: New York, NY, USA, 1995; pp. 33–57, ISBN 978-1-4615-1909-6.
2. Baiker, A. Crucial Aspects in the Design of Chirally Modified Noble Metal Catalysts for Asymmetric Hydrogenation of Activated Ketones. *Chem. Soc. Rev.* **2015**, *44*, 7449–7464. [CrossRef] [PubMed]
3. Meemken, F.; Baiker, A. Recent Progress in Heterogeneous Asymmetric Hydrogenation of C=O and C=C Bonds on Supported Noble Metal Catalysts. *Chem. Rev.* **2017**, *117*, 11522–11569. [CrossRef] [PubMed]
4. Klabunovskii, E.; Smith, G.V.; Zsigmond, Á. *Heterogeneous Enantioselective Hydrogenations-Theory and Practice*; Springer: Dordrecht, The Netherlands, 2006; pp. 161–170, ISBN 978-1-4020-4296-6.

5. Tungler, A.; Tarnai, T.; Mathe, T.; Petro, J. Enantioselective Hydrogenation of Acetophenone in the Presence of S-proline. *J. Mol. Catal.* **1991**, *67*, 277–282. [CrossRef]
6. Hess, R.; Vargas, A.; Mallat, T.; Bürgi, T.; Baiker, A. Inversion of Enantioselectivity in the Platinum-catalyzed Hydrogenation of Substituted Acetophenones. *J. Catal.* **2004**, *222*, 117–128. [CrossRef]
7. Vetere, V. New Approach toward the Synthesis of Asymmetric Heterogeneous Catalysts for Hydrogenation Reactions. *J. Catal.* **2004**, *226*, 457–461. [CrossRef]
8. Mills, P.L.; Ramachandran, P.A.; Chaudhari, R.V. Multiphase Reaction-engineering for Fine Chemicals and Pharmaceuticals. *Rev. Chem. Eng.* **1992**, *8*, 1–176. [CrossRef]
9. Perosa, A.; Tundo, P.; Selva, M. Multiphase Heterogeneous Catalytic Enantioselective Hydrogenation of Acetophenone over Cinchona-modified Pt/C. *J. Mol. Catal. A Chem.* **2002**, *180*, 169–175. [CrossRef]
10. Mhadgut, S.; Torok, M.; Esquibel, J.; Torok, B. Highly Asymmetric Heterogeneous Catalytic Hydrogenation of Isophorone on Proline Modified Base-supported Palladium Catalysts. *J. Catal.* **2006**, *238*, 441–448. [CrossRef]
11. Kubota, T.; Kubota, H.; Kubota, T.; Moriyasu, E.; Uchida, T.; Nitta, Y.; Sugimura, T.; Okamoto, Y. Enantioselective Hydrogenation of (E)-α-phenylcinnamic Acid over Cinchonidine-modified Pd Catalysts Supported on TiO_2 and CeO_2. *Catal. Lett.* **2009**, *129*, 387–393. [CrossRef]
12. Chen, Z.; Guan, Z.; Li, M.; Yang, Q.; Li, C. Enhancement of the Performance of a Platinum Nanocatalyst Confined within Carbon Nanotubes for Asymmetric Hydrogenation. *Angew. Chem. Int. Ed.* **2011**, *50*, 4913–4917. [CrossRef]
13. Guan, Z.H.; Lu, S.M.; Li, C. Enantioselective Hydrogenation of α,β-unsaturated Carboxylic Acid over Cinchonidine-modified Pd Nanoparticles Confined in Carbon Nanotubes. *J. Catal.* **2014**, *311*, 1–5. [CrossRef]
14. Bui Trung, T.S.; Kim, Y.; Kang, S.; Lee, H.; Kim, S. The Enantioselective Hydrogenation of (E)-α-phenylcinnamic Acid: Role of TiO_2 Coated on Al_2O_3 as a Novel Support for Cinchonidine-modified Pd Catalysts. *Catal. Commun.* **2015**, *66*, 21–24. [CrossRef]
15. Li, C. Chiral Synthesis on Catalysts Immobilized in Microporous and Mesoporous Materials. *Catal. Rev. Sci. Eng.* **2004**, *46*, 419–492. [CrossRef]
16. Hoxha, F.; Schimmoeller, B.; Cakl, Z.; Urakawa, A.; Mallat, T.; Pratsinis, S.E.; Baiker, A. Influence of Support Acid–base Properties on the Platinum-catalyzed Enantioselective Hydrogenation of Activated Ketones. *J. Catal.* **2010**, *271*, 115–124. [CrossRef]
17. Chaudhuri, R.G.; Paria, S. Core/Shell Nanoparticles: Classes, Properties, Synthesis Mechanisms, Characterization, and Applications. *Chem. Rev.* **2012**, *112*, 2373–2433. [CrossRef]
18. Zhang, L.; Zhou, N.; Wang, B.; Liu, C.; Zhu, G. Fabrication of Fe_3O_4/PAH/PSS@Pd Core-shell Microspheres by Layer-by-layer Assembly and Application in Catalysis. *J. Colloid Interface Sci.* **2014**, *421*, 1–5. [CrossRef] [PubMed]
19. Li, H.; Chen, G.X.; Yang, H.Y.; Wang, X.L.; Liang, J.H.; Liu, P.X.; Chen, M.; Zheng, N.F. Shape-controlled Synthesis of Surface-clean Ultrathin Palladium Nanosheets by Simply Mixing a Dinuclear Pd^I Carbonyl Chloride Complex with H_2O. *Angew. Chem. Int. Ed.* **2013**, *52*, 8368–8372. [CrossRef] [PubMed]
20. Huang, X.Q.; Tang, S.H.; Mu, X.L.; Dai, Y.; Chen, G.X.; Zhou, Z.Y.; Ruan, F.X.; Yang, Z.L.; Zheng, N.F. Freestanding Palladium Nanosheets with Plasmonic and Catalytic Properties. *Nat. Nanotechnol.* **2011**, *6*, 28–32. [CrossRef]
21. He, M.; Ji, J.; Liu, B.Y.; Huang, H.B. Reduced TiO_2 with Tunable Oxygen Vacancies for Catalytic Oxidation of Formaldehyde at Room Temperature. *Appl. Surf. Sci.* **2019**, *473*, 934–942. [CrossRef]
22. Tauster, S.J.; Fung, S.C.; Garten, R.L. Strong Metal-support Interactions. Group 8 Noble Metals Supported on TiO_2. *J. Am. Chem. Soc.* **1978**, *100*, 170–175. [CrossRef]
23. Kovtunov, K.V.; Barskiy, D.A.; Salnikov, O.G.; Burueva, D.B.; Khudorozhkov, A.K.; Bukhtiyarov, A.V.; Prosvirin, I.P.; Gerasimov, E.Y.; Bukhtiyarov, V.I.; Koptyug, I.V. Strong Metal-support Interactions for Palladium Supported on TiO_2 Catalysts in the Heterogeneous Hydrogenation with Parahydrogen. *ChemCatChem* **2015**, *7*, 2581–2584. [CrossRef]
24. Shen, W.J.; Okumura, M.; Matsumura, Y.; Haruta, M. The Influence of the Support on the Activity and Selectivity of Pd in CO Hydrogenation. *Appl. Catal. A Gen.* **2001**, *213*, 225–232. [CrossRef]
25. Tapin, B.; Epron, F.; Especel, C.; Ly, B.K.; Pinel, C.; Besson, M. Study of Monometallic Pd/TiO_2 Catalysts for the Hydrogenation of Succinic Acid in Aqueous Phase. *ACS Catal.* **2013**, *3*, 2327–2335. [CrossRef]
26. Lu, M.H.; Du, H.; Wei, B.; Zhu, J.; Li, M.S.; Shan, Y.H.; Song, C.S. Catalytic Hydrodeoxygenation of Guaiacol over Palladium Catalyst on Different Titania Supports. *Energy Fuels* **2017**, *31*, 10858–10865. [CrossRef]

27. Xiang, Q.; Lv, K.; Yu, J. Pivotal Role of Fluorine in Enhanced Photocatalytic Activity of Anatase TiO_2 Nanosheets with Dominant (001) Facets for the Photocatalytic Degradation of Acetone in Air. *Appl. Catal. B Environ.* **2010**, *96*, 557–564. [CrossRef]
28. Chen, X.B.; Liu, L.; Yu, P.Y.; Mao, S.S. Increasing Solar Absorption for Photocatalysis with Black Hydrogenated Titanium Dioxide Nanocrystals. *Science* **2011**, *331*, 746–750. [CrossRef]
29. Wang, G.; Wang, H.; Ling, Y.; Tang, Y.; Yang, X.; Fitzmorris, R.C.; Wang, C.; Zhang, J.Z.; Li, Y. Hydrogen-treated TiO_2 Nanowire Arrays for Photoelectrochemical Water Splitting. *Nano Lett.* **2011**, *11*, 3026–3333. [CrossRef]
30. Li, Y.B.; Zhang, C.B.; Ma, J.Z.; Chen, M.; Deng, H.; He, H. High Temperature Reduction Dramatically Promotes Pd/TiO_2 Catalyst for Ambient Formaldehyde Oxidation. *Appl. Catal. B Environ.* **2017**, *217*, 560–569. [CrossRef]
31. Masson, J.; Cividino, P.; Court, J. Selective Hydrogenation of Acetophenone on Chromium Promoted Raney Nickel Catalysts. III. The Influence of the Nature of the Solvent. *Appl. Catal. A Gen.* **1997**, *161*, 191–197. [CrossRef]
32. Bergault, I.; Fouilloux, P.; Joly-Vuillemin, C.; Delmas, H. Kinetics and Intraparticle Diffusion Modelling of a Complex Multistep Reaction: Hydrogenation of Acetophenone over a Rhodium Catalyst. *J. Catal.* **1998**, *175*, 328–337. [CrossRef]
33. Fujita, S.-I.; Onodera, Y.; Yoshida, H.; Arai, M. Selective Hydrogenation of Acetophenone with Supported Pd and Rh Catalysts in Water, Organic Solvents, and CO_2-dissolved Expanded Liquids. *Green Chem.* **2016**, *18*, 4934–4940. [CrossRef]
34. Su, N.; Gao, X.Y.; Chen, X.Y.; Yue, B.; He, H.Y. The Enantioselective Hydrogenation of Acetophenone over Pd Concave Tetrahedron Nanocrystals Affected by the Residual Adsorbed Capping Agent Polyvinylpyrrolidone (PVP). *J. Catal.* **2018**, *367*, 244–251. [CrossRef]
35. Aramendia, M.A.; Borau, V.; Gomez, J.F.; Herrera, A.; Jimenez, C.; Marinas, J.M. Reduction of Acetophenones over $Pd/AlPO_4$ Catalysts. Linear Free Energy Relationship (LFER). *J. Catal.* **1993**, *140*, 335–343. [CrossRef]
36. Drelinkiewicz, A.; Waksmundzka, A.; Makowski, W.; Sobczak, J.W.; Krol, A.; Zieba, A. Acetophenone Hydrogenation on Polymer-palladium Catalysts. The Effect of Polymer Matrix. *Catal. Lett.* **2004**, *94*, 143–156. [CrossRef]
37. Jun, S.; Joo, S.H.; Ryoo, R.; Kruk, M.; Jaroniec, M.; Liu, Z.; Ohsuna, T.; Terasaki, O. Synthesis of New, Nanoporous Carbon with Hexagonally Ordered Mesostructure. *J. Am. Chem. Soc.* **2000**, *122*, 10712–10713. [CrossRef]

© 2019 by the authors. Licensee MDPI, Basel, Switzerland. This article is an open access article distributed under the terms and conditions of the Creative Commons Attribution (CC BY) license (http://creativecommons.org/licenses/by/4.0/).

Article

Effect of Direct Reduction Treatment on Pt–Sn/Al$_2$O$_3$ Catalyst for Propane Dehydrogenation

Jae-Won Jung [1], Won-Il Kim [2], Jeong-Rang Kim [3], Kyeongseok Oh [4] and Hyoung Lim Koh [1],*

[1] Department of Chemical Engineering, RCCT, Hankyong National University, Anseong 456-749, Korea; rohyang@naver.com
[2] R&D Business Lab., Hyosung Co., Hoge-dong, Dongan-ku, Anyang, Gyeounggi 431-080, Korea; wikim@hyosung.com
[3] Carbon Resources Institute, Korea Research Institute of Chemical Technology, 141, Gajeong-ro, Yuseong-gu, Daejeon 34114, Korea; jrkim@krict.re.kr
[4] Chemical and Environmental Technology Department, Inha Technical College, Inha-ro 100, Michuhol-gu, Incheon 22212, Korea; kyeongseok.oh@inhatc.ac.kr
* Correspondence: hlkoh@hknu.ac.kr; Tel.: +82-31-670-5410

Received: 19 April 2019; Accepted: 10 May 2019; Published: 14 May 2019

Abstract: Pt–Sn/Al$_2$O$_3$ catalysts were prepared by the direct reduction method at temperatures from 450 to 900 °C, denoted as an SR series (SR450 to SR900 according to reduction temperature). Direct reduction was performed immediately after catalyst drying without a calcination step. The activity of SR catalysts and a conventionally prepared (Cal600) catalyst were compared to evaluate its effect on direct reduction. Among the SR catalysts, SR550 showed overall higher conversion of propane and propylene selectivity than Cal600. The nano-sized dispersion of metals on SR550 was verified by transmission electron microscopy (TEM) observation. The phases of the bimetallic Pt–Sn alloys were examined by X-ray diffraction, TEM, and energy dispersive X-ray spectroscopy (EDS). Two characteristic peaks of Pt$_3$Sn and PtSn alloys were observed in the XRD patterns, and these phases affected the catalytic performance. Moreover, EDS confirmed the formation of Pt$_3$Sn and PtSn alloys on the catalyst surface. In terms of catalytic activity, the Pt$_3$Sn alloy showed better performance than the PtSn alloy. Relationships between the intermetallic interactions and catalytic activity were investigated using X-ray photoelectron spectroscopy. Furthermore, qualitative analysis of coke formation was conducted after propane dehydrogenation using differential thermal analysis.

Keywords: direct reduction; propane dehydrogenation; Pt–Sn/Al$_2$O$_3$; Pt$_3$Sn alloy; PtSn alloy

1. Introduction

Light olefins such as ethylene and propylene are important chemical intermediates, and demand for these materials has increased continuously [1,2]. Propane dehydrogenation (PDH) is used to produce propylene and has drawn interest from researchers. As the price of propane has fallen, propylene supply is also becoming important. In addition to the steam cracking of light olefins or naphtha, the dehydrogenation of light alkanes is also popular nowadays. Since the 2010s, increasing shale gas exploitation has motivated the development of dehydrogenation processes because significant amounts of propane and butane are generated (although the main product of shale gas is methane). The commercial PDH process uses Pt/Al$_2$O$_3$ catalysts or Cr/Al$_2$O$_3$ catalysts [3–6], but the demand for highly active catalysts is still growing. In this paper, we refer to Al$_2$O$_3$ without differentiating its α-, γ-, and θ-phases because the effect of the Al$_2$O$_3$ phase is not the objective of this study.

The Pt catalyst used in the PDH process is often impregnated with Sn as a co-catalyst [7–11]. Because the Pt–alkene interaction is stronger than the Pt–alkane interaction, unwanted side reactions such as hydrogenolysis and isomerization often occur [12]. When Sn is added to a Pt/Al$_2$O$_3$ catalyst, the Pt–alkene interaction is weakened and side reactions are suppressed [13–17]. Thus, Sn may enhance the propylene selectivity of the PDH reaction [18,19]. It has also been reported that Sn plays important roles in preventing Pt sintering, lowering the acidity of the oxide support, and removing the coke generated during the PDH process [4,20]. Previously, Pt–Sn alloys that use SiO_2 as a support have been reported [10]. The formation of alloys and their catalytic activities depend upon the preparation conditions. Alloys of Pt–Sn have several forms, for example, PtSn, Pt_3Sn, and $PtSn_2$.

Deng and coworkers [10,21,22] used a direct reduction method to prepare PDH catalysts, which did not include a calcination step, unlike the conventional protocol for catalyst preparation involving drying, calcination, and reduction. They applied a hydrogen reduction stage at high temperature (800 °C) immediately after drying the catalyst without a calcination step. The direct reduction method yielded a catalyst with much higher performance than the conventionally calcined catalyst considering propane conversion, as well as propylene selectivity. They reported that the Pt_3Sn alloy formed by direct reduction was responsible for the observed enhancement in catalytic activity. However, the role of the Pt_3Sn alloy in promoting the PDH reaction is still controversial. For example, Vu [3] reported the reverse result, that is, the Pt_3Sn alloy may cause catalytic deactivation, but the PtSn alloy provides active sites.

In our previous study [23], we reported that oxychlorination treatment was effective in regenerating Pt–Sn/Al_2O_3 catalysts after the PDH reaction, and we observed that catalysts containing Pt_3Sn alloys after oxychlorination treatment showed improved PDH activity. When considering catalysts for commercial PDH applications, an Al_2O_3 support is more favorable than SiO_2 because of its superior thermal and mechanical stabilities. Thus, the evaluation of the effects of the direct reduction of a catalyst on an Al_2O_3 support for the PDH reaction is of significant value. In this study, direct reduction was employed to prepare Pt–Sn/Al_2O_3 catalysts. The Pt–Sn/Al_2O_3 catalysts were prepared at temperatures from 450 to 900 °C, and the catalytic performance was evaluated in terms of propane conversion and propylene selectivity. As a reference, a calcined catalyst was also prepared. The catalysts characteristics were examined by X-ray diffraction (XRD), transmission electron microscopy (TEM), energy dispersive spectroscopy (EDS), thermogravimetric analysis (TGA)/differential thermal analysis (DTA), and X-ray photoelectron spectroscopy (XPS).

2. Results

2.1. Catalyst Performance

Figure 1 shows the propane conversion for the PDH reaction over the SR catalysts prepared by the direct reduction method, as well as that for the Cal600 catalyst prepared by conventional calcination. The SR550 catalyst showed the best performance in propane conversion over 5 h. The conversion of SR550 was initially 41.2%, but this gradually decreased to 31.6% over the 5 h period. The second best catalyst was SR500, which showed good performance for 4 h. The SR500 catalyst showed 38.8% initial conversion, which decreased to 29.1% after 5 h. Meanwhile, the reference catalyst, Cal600, had a conversion of 32.3%, making it the third best concerning initial conversion. The conversion subsequently decreased to 28.5%, making it the second best after 5 h. The remainder of the catalysts showed conversion values in the order of SR600 > SR450 > SR800 > SR900. Of the SR catalysts, higher conversion values were achieved after direct reduction for 500 and 550 °C. It was observed that reduction temperatures higher than 600 °C had a negative effect on the propane conversion.

As shown in Figure 2, the propylene selectivity values determined after 1 h were in the order of SR500 > SR550 > SR600 > Cal600 > SR450 > SR800 > SR900. Similar to the trends in Figure 1, the selectivity values of SR500 and SR550 were similarly high until 180 min, after which SR550 showed slightly superior selectivity to SR500 until 5 h reaction time. As mentioned, the direct reduction method was first reported by Deng and coworkers [10,21]. They prepared Pt–Sn/SiO_2 catalysts and evaluated the effects of the direct reduction method. On comparison with the activity of a conventionally calcined catalyst, they observed higher activity after direct reduction at 800 °C. However, because we are considering commercial applications, in our study, an Al_2O_3 support was chosen. Al_2O_3 is more favorable than SiO_2 because Al_2O_3 is durable and more stable under the high temperature conditions of the PDH reaction. The main objective of this study was to verify whether Pt–Sn/Al_2O_3 prepared by a direct reduction will show similar superiority to the conventional catalyst for the PDH reaction.

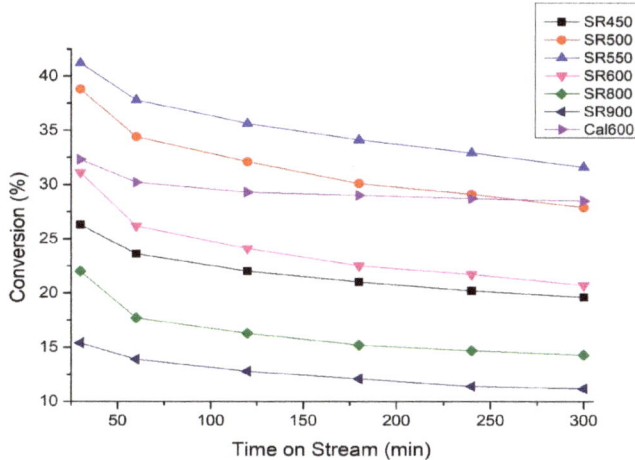

Figure 1. Propane conversion over Pt–Sn/Al_2O_3. Reaction conditions: catalyst weight, 0.1 g; reaction temperature, 600 °C; and flow rate, H_2:C_3H_8 = 32:32 (mL/min).

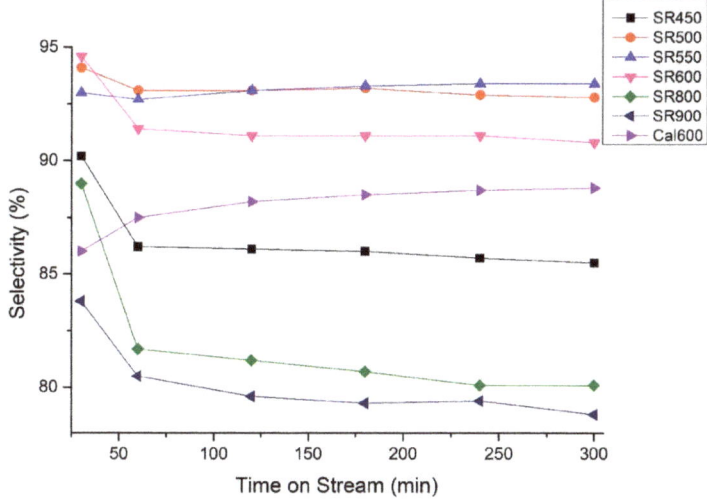

Figure 2. Propylene selectivity of Pt–Sn/Al_2O_3. Reaction conditions: catalyst weight, 0.1 g; reaction temperature, 600 °C; and flow rate, H_2:C_3H_8 = 32:32 (mL/min).

2.2. Catalyst Characterization

2.2.1. X-ray Diffraction Analysis

Figure 3 shows the XRD patterns of the five SR catalysts (SR450, SR500, SR550, SR600, and SR800) and the Cal600 catalyst. As shown in Figure 3, two different alloys, PtSn and Pt_3Sn, were detected between 38° and 46° in 2θ. The peaks located at 2θ values of 41.71° and 44.01° are related to the PtSn alloy, and the peaks located at 2θ values of 39.16° and 45.51° correspond to the Pt$_3$Sn alloy [10,21]. It has been reported that the Pt$_3$Sn alloy is only formed when precursor's molar ratios of Pt/Sn is higher than 1 [21]. In this study, both PtSn and Pt$_3$Sn are clearly indicated by the XRD patterns. The precursor ratio was fixed to 3 to 1.8 by weight (1:1 by mole). Interestingly, the SR catalysts showed a slight shift to a higher angle in the characteristic Pt$_3$Sn peak near 45°. Here, we infer that the Pt$_x$Sn$_y$ (x/y >3) alloy is possibly formed, and a small amount of Sn can be excluded from the existing Pt$_3$Sn alloy phase, as reported previously [10,21]. After reduction (Figure 4), the characteristic peaks of the PtSn alloy are not present in the XRD patterns of SR450, SR500, SR550, and SR600, although they were observed in the XRD pattern of SR800. As shown in Figure 4, the intensities of the characteristic peaks of PtSn alloy were reduced as the reduction temperature increased from 450 to 550 °C, and the peaks grew sharper as the reduction temperatures increased from 600 to 800 °C. Considering the results of the catalytic activity tests and the XRD analysis, we inferred the presence of PtSn alloy, which must affect the PDH reaction negatively. Meanwhile, the intensities of the characteristic peaks of the Pt$_3$Sn alloy did not change much with increasing direct reduction temperature. The characteristic Pt$_3$Sn peak shifted slightly to the characteristic peaks of Pt$_x$Sn$_y$ (x/y > 3) near 45.5°. Conceivably, the abundant Pt in Pt$_3$Sn or Pt$_x$Sn$_y$ (x/y > 3) provides more active sites for the PDH reaction. Thus, the activity for the PDH reaction was increased not by the increasing amount of PtSn alloy but the increasing amount of Pt$_3$Sn or Pt$_x$Sn$_y$ (x/y > 3) alloy. In addition, the presence of broad peaks in the XRD patterns implies that the Pt$_3$Sn alloy is more amorphous or dispersed and nano-sized [21]. In the case of Cal600, the characteristic peaks of both PtSn and Pt$_3$Sn alloys were very sharp, as shown in Figure 4f. Thus, both Pt$_3$Sn and PtSn alloys are highly crystalline in the Cal600 catalyst. Therefore, even though Pt$_3$Sn alloy in Cal600 was present as a highly crystalline phase with large particle sizes, the presence of the PtSn alloy negatively affected the PDH reaction. After the PDH reaction, we found that the XRD peaks maintained similar peak intensities, as shown in Figure 4.

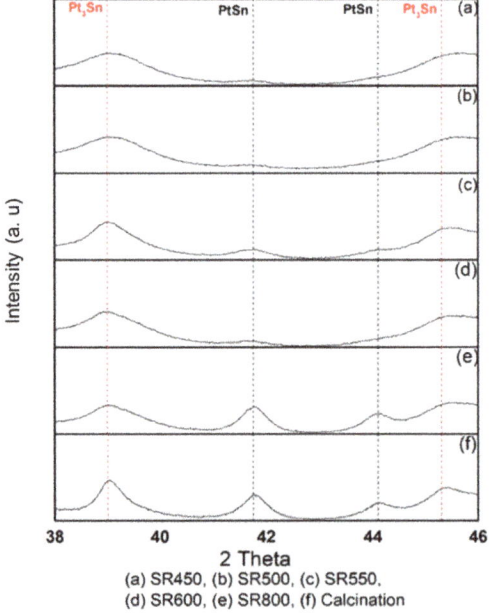

Figure 3. XRD patterns of Pt–Sn/Al$_2$O$_3$ catalysts after catalyst preparation.

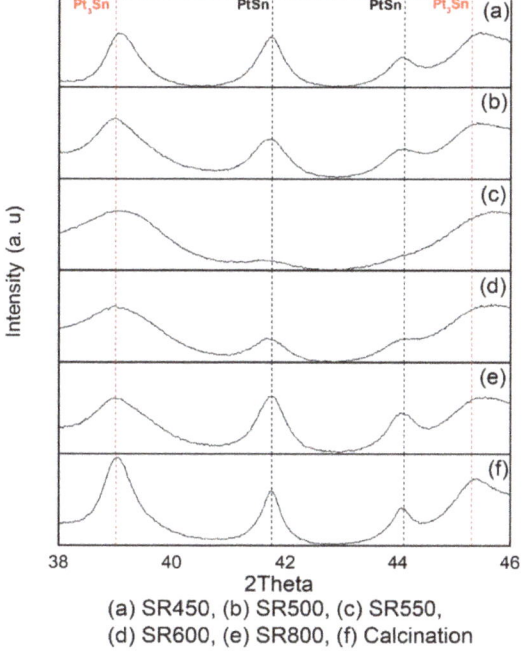

(a) SR450, (b) SR500, (c) SR550,
(d) SR600, (e) SR800, (f) Calcination

Figure 4. XRD patterns of Pt–Sn/Al$_2$O$_3$ catalysts after pre-reduction at 600 °C for 1 h under a hydrogen atmosphere.

2.2.2. TEM

Figure 5 shows TEM images of SR450, SR500, SR600, and Cal600. Images of each catalyst are shown at four magnifications. Compared to Cal600, the SR catalysts contain much smaller alloy particles. In particular, the Pt–Sn particles of SR550 are the smallest and also distributed rather uniformly. This is consistent with the XRD results shown in Figure 3. Thus, the SR550 catalyst has well-dispersed Pt–Sn alloys, which are responsible for the high activity for the PDH reaction. In addition, large particles of Pt–Sn alloys were observed in Cal600.

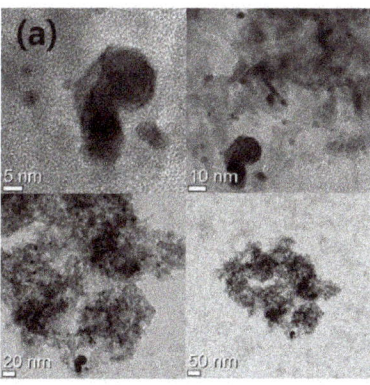
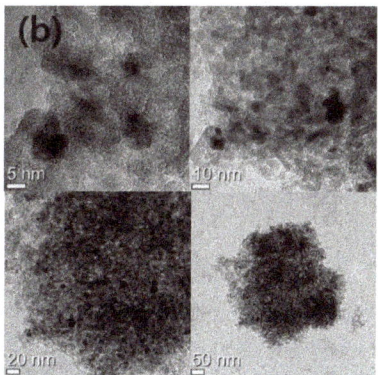

Figure 5. Cont.

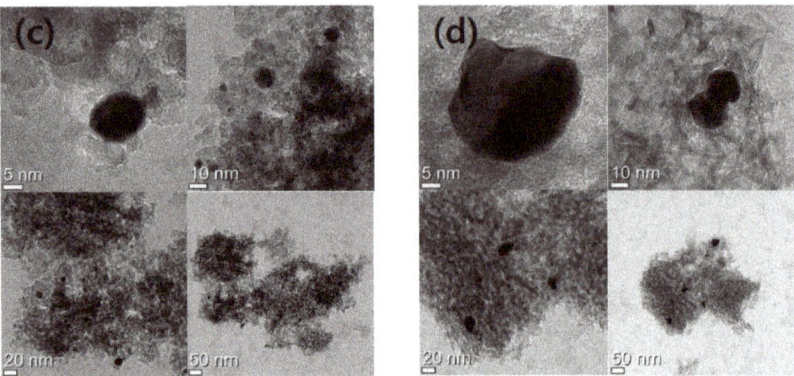

Figure 5. TEM images of (**a**) SR450, (**b**) SR550, (**c**) SR900, and (**d**) Cal600.

In Figure 6, EDS analyses of the SR550 and Cal600 catalysts are shown. Figure 6a shows the atomic percentages of Pt and Sn in SR550 catalyst (77.20 at.% Pt and 22.79 at.% Sn), which are close to the percentages of Pt and Sn in Pt_3Sn alloy. In the case of the Cal600 catalyst, the atomic percentages (56.05 at.% Pt and 43.94 at.% Sn) are close to those of the PtSn alloy. Interestingly, different amounts of Pt in SR550 were observed at the edge as well as in the core of a metal alloy particle as shown in Figure 6b. Pt contents were 50.45 at.% at the edge and 58.26 at.% in the core of a metal (alloy) particle.

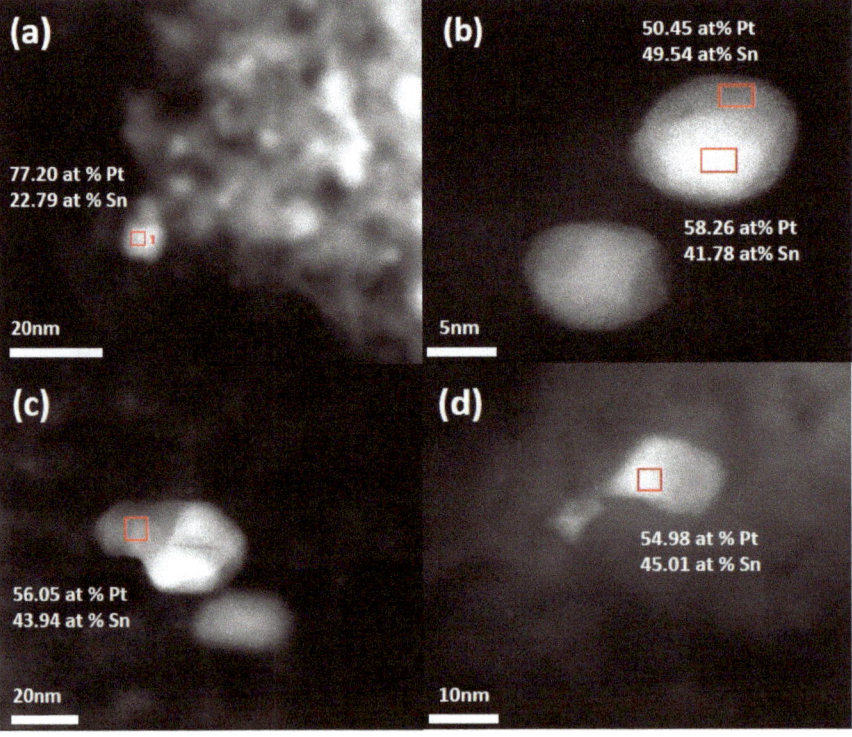

Figure 6. EDS analysis of (**a**,**b**) SR550 and (**c**,**d**) Cal600.

In the case of the SR550 catalyst, the difference in Pt contents was 50.45 at.% at the edge and 58.26 at.% in the core of a metal particle.

2.2.3. DTA Analysis

After the PDH reaction, coke is deposited on each catalyst, and this was characterized by DTA analysis. The amount of coke is summarized in Table 1 (in the Discussion section). As the direct reduction temperature increased from 450 to 550 °C, the amount of coke increased and then decreased again at 600 °C. The amount of coke gradually increased with increasing catalyst activity. As shown in Figure 7, the SR catalysts showed the first peak at 417–454 °C, and the second peak occurred at 482–503 °C during coke oxidation.

Table 1. Amount of coke and XRD peak ratios in the Pt_3Sn and PtSn alloys after catalyst preparation, direct reduction, and PDH reaction.

Catalyst	XRD (I_{Pt3Sn}/I_{PtSn}) After Preparation	XRD (I_{Pt3Sn}/I_{PtSn}) After Reduction	XRD (I_{Pt3Sn}/I_{PtSn}) After Reaction	Coke (wt.%)
SR450	9.15	1.08	1.58	5.85
SR500	5.16	1.55	1.58	7.86
SR550	4.12	7.79	2.33	8.60
SR600	6.97	2.48	3.21	7.04
Cal600	1.62	1.63	1.54	4.00

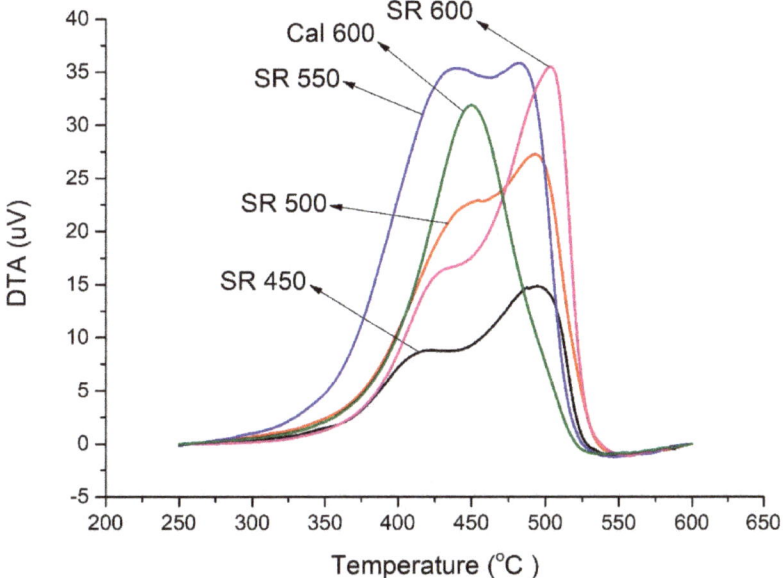

Figure 7. DTA curves of the Pt–Sn/Al_2O_3 catalysts. Inflection points of SR450 were observed at 417.62 and 494.38 °C, SR500 = 454.29 and 492.97 °C, SR550 = 439.61 and 482.68 °C, SR600 = 429.71 and 503.27 °C, and Cal600 = 450.16 °C.

It has been reported [24] that the first peak near 417 °C can be attributed to coke combustion on either Pt or Pt–Sn alloy sites, whereas the second peak near 494.4 °C is due to the "drain-off" effect. Here, drain-off is the case of the coke on metal sites migrating to the oxide support. The drain-off effect is known to be greater when Sn is added to Pt [5]. Comparing the results for the SR catalysts to that of Cal600, the second peak of the SR catalysts was larger. This is indirect evidence that the bimetallic catalysts induced the cokes to move from the metal sites to the surface of the catalyst support when coke accumulates. In particular, there are two inflection points of equal magnitude in the DTA curve of SR550. The first peak in the DTA curve of SR550 is relatively strong and is related to the large amount of coke on metal sites. Note that SR550 has the highest activity of all the catalysts for the PDH reaction. We suspect that the interaction between propane and the Pt–Sn alloys is greater for the Pt_3Sn alloy than the PtSn alloy. Thus, the greater number of interactions enhances the PDH reaction and more coke is formed. When coke accumulates, the metal sites become covered, which could inhibit the PDH reaction. If drain-off occurs, the active sites of catalysts are recovered. Unlike that of SR550, the DTA curve of Cal600 showed a single peak at 450.2 °C. Thus, no drain-off effects occurred in the Cal600 catalyst.

2.2.4. XPS Analysis

XPS spectra of the Pt 4f and Sn 3d regions of SR550 and Cal600 are shown in Figure 8. Peak deconvolution was conducted, and the overlap of the Pt 4f and Al 2p peaks was considered. The dashed curves show the Al 2p and Pt 4f peaks in Figure 8. The binding energies of Pt $4f_{7/2}$ and $4f_{5/2}$ peaks were 71.6 and 74.8 eV, respectively. On the basis of this binding energy, it can be confirmed that the state of Pt in both SR550 and Cal600 catalysts is zero valent (metallic Pt). A comparison of the Pt 4f peak areas between the catalysts reveals that Pt was distributed less on the surface of SR550 than that of Cal600. As shown in Figure 8c,d, the binding energies of the Sn $3d_{5/2}$ peaks were 487 and 485 eV, and deconvolution was performed to compare the contents of Sn(II, IV) and Sn(0). Because the relative intensity of the Sn(II, IV) of Cal600 is greater than that of SR550, Sn may cover the surface of the Pt or Pt–Sn alloy in the SR550 catalyst

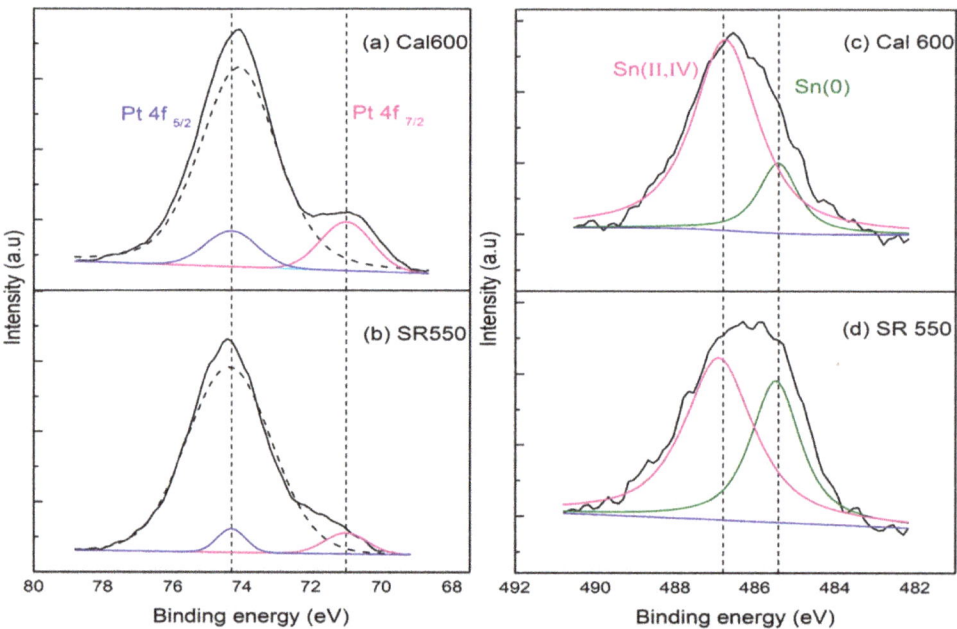

Figure 8. XPS Pt 4f spectra of (**a**) SR550, (**b**) Cal 600 and Sn 3d spectra of (**c**) SR550, and (**d**) Cal 600.

XPS spectra after depth-profiling sputtering with Ar ions for SR550 and Cal600 catalysts are shown in Figure 9. The results support the hypothesis that the surfaces of both metallic Pt or Pt–Sn alloys in the SR550 catalyst are covered with Sn. There was no change in the peak area of the Pt 4d for Cal600 catalyst after Ar ion sputtering for 6 and 18 s. On the other hand, the Pt 4d peak area of the SR550 catalyst after 18 s of Ar ion sputtering is larger than that after 6 s of Ar ion sputtering. The effect of Sn enrichment on the Pt particle surface has been reported elsewhere [9,22]. Zhu et al. [9] reported a Sn-surface-enriched Pt–Sn bimetallic nanoparticle catalyst prepared on a $MgAl_2O_4$ support using an organometallic chemistry concept, and the catalyst exhibited high selectivity and stability during the PDH reaction. Experiments by Deng and colleagues have shown that highly dispersed Sn-surface-enriched Pt–Sn alloy nanoparticles are formed when Pt–Sn/SiO_2 catalysts are prepared by direct reduction [22]. In addition, when Pt is in a very electron-rich state, the dehydrogenation activity of n-butane could be greatly improved by the strong interaction of Pt and SnO_2 (SMSI). Therefore, we can propose that the synergistic effect of the interaction of Pt and Sn covering the surface of Pt, as well as the accessible Pt sites, could improve the dehydrogenation performance of the catalyst prepared by the direct reduction method.

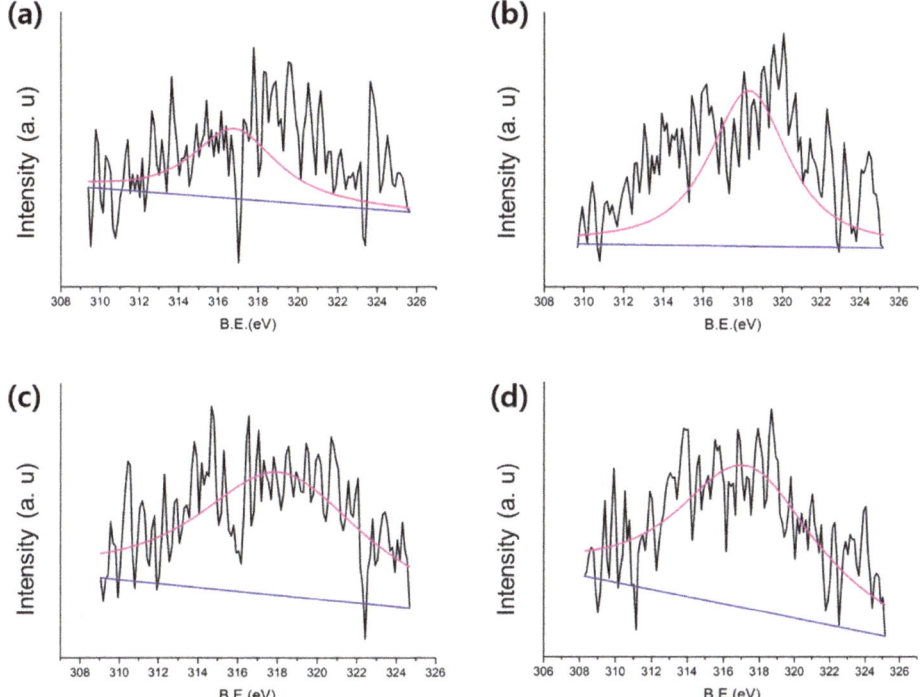

Figure 9. XPS Ar sputter profile Pt 4d spectra of (**a**) SR550 6s, (**b**) SR550 18s, (**c**) Cal600 6s, and (**d**) Cal600 18s (B.E. = binding energy).

3. Discussion

The reduction time in the direct reduction method is less than the typical time required for calcination. In addition, the catalyst might be affected by the Pt and Sn precursors left behind because of the omission of a calcination step. According to Arteaga [25], the dispersion of Pt particles is due to oxychlorination with chlorine compounds followed by reduction. It was hypothesized that the calcination method may remove a considerable amount of chlorine originating from the Pt and Sn precursors. This can cause oxychlorination before the reduction treatment. However, in the direct reduction method, oxychlorination and reduction occur simultaneously in the presence of chlorine from the precursors, which enhances the Pt dispersion and results in the formation of the Pt_3Sn alloy.

The XRD intensity ratios are listed in Table 1. These values were determined from the peak area of reflections at 39.2° for the Pt_3Sn alloy and 41.7° for the PtSn alloy. The peak areas were measured after baseline adjustment for the individual XRD patterns. The effects of catalyst preparation, direct reduction, and the PDH reaction were considered and compared. It should be noted that the XRD intensity ratio values for SR catalysts changed significantly but showed a mixed trend for each stage, unlike the Cal600 catalyst. In the case of the SR450 catalyst, the initial XRD intensity ratio value was 9.15, the highest value, but this rapidly decreased to 1.08, the lowest value, after direct reduction. As shown in Figure 4a, the characteristic XRD peaks of the Pt_3Sn and PtSn alloys were all sharp, which can be attributed to an increase in size of both alloy particles. Larger metallic sites would result in reduced catalytic activity. This trend is similar to that of the SR500 catalyst. In the case of the SR550 catalyst, the highest value of the XRD peak ratio was obtained after direct reduction. In Figure 4c, XRD characteristic peaks of Pt_3Sn and PtSn alloys are not sharp but broad, indicating an amorphous-like state. The dominance of Pt_3Sn alloy over PtSn alloy also induced the greatest production of coke after the PDH reaction. This result is consistent with the high propane conversion and propylene selectivity of the SR550 catalyst. Lower catalytic performance was observed for the SR600 catalyst. The XRD peak ratio values were decreased significantly in the SR600 catalyst.

In Figure 10, hypothetical models of the catalyst dispersion are shown. In the case of the Cal600 catalyst, the large metal alloy particles are dispersed over the support, which may be covered with Sn and/or SnO_2. Meanwhile, smaller alloy particles and a support only partly covered with SnO_2 is shown for the SR550 catalyst. Based on Figure 8, the population of Pt_3Sn alloy particles may be greater in the SR550 catalyst. In addition, metallic Sn could be present on the alloy surface. We have interpreted the presence of zero-valent Sn in Figure 8B as evidence of both Sn in alloys and isolated Sn particles.

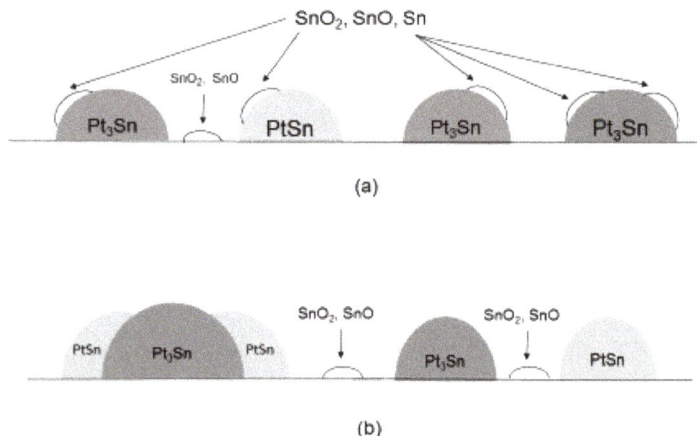

Figure 10. Hypothetical models of dispersed metals on Al_2O_3 support. (**a**) Conceptual metal dispersion after direct-reduction for SR550 catalyst. (**b**) Conceptual metal dispersion for Cal600 catalyst.

4. Materials and Methods

4.1. Catalyst Preparation

Hydrogen hexachloroplatinate, ($H_2PtCl_6 \cdot 5.5H_2O$, Kojima Chemicals, Saitama, Japan) and stannous chloride ($SnCl_2$, Sigma Aldrich, St. Louis, MO, USA) were used as metal precursors for the supported Pt–Sn catalysts. Al_2O_3 (γ-Al_2O_3, PURALOX® supplied by SASOL, Hamburg, Germany) was used as a catalyst support. $H_2PtCl_6 \cdot 5.5H_2O$ and $SnCl_2$ were dissolved in ethanol (C_2H_5OH, 99.5%, Daejung, Seoul, Korea), and Al_2O_3 was co-impregnated with these precursors. The co-impregnated Pt–Sn/Al_2O_3 catalyst was dried at 110 °C for 12 h. Subsequently, the Pt–Sn/Al_2O_3 catalysts were prepared by direct reduction at temperatures from 450 to 900 °C under H_2:N_2 flow (90 mL/min). The Pt content was controlled to be 3 wt.%, and the Sn content was 1.8 wt.%. It should be noted that direct reduction was performed immediately after drying the catalyst without calcination. Each SR sample was kept at the final reduction temperature for 30 min. Then, the sample was cooled or heated to 500 °C (10 °C/min) under H_2, N_2 flow (90 mL/min). Then, the catalyst samples were kept at 500 °C for 60 min again. For comparison, a calcined catalyst was also prepared (Cal600). Calcination was performed at 600 °C for 240 min in air.

4.2. Catalytic Activity Measurements

The catalytic activity during the PDH reaction was evaluated in a fixed-bed reactor (quartz, inner diameter 18 mm) using 0.1 g of each catalyst (20–40 mesh). The PDH reaction was carried out at 600 °C for 5 h at atmospheric pressure in the presence of C_3H_8 (30 mL/min), H_2 (30 mL/min), and N_2 (70 mL/min). Before the PDH reaction, the samples were heated to 600 °C at 10 °C/min in the presence of H_2 (30 mL/min) and N_2 (100 mL/min). The reaction product was collected at various time intervals, subsequently being analyzed by gas chromatography (flame ionization detector, 5890 Series2 Plus, Hewlett Packard, Wilmington, DE, USA). A 50 m × 0.53 mm GS-Alumina capillary column was used.

4.3. Characterization

XRD patterns were measured using a SMART LAB X-ray diffractometer (Rigaku, Tokyo, Japan) with Cu-$K\alpha$ radiation. The X-ray tube was operated at 40 kV and 200 mA. XRD patterns were obtained from $2\theta = 34°$ to $48°$ at a scanning speed of 4°/min. TEM images were obtained using an FEI TEM (Titan^{TM} 80–300, Hillsboro, OR, USA) operating at an accelerating voltage of 300 kV. In addition, XPS spectra were obtained with a PHI 5000 Versa Probe spectrometer (Ulvac-PHI, Kanagawa, Japan) equipped with a monochromatic electro analyzer and a monochromatic Al-$K\alpha$ 150 W X-ray source.

5. Conclusions

Pt–Sn/Al_2O_3 catalysts were prepared by direct reduction at temperatures from 450 to 900 °C, denoted SR450 to SR900 according to the reduction temperature. Direct reduction was performed by reduction treatment of the catalysts immediately after drying without conventional calcination. The effect of direct reduction was analyzed by comparison of the activity with that of a conventionally prepared catalyst (Cal600). Concerning catalytic performance, the SR catalysts showed overall higher values for the conversion of propane, as well as propylene selectivity, compared to Cal600. Of the SR catalysts, SR550 showed the highest activity. The direct reduction method results in the formation of different Pt–Sn alloys, and PtSn and Pt_3Sn alloys were identified. A larger amount of Pt_3Sn and a smaller amount of PtSn alloy was formed after direct reduction when the precursor ratio of Pt/Sn was 3 to 1.8 by weight (1:1 by mole), which is consistent with the literature [21]. This might be attributed to the difference in catalyst support. In our case, Al_2O_3 may provide more favorable conditions to form the Pt_3Sn alloy than the SiO_2 support during direct reduction. However, we agree that PtSn alloy can be formed on both supports, but the rearrangement of different alloys may occur when a direct reduction process is employed. Cal600 was found to contain larger metal particles, whereas the SR catalysts have nano-sized metal particles. Possibly, metal redistribution occurred during hydrogen reduction. The formation of smaller metal particles formed during reduction can result in increased interactions between nano-sized Pt and nano-size Sn particles. From a stoichiometric point of view, more Pt particles than Sn particles can interact to form a Pt_3Sn alloy. In this study, the highest activity was obtained when a reduction temperature of 550 °C was used. In addition, the coke behavior was analyzed by DTA. The DTA curve for Cal600 has one inflection point. However, the DTA plots of the SR catalysts contained two inflection points, and the second peak is related to the coke "drain-off" effect. When the PDH reaction proceeds, coke is generated. High activity catalysts could generate more coke, which would accumulate and cover the catalyst active sites. The well-dispersed Pt_3Sn alloy may accelerate the PDH reaction and generate more coke. However, the coke can migrate to the edge of Pt_3Sn alloy and/or Pt_xSn_y ($x/y > 3$) alloy, resulting in the preservation of active sites, allowing the PDH reaction to proceed for longer.

Author Contributions: The experimental work was designed and performed by J.-W.J.; W.-I.K. and J.-R.K. analyzed the data; writing, review, and editing was done by K.S.O.; and writing and original draft preparation was done by H.L.K.

Funding: This work was financed by Basic Science Research Program through the National Research Foundation of Korea (NRF) funded by the Ministry of Education (2017R1D1A1B03034244).

Conflicts of Interest: The authors declare no conflicts of interest.

References

1. Sattler, J.J.H.B.; Ruiz-Martinez, J.; Santillan-Jimenez, E.; Weckhuysen, B.M. Catalytic dehydrogenation of light alkanes on metals and metal oxides. *Chem. Rev.* **2014**, *114*, 10613–10653. [CrossRef] [PubMed]
2. Nawaz, Z. Light alkane dehydrogenation to light olefin technologies: A comprehensive review. *Rev. Chem. Eng.* **2015**, *31*, 413–436. [CrossRef]
3. Vu, B.K.; Song, M.B.; Ahn, I.Y.; Suh, Y.W.; Suh, D.J.; Kim, W.I.; Koh, H.L.; Choi, Y.G.; Shin, E.W. Pt–Sn alloy phases and coke mobility over Pt–Sn/Al_2O_3 and Pt–Sn/$ZnAl_2O_4$ catalysts for propane dehydrogenation. *Appl. Catal. A Gen.* **2011**, *400*, 25–33. [CrossRef]
4. Seo, H.; Kwon, J.; Gi, U.; Park, G.; Yoo, Y.; Lee, J.; Chang, H.; Kyu, I. Direct dehydrogenation of n-butane over Pt/Sn/M/γ-Al_2O_3 catalysts: Effect of third metal (M) addition. *CATCOM* **2014**, *47*, 22–27. [CrossRef]
5. Iglesias-Juez, A.; Beale, A.M.; Maaijen, K.; Weng, T.C.; Glatzel, P.; Weckhuysen, B.M. A combined in situ time-resolved UV-Vis, Raman and high-energy resolution X-ray absorption spectroscopy study on the deactivation behavior of Pt and PtSn propane dehydrogenation catalysts under industrial reaction conditions. *J. Catal.* **2010**, *276*, 268–279. [CrossRef]
6. Lee, H.; Kim, W.I.; Jung, K.D.; Koh, H. Effect of Cu promoter and alumina phases on Pt/Al_2O_3 for propane dehydrogenation. *Korean J. Chem. Eng.* **2017**, *34*, 1337–1345. [CrossRef]
7. Shan, Y.; Sui, Z.; Zhu, Y.; Chen, D.; Zhou, X. Effect of steam addition on the structure and activity of Pt–Sn catalysts in propane dehydrogenation. *Chem. Eng. J.* **2015**, *278*, 240–248. [CrossRef]
8. Hauser, A.W.; Gomes, J.; Bajdich, M.; Head-Gordon, M.; Bell, A.T. Subnanometer-sized Pt/Sn alloy cluster catalysts for the dehydrogenation of linear alkanes. *Phys. Chem. Chem. Phys.* **2013**, *15*, 20727–20734. [CrossRef]
9. Zhu, H.; Anjum, D.H.; Wang, Q.; Abou-Hamad, E.; Emsley, L.; Dong, H.; Laveille, P.; Li, L.; Samal, A.K.; Basset, J. Sn surface-enriched Pt – Sn bimetallic nanoparticles as a selective and stable catalyst for propane dehydrogenation. *J. Catal.* **2014**, *320*, 52–62. [CrossRef]
10. Deng, L.; Shishido, T.; Teramura, K.; Tanaka, T. Effect of reduction method on the activity of Pt-Sn/SiO_2 for dehydrogenation of propane. *Catal. Today* **2014**, *232*, 33–39. [CrossRef]
11. Pham, H.N.; Anderson, A.E.; Johnson, R.L.; Schwartz, T.J.; O'Neill, B.J.; Duan, P.; Schmidt-Rohr, K.; Dumesic, J.A.; Datye, A.K. Carbon Overcoating of Supported Metal Catalysts for Improved Hydrothermal Stability. *ACS Catal.* **2015**, *5*, 4546–4555. [CrossRef]
12. Bhasin, M.M.; McCain, J.H.; Vora, B.V.; Imai, T.; Pujadó, P.R. Dehydrogenation and oxydehydrogenation of paraffins to olefins. *Appl. Catal. A Gen.* **2001**, *221*, 397–419. [CrossRef]
13. Larese, C.; Campos-Martin, J.M.; Calvino, J.J.; Blanco, G.; Fierro, J.L.G.; Kang, Z.C. Alumina- and Alumina–Zirconia-Supported PtSn Bimetallics: Microstructure and Performance for the n-Butane ODH Reaction. *J. Catal.* **2002**, *208*, 467–478. [CrossRef]
14. Zhang, Y.; Zhou, Y.; Yang, K.; Li, Y.; Wang, Y.; Xu, Y.; Wu, P. Effect of hydrothermal treatment on catalytic properties of PtSnNa/ZSM-5 catalyst for propane dehydrogenation. *Microporous Mesoporous Mater.* **2006**, *96*, 245–254. [CrossRef]
15. Huang, L.; Xu, B.; Yang, L.; Fan, Y. Propane dehydrogenation over the PtSn catalyst supported on alumina-modified SBA-15. *Catal. Commun.* **2008**, *9*, 2593–2597. [CrossRef]
16. Del Angel, G.; Bonilla, A.; Peña, Y.; Navarrete, J.; Fierro, J.L.G.; Acosta, D.R. Effect of lanthanum on the catalytic properties of PtSn/γ-Al_2O_3 bimetallic catalysts prepared by successive impregnation and controlled surface reaction. *J. Catal.* **2003**, *219*, 63–73. [CrossRef]
17. Zangeneh, F.T.; Mehrazma, S.; Sahebdelfar, S. The influence of solvent on the performance of Pt–Sn/θ-Al2O3 propane dehydrogenation catalyst prepared by co-impregnation method. *Fuel Process. Technol.* **2013**, *109*, 118–123. [CrossRef]

18. Vu, B.K.; Song, M.B.; Ahn, I.Y.; Suh, Y.W.; Suh, D.J.; Kim, W. Il; Koh, H.L.; Choi, Y.G.; Shin, E.W. Propane dehydrogenation over Pt–Sn/Rare-earth-doped Al_2O_3: Influence of La, Ce, or Y on the formation and stability of Pt–Sn alloys. *Catal. Today* **2011**, *164*, 214–220. [CrossRef]
19. Kikuchi, I.; Ohshima, M.; Kurokawa, H.; Miura, H. Effect of Sn Addition on *n*-Butane Dehydrogenation over Alumina-supported Pt Catalysts Prepared by Co-impregnation and Sol-gel Methods. *J. Jpn Pet. Inst.* **2012**, *55*, 206–213. [CrossRef]
20. Praserthdam, P.; Grisdanurak, N.; Yuangsawatdikul, W. Coke formation over Pt–Sn-K/Al_2O_3 in C3, C5-C8 alkane dehydrogenation. *Chem. Eng. J.* **2000**, *77*, 215–219. [CrossRef]
21. Deng, L.; Miura, H.; Shishido, T.; Hosokawa, S.; Teramura, K.; Tanaka, T. Dehydrogenation of propane over silica-supported platinum-tin catalysts prepared by direct reduction: Effects of tin/platinum ratio and reduction temperature. *ChemCatChem* **2014**, *6*, 2680–2691. [CrossRef]
22. Deng, L.; Miura, H.; Shishido, T.; Wang, Z.; Hosokawa, S.; Teramura, K.; Tanaka, T. Elucidating strong metal-support interactions in Pt–Sn/SiO_2 catalyst and its consequences for dehydrogenation of lower alkanes. *J. Catal.* **2018**, *365*, 277–291. [CrossRef]
23. Hee Kim, G.; Jung, K.-D.; Kim, W.-I.; Um, B.-H.; Shin, C.-H.; Oh, K.; Lim Koh, H. Effect of oxychlorination treatment on the regeneration of Pt–Sn/Al_2O_3 catalyst for propane dehydrogenation. *Res. Chem. Intermed.* **2015**, *42*. [CrossRef]
24. Deng, L.; Arakawa, T.; Ohkubo, T.; Miura, H.; Shishido, T.; Hosokawa, S.; Teramura, K.; Tanaka, T. Highly Active and Stable Pt–Sn/SBA-15 Catalyst Prepared by Direct Reduction for Ethylbenzene Dehydrogenation: Effects of Sn Addition. *Ind. Eng. Chem. Res.* **2017**, *56*, 7160–7172. [CrossRef]
25. Arteaga, G.J.; Anderson, J.A.; Becker, S.M.; Rochester, C.H. Influence of oxychlorination treatment on the surface and bulk properties of a Pt–Sn/Al_2O_3 catalyst. *J. Mol. Catal. A Chem.* **1999**, *145*, 183–201. [CrossRef]

© 2019 by the authors. Licensee MDPI, Basel, Switzerland. This article is an open access article distributed under the terms and conditions of the Creative Commons Attribution (CC BY) license (http://creativecommons.org/licenses/by/4.0/).

Article

Energy Efficient and Intermittently Variable Ammonia Synthesis over Mesoporous Carbon-Supported Cs-Ru Nanocatalysts

Masayasu Nishi *, Shih-Yuan Chen and Hideyuki Takagi

Energy Catalyst Technology Group, Research Institute of Energy Frontier (RIEF), Department of Energy and Environment, National Institute of Advanced Industrial Science and Technology (AIST), 16-1 Onogawa, Tsukuba, Ibaraki 305-8589, Japan; sy-chen@aist.go.jp (S.-Y.C.); hide-takagi@aist.go.jp (H.T.)
* Correspondence: m.nishi@aist.go.jp; Tel.: +81-29-861-8261

Received: 9 April 2019; Accepted: 26 April 2019; Published: 30 April 2019

Abstract: The Cs-promoted Ru nanocatalysts supported on mesoporous carbon materials (denoted as Cs-Ru/MPC) and microporous activated carbon materials (denoted as Cs-Ru/AC) were prepared for the sustainable synthesis of ammonia under mild reaction conditions (<500 °C, 1 MPa). Both Ru and Cs species were homogeneously impregnated into the mesostructures of three commercial available mesoporous carbon materials annealed at 1500, 1800 and 2100 °C (termed MPC-15, MPC-18 and MPC-21, respectively), resulting in a series of Cs-Ru/MPC catalysts with Ru loadings of 2.5–10 wt % and a fixed Cs loading of 33 wt %, corresponding to Cs/Ru molar ratios of 2.5–10. However, the Ru and Cs species are larger than the pore mouths of microporous activated carbon (shortly termed AC) and, as a consequence, were mostly aggregated on the outer surface of the Cs-Ru/AC catalysts. The Cs-Ru/MPC catalysts are superior to the Cs-Ru/AC catalyst in catalysing mild ammonia synthesis, especially for the 2.5Cs-10Ru/MPC-18 catalyst with a Ru loading of 10 wt % and a Cs/Ru ratio of 2.5, which exhibited the highest activity across a wide SV range. It also showed an excellent response and stability during cycling tests over a severe temperature jump in a short time, presumably due to the open mesoporous carbon framework and suitable surface concentrations of CsOH and metallic Ru species at the catalytically active sites. This 2.5Cs-10Ru/MPC-18 catalyst with high activity, fast responsibility and good stability has potential application in intermittently variable ammonia synthesis using CO_2-free hydrogen derived from electrolysis of water using renewable energy with fast variability.

Keywords: sustainable ammonia synthesis; ruthenium; caesium; porous carbons; renewable hydrogen

---

1. Introduction

The latest U.S. Energy Information Administration (EIA) report has projected that the world energy consumption will grow by 28% between 2015 and 2040 due to a continuous increase in human population and improvement in living standards [1]. Fossil fuels will contribute to more than 75% of the energy required through 2040; the inevitable consequence is that large amounts of greenhouse gases (GHGs), especially carbon dioxide (CO_2) and particulate matter, would be produced, causing global warming, air pollution and extreme climates. To create a sustainable society with a low carbon economy for future generations, the Paris agreement was adopted in 2015. Its major objective is to limit the increase of global temperature below 2 °C above the pre-industrial levels or if possible, to less than 1.5 °C [2]. Following up on the Paris agreement, Japan promised to reduce 26% of its CO_2 emissions by 2030 as compared to those in 2013 and further reduce them by 80% by 2050 as part of its long-term plan [3]. This ambitious goal has motivated scientific research and industrial development in the fields of new energy resources, such as carbon dioxide (CO_2)-free hydrogen and energy-effective processes to build a low-carbon society [4]. CO_2-free hydrogen can be synthesized by the electrolysis of water using renewable electricity, decomposition of methane combined with carbon capture and storage (CCS) techniques or other well-known methods [5,6]. However, hydrogen is difficult to store, transport and utilize due to its very low boiling point (−252.8 °C), high flammability and price (particularly when hydrogen is produced by renewable energy). These technical problems have hindered the extensive use of hydrogen, especially renewable hydrogen, as a primary energy source.

The incorporation of hydrogen in chemical compounds that can be easily stored, transported and utilized, the so-called hydrogen carriers, is an alternative method to preserve and utilize hydrogen with a high level of safety and security. Ammonia of high hydrogen content (17.6 wt %) and a relatively high boiling point (−33 °C at the standard condition) is a promising hydrogen carrier [7–10]. The ammonia industry, which had a production capacity of 140 million tons in 2018, is a mature industry with vast infrastructure for the production, storage, transportation and utilization, especially in agriculture and fine chemical chains [11]. However, ammonia is conventionally synthesized by the Haber-Bosch process using Fe_3O_4-K-Al_2O_3-based catalysts under severe reaction conditions (>450 °C and 20 MPa), which consumes around 1%–2% of the global energy and releases a massive amount of CO_2 (1.2 ton CO_2 per ton of NH_3) [12–14]. Recent studies mainly aim at sustainable ammonia synthesis, catalysed by novel nanostructured materials with enhanced energy efficiency and durability and a reduced carbon footprint under mild conditions using renewable hydrogen as a feedstock. Pioneering works by the Ozaki and Aika research groups and other renowned research groups demonstrated that activated carbon-supported Ru catalysts were superior to conventional Fe-based catalysts in catalysing ammonia synthesis, especially when mild reaction conditions were used [15–21]. An advanced ammonia synthesis reaction with enhanced energy efficient catalysed by a graphitized carbon-supported Ru-based catalyst was commercialized by Kellogg Brown Root (KBR) in the 1990s; this is the so-called Kellogg Advanced Ammonia Process (KAAP) [22]. The activity of Ru-based catalysts for mild ammonia synthesis can be further improved by the addition of promoters, such as Ce and Ba, or by using new supporting materials with tuneable electronic and structural properties such as the oxides of alkaline, rare-earth elements and transition metals, novel electrolytes and mesoporous carbons with open structures [23–32]. For example, Ru catalysts supported on Pr_2O_3 [23] and a composite material of $La_{0.5}Ce_{0.5}O_{1.75}$ [24] exhibited a higher activity for mild ammonia synthesis than conventional Ru-based catalysts. Novel materials of $12CaO·7Al_2O_3$ and Ba-doped $Ca(NH_2)_2$ could be used as supporting materials to fabricate next-generation Ru catalysts for mild ammonia synthesis [25,26]. The recent study further demonstrated that the composite materials of K/Ru/$TiO_{2-x}H_x$ were effective in sustainable ammonia synthesis using a solar thermal approach under atmospheric pressure [27]. On the other hand, Cs- and Ba-promoted Ru catalysts supported on microporous carbons with enhanced activity have also been reported for the mild synthesis of ammonia [28–31]. The microporous carbon-supported Cs-promoted Ru catalysts might be suitable for sustainable ammonia synthesis using a water-rich hydrogen feedstock derived from electrolysis of water powered by renewable energy whereas the deactivation was observed for the Ba-promoted counterparts [32]. However, homogeneous impregnation of nanosized Ru particles into microporous carbons with small pore mouths is a difficult task. The recent study found that the activated carbon-supported Ru catalyst was deactivated seriously during ammonia synthesis due to the sintering of surface Ru particles [33]. This deactivation process can presumably be avoided by replacing activated carbon with mesoporous carbon materials with an open-pore structure at the nanoscale; these mesopores are expected to firmly confine nanosized Ru particles. In addition, mesoporous carbon materials are superior to the analogues of silica and alumina in ammonia synthesis due to their higher electronic properties and structural stability under the reaction conditions [34–43]. In this study, a series of mesoporous carbon-supported Cs-Ru nanocatalysts with different pore structures and graphite crystallinities were prepared for energy efficient and intermittently variable ammonia synthesis under mild conditions (280–450 °C and <1 MPa); their performance was contrasted with that of a microporous carbon-supported Cs-Ru catalyst. We paid special attention to the influence of the structures of mesoporous carbon supports and the catalytically active sites on the performance of the prepared Cs-Ru catalysts with various Ru loadings and Cs/Ru ratios for mild ammonia synthesis. In addition, ammonia synthesis with rapid changes in the reaction conditions, such as reaction temperature and a wide space velocity (SV) range, was analysed to evaluate the catalytic performance of the prepared Cs-Ru catalysts. Such study is crucial for analysing their potential in intermittently variable ammonia synthesis, where CO_2-free hydrogen derived from water electrolysis powered by renewable energy can be used.

2. Results and Discussion

2.1. Characterizations

Figures 1 and 2 show the X-ray diffraction (XRD) patterns and N_2 adsorption-desorption isotherms of the carbon supports and prepared Cs-Ru catalysts, respectively. In the diffraction patterns of the three MPC supports, several peaks were observed at 25.8°, 26.4°, 42.5°, 53.5° and 77.6°, corresponding to graphite (PDF card number: 9008568). With an increase in annealing temperature, all the diffraction peaks increased to higher values and that of the (002) plane split into two peaks at 25.8° and 26.4°, owing to the growth and crystallization of the mesoporous carbon framework through the heterogeneous graphitization of turbostratic and graphitic structures [44]. In contrast, the AC support exhibits no diffraction peaks in the wide-angle region, suggesting that the carbon framework is amorphous in nature. The N_2 physisorption data of the three MPC supports exhibit a classical type IV isotherm with a wide hysteresis loop in a relatively large P/P_0 range of 0.5–0.9, corresponding to large mesoporous structures. Meanwhile, the AC support exhibits a classical type I isotherm with no hysteresis loop at relatively high P/P_0 regions, indicating the presence of microporous structures with pores smaller than 2 nm.

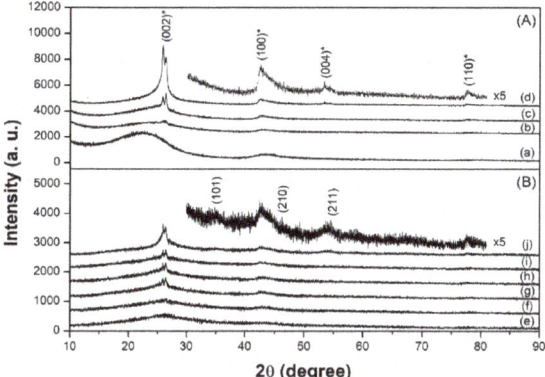

Figure 1. Wide-angle X-ray diffraction (XRD) patterns of (**A**) carbon materials—(a) AC, (b) MPC-15, (c) MPC-18 and (d) MPC-21 and (**B**) the prepared Cs-Ru catalysts—(e) 2.5Cs-10Ru/AC, (f) 2.5Cs-10Ru/MPC-15, (g) 10Cs-2.5Ru/MPC-18, (h) 5Cs-5Ru/MPC-18, (i) 2.5Cs-10Ru/MPC-18 and (j) 2.5Cs-10Ru/MPC-21. The "asterisk" peaks are associated with the carbon materials. The peaks of samples (d) and (j) are enlarged and inserted in Figure 1A,B, respectively.

In the case of the prepared Cs-Ru catalysts, the amount of nitrogen uptake decreased, and the diffraction peaks associated with the graphite structure weakened, indicating that the mesoporous carbon framework is slightly influenced by the thermal treatment used for impregnation. A series of diffraction peaks gradually appear with an increase in the Ru loading, which is associated with the formation of crystalline RuO_2 species (PDF card number: 1,000,058). However, no diffraction signals corresponding to the Cs species can be found in the wide-angle XRD patterns. The size of the RuO_2 crystallites is too small to be estimated from their diffraction peaks using Scherrer's equation.

This suggests that the Ru and Cs precursors thermally decomposed into nanosized RuO_2 particles with low crystallinity (<ca. 5 nm) and amorphous Cs species, respectively, on the carbon materials.

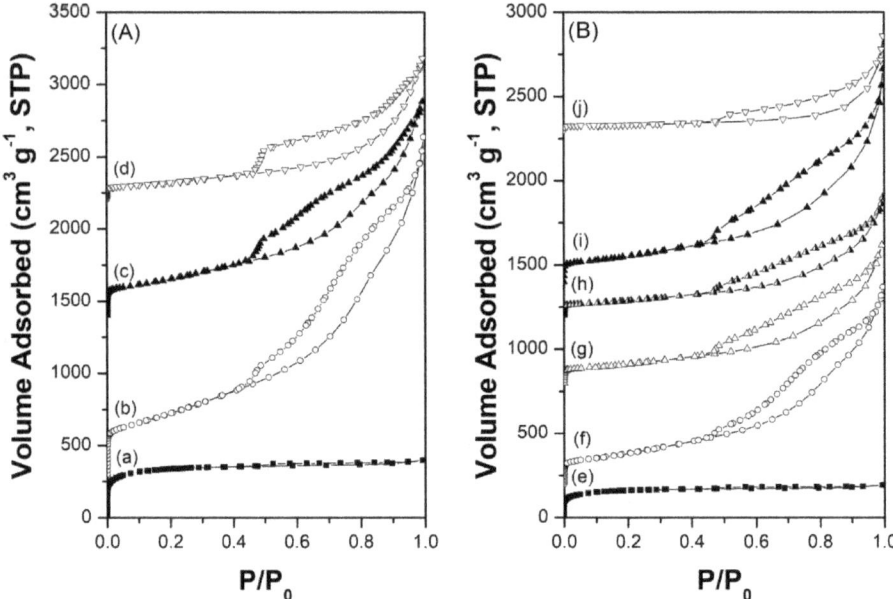

Figure 2. N_2 adsorption-desorption isotherms of (**A**) carbon materials—(a) AC, (b) MPC-15, (c) MPC-18 and (d) MPC-21 and (**B**) the prepared Cs-Ru catalysts—(e) 2.5Cs-10Ru/AC, (f) 2.5Cs-10Ru/MPC-15, (g) 10Cs-2.5Ru/MPC-18, (h) 5Cs-5Ru/MPC-18, (i) 2.5Cs-10Ru/MPC-18 and (j) 2.5Cs-10Ru/MPC-21.

High-resolution transmission electron microscope (HRTEM) and high-angle annular dark-field scanning transmission electron microscopy-scanning transmission electron microscopy (HAADF-STEM) analyses were performed to understand the microstructure of the prepared Cs-Ru catalysts and the sizes and size distributions of Ru and Cs species; the results are compared with those of carbon materials. The MPC supports contain many graphite grains which aggregate to form mesoporous carbon framework with open-pore structure, whereas the AC support contained large aggregates with an amorphous structure and no visible mesopores (Figure S1, electronic supplementary information (ESI)). In the case of the prepared Cs-Ru/MPC catalysts, the HRTEM images in Figure 3b–d show that the dark spots of nanosized RuO_2 particles—ca. 2–3 nm in size—are uniformly impregnated into the mesoporous carbon framework with open-pore structure and the influence of the impregnation is hardly found. It should be noted that the dark spots observed in the HRTEM images are mostly Ru particles as supported by the HAADF images (the bright spots in Figures S2–S4, ESI). The size and size distribution of nanosized RuO_2 particles increased slightly in the 2.5Cs-10Ru/MPC-21 catalyst, probably due to the decrease in surface area and porosity at high annealing temperatures. The HAADF-STEM images further show that the Cs species are also impregnated into the mesoporous carbon framework and they are presumably accumulated near to nanosized Ru particles (Figures S2–S4, ESI). Microporous carbon-supported Cs-Ru catalysts also contain nanosized RuO_2 particles with a narrow size distribution; these particles are surrounded by the Cs species (Figure 3a and Figure S5 (ESI)). However, the Ru particles on the 2.5Cs-10Ru/AC catalyst are close to each other, suggesting that they are presumably impregnated on the outer surface of the AC support.

Table 1. Structural properties of carbon materials and the as-prepared Cs-Ru catalysts.

Samples	Cs/Ru Molar Ratio		Ru Loading (wt %)		S_{BET} ($m^2\,g^{-1}$)	V_{Total} ($cm^3\,g^{-1}$)	$V_{Micro.}$ ($cm^3\,g^{-1}$) [3]	$V_{Meso.}$ ($cm^3\,g^{-1}$) [4]	Pore Size (nm)
	(Expected) [1]	(Solid) [2]	(Expected) [1]	(Solid) [2]					
AC	-	-	-	-	1260	0.62	0.51	0.11	1.6
MPC-15	-	-	-	-	1180	2.94	0.55	2.39	5.8
MPC-18	-	-	-	-	930	2.29	0.37	1.92	5.1
MPC-21	-	-	-	-	270	1.28	0.11	1.17	6.8
2.5Cs-10Ru/AC	2.5	1.3	10	13.4	580	0.30	0.24	0.06	1.6
2.5Cs-10Ru/MPC-15	2.5	1.1	10	13.7	680	1.81	0.27	1.54	5.8
2.5Cs-10Ru/MPC-18	2.5	1.1	10	13.8	430	1.33	0.17	1.16	5.8
5Cs-5Ru/MPC-18	5.0	2.1	5.0	8.1	440	1.40	0.18	1.22	5.8
10Cs-2.5Ru/MPC-18	10	3.7	2.5	4.8	500	1.52	0.20	1.32	5.8
2.5Cs/MPC-18	-	.5	-	-	580	1.71	0.23	1.48	5.8
10Ru/MPC-18	-	-	10	11.3	800	1.84	0.32	1.52	5.1
2.5Cs-10Ru/MPC-21	2.5	1.1	10	13.4	110	0.58	0.04	0.54	7.3

[1] Theoretical Cs/Ru molar ratios and Ru loadings, which were based on the carbon content. [2] The Cs and Ru loadings in solids were measured by EA and XRF methods. [3] Microporous pore volume (V_{Micro}) was calculated using the Dubinin-Astakhov (DA) plot and the α_s-plot method. [4] Mesoporous pore volume (V_{Meso}) was calculated as $V_{Total} - V_{Micro}$. [5] The Cs loading was 32.4 wt % according to the EA and XRF method.

The structural properties of the carbon materials and the corresponding Cs-Ru catalysts are listed in Table 1. The AC support has a high surface area (S_{BET} = 1260 m^2 g^{-1}) and a moderate pore volume (V_{Total} = 0.62 cm^3 g^{-1}), which is mostly related to its microporous structure. In contrast, the MPC supports with mesopores 5–7 nm in size have surface areas of 270–1180 m^2 g^{-1} and pore volumes of 1.28–2.94 cm^3 g^{-1}, respectively; such reduced values can be attributed to the growth of graphite structures as the annealing temperature increased. In the prepared Cs-Ru/MPC catalysts, the overall surface area and pore volume decreased upon impregnation; further, the size of the mesopores also varied. In contrast, the pore sizes of AC and 2.5Cs-10Ru/AC are nearly unchanged, which suggests that the Cs and Ru species are mainly impregnated on the outer surface of the AC support. The elemental analysis of carbon, hydrogen and nitrogen (CHN) and the X-ray Fluorescence (XRF) data show that Cs and Ru are indeed impregnated on the MPC and AC supports; however, the Ru loadings and Cs/Ru molar ratios in the fresh samples are slightly different from those of the recipes. It indicates that the compositions and chemical environments of the prepared Cs-Ru catalysts are presumably affected by their nature upon impregnation; however, they are hardly estimated by conventional techniques of CHN elemental and XRF analyses, which are similar to the literature reports [45,46].

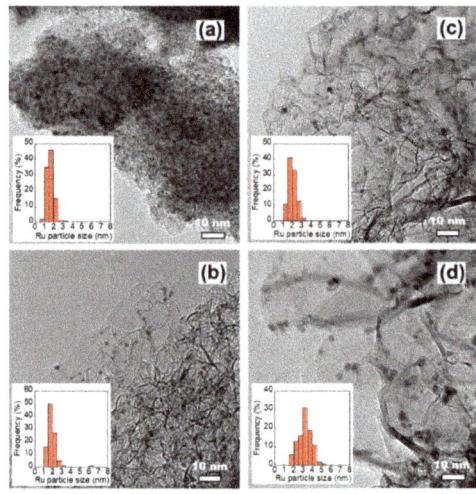

Figure 3. HRTEM images and Ru particle size distributions of freshly prepared catalysts. (**a**) 2.5Cs-10Ru/AC, (**b**) 2.5Cs-10Ru/MPC-15, (**c**) 2.5Cs-10Ru/MPC-18 and (**d**) 2.5Cs-10Ru/MPC-21.

By analysing more than 100 particles in the HRTEM images, the RuO$_2$ size and size distribution in 2.5Cs-10Ru/AC, 2.5Cs-10Ru/MPC-15, 2.5Cs-10Ru/MPC-18 and 2.5Cs-10Ru/MPC-21 were found to be (1.6 ± 0.4) nm, (1.9 ± 0.6) nm, (2.1 ± 0.4) nm and (3.2 ± 0.8) nm, respectively; they were found to be inversely proportional to the surface area of the carbon support and Ru loading (Table 2). The HRTEM analysis is consistent with the XRD study, which previously indicated that the RuO$_2$ sizes of the prepared Cs-Ru catalysts should be smaller than 5 nm. The prepared Cs-Ru catalysts with reduction pre-treatment at 450 °C were further studied using the CO chemisorption and CO$_2$-TPD methods. Noted that the reduction pre-treatment was carried out by the same procedure as described in the mild ammonia synthesis. The sizes of metallic Ru nanoparticles calculated by the CO chemisorption method are similar to those by the HRTEM analysis, except the 2.5Cs-10Ru/AC catalyst. The uptakes of CO$_2$ over the prepared Cs-Ru/MPC-18 catalysts are around 2.4–2.7 mmol g^{-1}, which values are smaller than that of the 2.5Cs/MPC-18 catalyst and larger than that of 2.5Cs-10Ru/AC catalyst. In combination with CO chemisorption, CO$_2$-TPD analysis and several characterizations as aforementioned, it can be suggested that nanosized RuO$_2$ particles partially laid on the Cs species can be homogeneously impregnated into the mesoporous carbon framework of MPC supports and they can be reduced to corresponding

Ru metals with similar sizes after the reduction pre-treatment. In contrast, the sintering of Ru and Cs species is presumably occurred for those supported on the AC support with microporous carbon framework, suggesting that the Ru and Cs species in the 2.5Cs-10Ru/AC catalyst with microporous carbon framework are relatively unstable.

2.2. Temperature-Programmed Studies

The TPR technique was employed to analyse the compositions and chemical environments of the prepared Cs-Ru catalysts and their nature (i.e. reducibility, activation, stability, etc.) in the reduction atmosphere at ambient pressure. The TPR profiles recorded by TCD are shown in Figure 4A and those recorded by MS are shown in Figure 4B–F) and Figures S10 and S11 (ESI). The TPR profiles of ruthenium oxide (RuO_2), MPC-18, 10Ru/MPC-18 and 2.5Cs/MPC-18 as the reference materials were also measured using the same procedures (Figures S6–S9, ESI). No signals could be observed in the MPC-18 sample, suggesting that the unimpregnated mesoporous carbon framework does not react in the reduced environment [47]. A one-step reduction of RuO_2 to metallic Ru (Equation (1)) on the RuO_2 standard and the 10Ru/MPC-18 sample was observed at temperature below 200 °C [19].

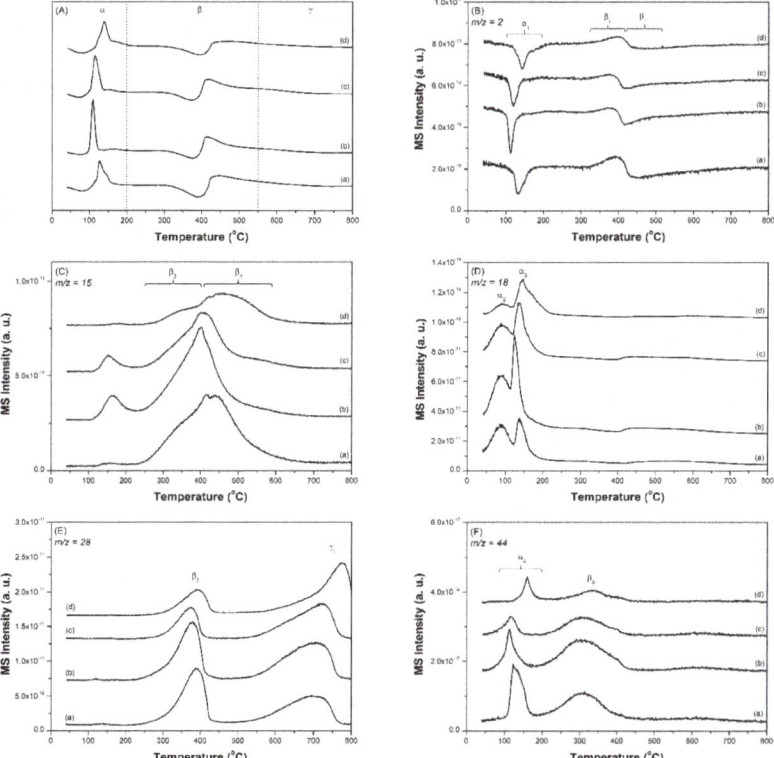

Figure 4. (**A**) Temperature-programmed reduction equipped with a thermal conductivity detector (TPR-TCD) and (**B–F**) temperature-programmed reduction equipped with a mass spectrometer (TPR-MS) profiles of freshly prepared catalysts. (a) 2.5Cs-10Ru/AC, (b) 2.5Cs-10Ru/MPC-15, (c) 2.5Cs-10Ru/MPC-18 and (d) 2.5Cs-10Ru/MPC-21.

This result once again suggests that RuO_2 is formed over the 10Ru/MPC-18 sample by the thermal decomposition of the Ru precursor in an N_2 atmosphere, which is consistent with the XRD and HRTEM studies. It should also be noted that the reduction of RuO_2 over the 10Ru/MPC-18 sample was observed at relatively low temperatures (ca. 100 °C), suggesting that the nanosized RuO_2 species can be easily dispersed on the MPC-18 support and its size is smaller than that of the bulk RuO_2 standard. On the other hand, small amounts of CH_4 and H_2O are gradually formed over the 10Ru/MPC-18 sample in the temperature range of 200–700 °C; this is accompanied by a continuous consumption of small amounts of hydrogen. This observation might be attributed to the methanation of surface oxygenated groups, such as carboxylic acid (-COOH) or carbonyl (-CO) catalysed by the metallic Ru species in the reduction atmosphere to form the clean surface of carbon materials (Equation (2)) [19]. Noted that those superficial oxygenated compounds are presumably formed by surface reaction of nitrate ions and carbon species during the impregnation of $Ru(NO)(NO_3)_3$. These gas molecules were only observed by reduction of freshly prepared Cs-Ru catalysts in our study and they were undetectable in the subsequent ammonia synthesis, which will be discussed hereafter. In the industry, the Ru-based catalysts are usually activated at 400–500 °C, which is close to the temperature region of the TPR study as mentioned previously. It speculates that the activation of the Ru-based catalysts used in the ammonia synthesis is not only to form metallic Ru species as the catalytically active sites but also to make metallic Ru species contact with graphite structure of clean surface more. When the temperature is higher than 600 °C, thermal decomposition of the mesoporous carbon framework occurs over 10Ru/MPC-18 and consequently a large amount of CO is formed (Equation (3)) [48]. Similar phenomena are observed for other Cs- and Ru-containing samples, suggesting that the stability of carbon supports in the presences of Cs and Ru species is up to ca. 600 °C under a reduced atmosphere.

$$RuO_2(s) + 2H_2(g) \rightarrow Ru(s) + 2H_2O(g) \tag{1}$$

$$C(s)\text{-COOH} + 4H_2(g) \rightarrow C(s)\text{-H} + CH_4(s) + 2H_2O(g) \tag{2}$$

$$C(s)\text{-C=O} + 0.5H_2(g) \rightarrow C(s)\text{-H} + CO(g) \tag{3}$$

In the 2.5Cs/MPC-18 sample, desorption of physically adsorbed H_2O and CO_2 was observed at ca. 100 and 160 °C, respectively. The decomposition of $Cs_2(CO_3)$ in the presence of H_2 to form CsOH and CO_2 (Equation (4)) occurred at 350–550 °C [49]. When the temperature was higher than ca. 420 °C, CO and H_2O were formed accompanied by a continuous decrease in hydrogen; however, CH_4 was not present. This result might be attributed to the decomposition of surface carboxylic groups on the mesoporous carbon framework. The reaction is catalysed by CsOH species in the reduced atmosphere (Equation (5)). It implies that the Cs species might take part in making clean surface of carbon supports, which are able to contact firmly with metallic Ru species after the activation process as mentioned above. Similarly, CO is formed due to the thermal decomposition of mesoporous carbon framework at higher temperatures (Equation (3)).

$$Cs_2(CO_3)(s) + H_2(g) \rightarrow 2CsOH(s) + CO_2(s) \tag{4}$$

$$C(s)\text{-COOH} + H_2(g) \rightarrow C(s)\text{-H} + CO(g) + H_2O(g) \tag{5}$$

The TPR-TCD and TPR-MS profiles of the prepared Cs-Ru catalysts can be divided into three parts—α, β and γ regions in the temperature ranges of 50–200 °C, 200–550 °C and 550–800 °C, respectively. The positive TCD signal centred at ca. 120–130 °C is due to the reduction of RuO_2 to metallic Ru species (Equation (1)); this observation is also supported by the MS signals (α_1 in Figure 4B and α_3 in Figure 4D corresponding to $RuO_2(s) + 2H_2(g) \rightarrow Ru(s) + 2H_2O(g)$ (Equation (1), also see Table S1, ESI)). Compared with the 10Ru/MPC-18 sample, the reduction temperatures of RuO_2 over the prepared Cs-Ru catalysts shift to higher values, suggesting that the Ru and Cs species co-existed in the mesoporous carbon framework are strongly interacted. Note that the baselines of the TPR-TCD profiles in the α region are slightly different. The MS signals marked as α_2 (ca. 90 °C) in Figure 4D and α_4 (120–160 °C) in Figure 4F indicate that the variation in baselines is due to the desorption of physically adsorbed H_2O and CO_2 from the prepared Cs-Ru catalysts, respectively (also see Table S1, ESI). On the other hand, CH_4 is visible in the case of the 2.5Cs-10Ru/MPC-15 and 2.5Cs-10Ru/MPC-18 catalysts with relatively high surface area and porosity, presumably due to the decomposition of ethanol residues trapped in the mesopores.

In the β region, S-shaped curves (Figure 4A,B), associated with a balance between several surface reactions involving H_2 production ($β_1$ in Figure 4B) and consumption ($β_2$ in Figure 4B), can be observed. When the temperature is lower than ca. 420 °C, corresponding to a MS signal of H_2 production in the $β_1$ region, CH_4 (marked as $β_3$ in Figure 4C), CO (marked as $β_5$ in Figure 4E) and CO_2 (marked as $β_6$ in Figure 4F) are formed; at the same time, a small amount of H_2O can be observed (Figure S11, ESI). The increase in H_2 concentration in the downstream is presumably due to the desorption of H_2 molecules previously adsorbed on metallic Ru particles during the TPR process (Equation (6)) [50]. A large amount of CH_4 is formed over the prepared Cs-Ru catalysts, indicating that the dissociation of H_2 and subsequent methanation of surface oxygenated groups can be facilitated in the presences of Cs and Ru species (Equation (2)). Similarly, CO is largely formed due to the decomposition of surface carboxylic groups before methanation (Equation (5)). It is to be noted that the amounts of CH_4 and CO decreased upon increasing the annealing temperature whereas their signals shift to higher temperature regions, particularly for the 2.5Cs-10Ru/MPC-21 catalyst. It is a fact that the amount of surface oxygenated groups on the MPC-21 support is low due to its high annealing temperature. On the other hand, CO_2, which is a co-product of the decomposition reaction of $Cs_2(CO_3)$ to form CsOH in the presence of H_2 (Equation (4)), gradually moved towards lower temperature regions for the 2.5Cs-10Ru/MPC-18 and 2.5Cs-10Ru/MPC-21 catalysts. It is presumable that this reaction was facilitated by the Cs and Ru species impregnated on mesoporous carbons with a highly crystalline graphite structure. These observations further assume that the activation of the prepared Cs-Ru catalysts in ammonia synthesis is associated with adsorption of H_2 molecules on the metallic Ru species, spillover to the interfaces of Ru, Cs and C species and consequently to form active phases of metallic Ru species and CsOH species close to each other on the clean surface of graphite structure. We are currently conducting more surface characterization research using the diffuse reflectance infrared Fourier transform (DRIFT) and extended X-ray absorption fine structure (EXAFS) techniques to prove this assumption and the results will be discussed in our future reports. On the other hand, CH_4 (marked as $β_4$ in Figure 4C), corresponding to the $β_2$ region in Figure 4B gradually moved to higher temperature regions upon increasing the annealing temperature, especially in the case of the 2.5Cs-10Ru/MPC-21 catalyst. This is another indication that the surface oxygenated groups on MPC-21 hardly converted to CH_4 due to the high annealing temperatures it is subjected to.

$$Ru\text{-}* + H_2 \leftrightarrows 2Ru\text{-}H \qquad (6)$$

In the γ region, an intense MS signal corresponding to CO could only be observed up to ca. 800 °C (Figure 4E), which temperature shifts to higher regions by increasing the annealing temperatures of MPC samples. This is another indication that the mesoporous carbon framework with its higher annealing temperature contain relatively low amounts of surface oxygenated groups and thus it is highly stable when subjected to thermal treatment in the reduced atmosphere [48].

Ammonia synthesis over the prepared Cs-Ru catalysts was further examined by the temperature-programmed method at ambient pressure. A mixed gas of N_2 and H_2 (H_2/N_2 ratio = 3, flow rate = 30 mL min^{-1}) was used as a feedstock. The ramp rate was kept at 5 °C min^{-1} up to 800 °C. Prior to the temperature-programmed measurement, freshly prepared samples (around 50 mg) were reduced by a H_2 flow (50 mL min^{-1}) at 450 °C, followed by cooling to 100 °C under an atmosphere of Ar (50 mL min^{-1}). The results are shown in Figure 5, in comparison to those of the 10Ru/MPC-18 and 2.5Cs/MPC-18 catalysts. For the Cs-Ru/MPC catalysts, ammonia could be synthesized in the temperature range of 300–500 °C with maxima MS signals at around 380–390 °C and methane was formed at the higher temperature region (>540 °C). The other gases, such as H_2O, CO and CO_2, were not detectable under N_2 and H_2 atmosphere using in the temperature-programmed measurement. In contrast, ammonia synthesis over the 10Ru/MPC-18 catalyst was observed at high temperature region (>450 °C) whereas no ammonia was detectable for the 2.5Cs/MPC-18 catalyst. The results speculate that nanosized Ru metals impregnated in the mesoporous carbon materials are active in ammonia synthesis and the addition of Cs as the promoter was necessary to carry out mild ammonia synthesis. The 2.5Cs-10Ru/AC catalyst could only catalyse ammonia synthesis at high temperature region (350–550 °C) and the signal of ammonia was significantly weakened, indicating that the 2.5Cs-10Ru/AC catalyst with microporous carbon framework was inefficient in mild ammonia synthesis. Besides, methane over the 2.5Cs-10Ru/AC catalyst was formed at relatively low temperature region (<390 °C), suggesting that ammonia synthesis and methane formation over the 2.5Cs-10Ru/AC catalyst compete each other.

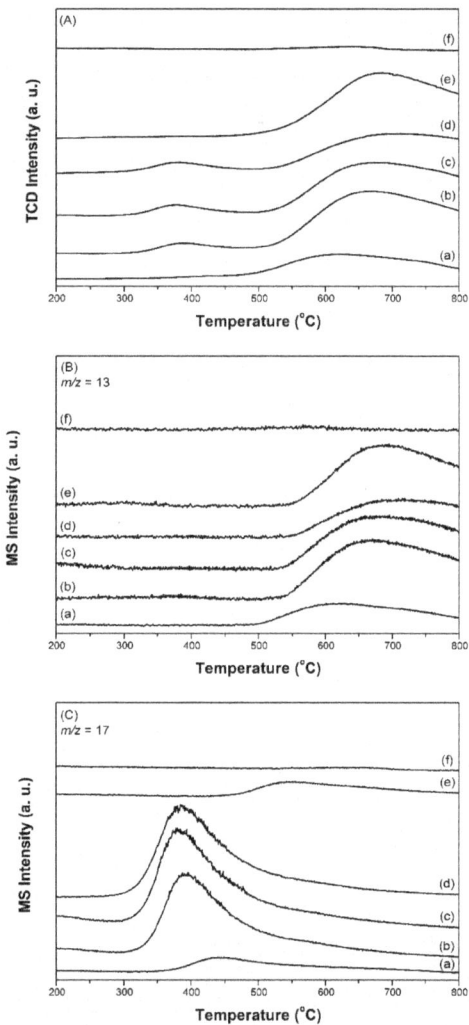

Figure 5. (**A**) TPR-TCD profile, (**B**) TPR-MS profile recorded by a m/z ratio of 13, and (**C**) TPR-MS profile recorded by a m/z ratio of 17 for the ammonia synthesis over freshly prepared catalysts. (a) 2.5Cs-10Ru/AC, (b) 2.5Cs-10Ru/MPC-15, (c) 2.5Cs-10Ru/MPC-18, (d) 2.5Cs-10Ru/MPC-21, (e) 10Ru/MPC-18 and (f) 2.5Cs/MPC-18.

2.3. Mild Ammonia Synthesis

Ammonia synthesis on the prepared 2.5Cs-10Ru catalysts with different porosities and graphite-structure crystallinities was carried out in a stainless-steel fixed-bed reactor under mild conditions (280–450 °C and <1 MPa) at an SV value of 9000 h^{-1}. The Ru loading and Cs/Ru molar ratio were kept at 10 wt % and 2.5, respectively. The downstream flow was analysed using an online GC-TCD instrument after a specific reaction time, where ammonia as a product was detected in addition to N_2 and H_2 molecules. Noted that the prepared Cs-Ru catalysts were reduced at 450 °C for 2 h at an SV value of 10,000 h^{-1} using a pure H_2 flow prior to the ammonia synthesis. Similar to the TPR study, several gas molecules (H_2O, CO, CO_2, CH_4) were present in the downstream of the catalyst bed during this activation process but they were undetectable during the ammonia synthesis. It suggests once again that the activation process is to produce metallic Ru and CsOH species as the catalytically active sites on the clean surface of carbon supports, which should be stable under the reaction conditions as aforementioned, through several surface reactions as discussed in the TPR study. Figure 6 shows that the rate of ammonia synthesis over the Cs-Ru catalysts was influenced by the reaction temperature and the type of carbon material used as the supporting material. The rates of ammonia synthesis over 2.5Cs-10Ru/AC, 2.5Cs-10Ru/MPC-15, 2.5Cs-10Ru/MPC-18 and 2.5Cs-10Ru/MPC-21 reached their maximum values of 2.2 mmol g^{-1} h^{-1} at 400 °C, 8.1 mmol g^{-1} h^{-1} at 370 °C, 10.2 mmol g^{-1} h^{-1} at 360 °C and 7.3 mmol g^{-1} h^{-1} at 360 °C, respectively; however, the rates decreased beyond these temperatures as the reverse reaction of ammonia decomposition can occur rapidly [51]. The three mesoporous 2.5Cs-10Ru/MPC catalysts yielded higher rates for ammonia synthesis at lower reaction temperatures, in comparison to the microporous 2.5Cs-10Ru/AC catalyst. The HRTEM images show that the Ru particle sizes and size distributions in the used 2.5Cs-10Ru/MPC catalysts are akin to those of fresh samples, whereas the 2.5Cs-10Ru/AC catalyst is unstable and large Ru crystallites (>10 nm) can be seen (Figure S12, ESI), which is consistent with the study of CO chemisorption. Previous studies demonstrated that Ru clusters, 1.8–3.5 nm in diameter, are rich in surface steps or B5 sites, which are defined as highly active structures for ammonia synthesis [52]. Moreover, the promoter of Cs preferable in the form of the CsOH species, is presumably present at the vicinity of the Ru surface, at which the N_2 dissociation as the rate determining step of ammonia synthesis can be facilitated [53]. The present study further demonstrates that mesoporous carbon materials are suitable supporting materials for the homogeneous dispersion of nanosized Cs and Ru species, which give strong synergetic properties and stability in ammonia synthesis and the molecular diffusion through these open mesoporous structures can be facilitated. In contrast, sintering of Ru particles over the 2.5Cs-10Ru/AC catalyst is observed (Figures S5 and S12), indicating that the Cs and Ru species are too big to be impregnated inside the microporous framework and their synergetic effect is suppressed due to serious deactivation by aggregation. This result is supported by the XRD pattern, which shows that large Ru^0 particles (ca. 20 nm) were formed in the used 2.5Cs-10Ru/AC catalyst (Figure S13).

The influence of Ru loading (2.5–10 wt %) on the activity of the prepared Cs-Ru/MPC-18 catalysts with a Cs loading of 33 wt % and Cs/Ru molar ratio in the range of 2.5–10 during ammonia synthesis was studied. The reaction conditions were 280–450 °C and 0.99 MPa at 9000 h^{-1}. Figure 7 shows that the rate of ammonia synthesis over Cs-Ru/MPC-18 catalyst is negatively related to the Ru loading whereas the reverse is true for the corresponding reaction temperature. Once again, the 2.5Cs-10Ru/MPC-18 catalyst with a Ru particle size of 2.4 nm calculated by the CO chemisorption method results in the highest activity. The other two catalysts, 5Cs-5Ru/MPC-18 and 10Cs-2.5Ru/MPC-18, with their smaller Ru particles result in relatively low rates of ammonia synthesis at relatively high temperatures, probably due to a decrease in the number of B5 sites when the Ru size is lower than 2 nm [54].

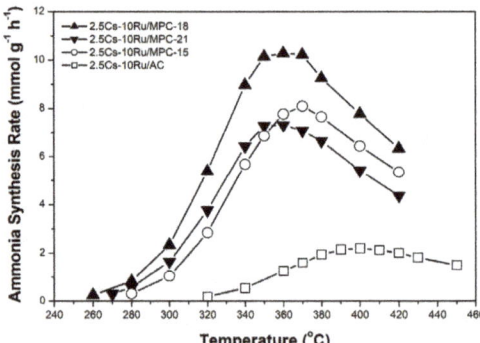

Figure 6. Rate of ammonia synthesis as a function of reaction temperature over the prepared Cs-Ru catalysts at an SV value of 9000 h^{-1}.

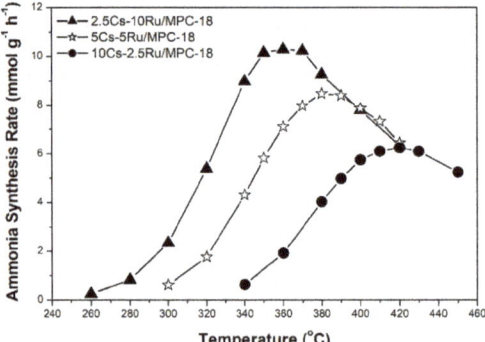

Figure 7. Rate of ammonia synthesis as a function of Ru loading over Cs-Ru/MPC-18 catalysts.

The correlation between the Ru size, the rate of ammonia synthesis and the TOF value as a function of surface Ru concentration over the prepared Cs-Ru/MPC catalysts is further discussed. The Ru sizes in the metallic state were determined by the CO chemisorption method. Figure 8a shows that the Ru sizes of the Cs-Ru/MPC catalysts are increased almost linearly from 1.3 nm to 2.5 nm by increasing the surface Ru concentration to 3.2 µmol m^{-2} and slightly increased to 3.7 nm at a high Ru concentration of 12 µmol m^{-2}. Nano-sized Ru particles in the 1–4 nm region can be easily impregnated on the MPC supports in a wide range of surface Ru concentration and its sizes are highly associated with the Ru loading and the structural property of mesoporous carbon framework. However, the surface Cs concentration has no significant influence on the Ru sizes of prepared catalysts, particularly for 10Ru/MPC-18 and 2.5Cs-10Ru/MPC-18. The correlation between the rate of ammonia synthesis and the surface Ru concentration forms a volcano-shape curve for the prepared Cs-Ru/MPC catalysts (Figure 8b). The 2.5Cs-10Ru/MPC-18 catalyst with a surface Ru concentration of 3.2 µmol m^{-2} and a Ru size of 2.4 nm gives the highest rate of ammonia synthesis, similar to the discussion aforementioned. The rate of ammonia synthesis over the 10Ru/MPC-18 catalyst with a surface Ru concentration of 1.4 µmol m^{-2} and a Ru size of 2.1 nm is significantly reduced by ca. 80% and the 2.5Cs/MPC-18 catalyst is inactive in ammonia synthesis (Figure S14, ESI). It is another evidence that the co-existing Ru and CsOH impregnated on the mesoporous carbon framework give the synergetic effect in ammonia synthesis, which can be maximized by optimizing the structural parameters of surface Ru concentration (~3.2 µmol m^{-2}), Ru size (2.4 nm) and surface Cs/Ru ratio (~1). Regarding to the other Cs-Ru/MPC catalysts, the rates of ammonia synthesis are reduced, presumably due to an improper combination of Ru size, surface Ru concentration and surface Cs/Ru ratio. Nevertheless, the TOF values over the prepared Cs-Ru/MPC catalysts remain nearly unchanged (Figure 8c), presuming that the B5 sites of nanosized Ru particles impregnated on the CsOH-containing mesoporous carbon frameworks are fully accessible. In contrast, the 2.5Cs-10Ru/AC catalyst gives a low rate of ammonia synthesis and a small TOF value because the B5 sites were lost by the sintering of nanosized Ru particles on the outer surface of CsOH-containing AC support and the molecular diffusion through the microporous carbon framework is hindered.

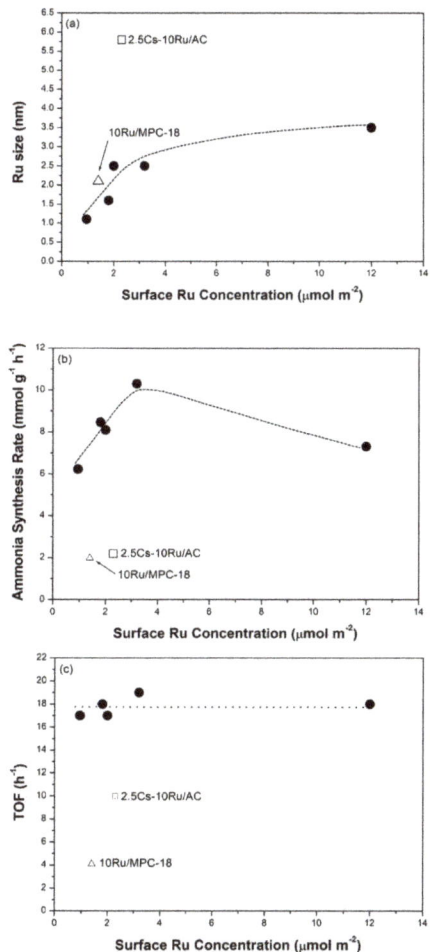

Figure 8. (a) The Ru size; (b) ammonia synthesis rate; and (c) TOF as a function of surface Ru concentration over the prepared Cs-Ru/MPC catalysts, in comparison to those of 10Ru/MPC-18 and 2.5Cs-10Ru/AC catalysts.

For sustainable ammonia synthesis using CO_2-free hydrogen as a feedstock, Ru-based catalysts must be subjected to a short warm-up and shut-down period to cooperate with the variable production rates of renewable hydrogen from electrolysis of water through intermittently available electricity, such as wind and solar powers. The potential of the 2.5Cs-10Ru/MPC-18 catalyst for sustainable ammonia synthesis was examined across a wide SV range and its performance was compared with that of the 2.5Cs-10Ru/AC catalyst. Figure 9a shows that the rate of ammonia synthesis over the 2.5Cs-10Ru/MCP-18 catalyst was high and stable across an SV range of 3000–20,000 h^{-1} and reaction temperatures in the range of 320–360 °C. Figure 10 further shows that the rate of ammonia synthesis over the 2.5Cs-10Ru/MPC-18 catalyst can be quickly tuned within a short response time period (<30 min) at a temperature jump of 60 °C for around 15 cycles. It can be said that the 2.5Cs-10Ru/MPC-18 catalyst has a high potential in sustainable ammonia synthesis using CO_2-free hydrogen generated from renewable energy resources. In comparison, the 2.5Cs-10Ru/AC catalyst exhibits lower activity and leads to a larger variation in ammonia synthesis when the SV values change (Figure 9b). It also suffers from a slow response time and low rate of ammonia synthesis when the reaction temperature is quickly varied. The rate of ammonia synthesis varied after each temperature jump, implying that the structure of the 2.5Cs-10Ru/AC catalyst might have changed during the reaction and it requires a long activation time for ammonia synthesis.

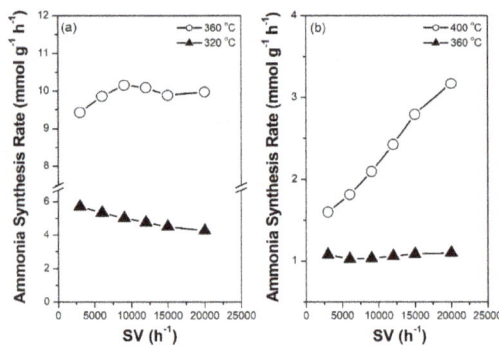

Figure 9. Rate of ammonia synthesis as a function of space velocity (SV) over the prepared Cs-Ru catalysts: (**a**) 2.5Cs-10Ru/MPC-18 and (**b**) 2.5Cs-10Ru/AC.

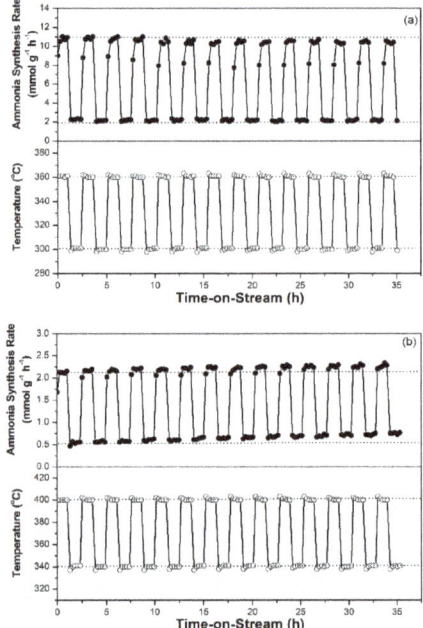

Figure 10. Intermittently variable ammonia synthesis over the prepared Ru-Cs catalysts: (**a**) 2.5Cs-10Ru/MPC-18 and (**b**) 2.5Cs-10Ru/AC.

Table 2. HRTEM image and gas chemisorption studied on the Ru and Cs species of the prepared Cs-Ru catalysts and their catalytic performances in ammonia synthesis.

Samples	Ru Conc. (μmol m^{-2})	Ru Size (nm)		Cs Conc. (μmol m^{-2})	CO_2 Uptake (mmol g^{-1}) [3]	Ammonia Synthesis Activity	
		HRTEM [1]	CO Chem. [2]			Maximum Rate (mmol g^{-1} h^{-1}) [4]	TOF (h^{-1})
2.5Cs-10Ru/AC	2.3	1.6 ± 0.4	5.8 (16%)	3.0	1.6	2.2 (400)	10
2.5Cs-10Ru/MPC-15	2.0	1.9 ± 0.6	2.5 (36%)	2.2	2.4	8.1 (370)	17
2.5Cs-10Ru/MPC-18	3.2	2.1 ± 0.4	2.4 (40%)	3.5	2.4	10 (360)	19
5Cs-5Ru/MPC-18	1.8	1.6 ± 0.3	1.6 (59%)	3.8	2.6	8.5 (380)	18
10Cs-2.5Ru/MPC-18	0.95	1.2 ± 0.3	1.3 (79%)	3.5	2.5	6.2 (420)	17
2.5Cs/MPC-18	-	-	-	4.2	3.5	0	-
10Ru/MPC-18	1.4	2.1 ± 0.6	2.1 (44%)	-	0	2.0 (510)	4.0
2.5Cs-10Ru/MPC-21	12	3.2 ± 0.8	3.7 (31%)	13	2.7	7.3 (360)	18

[1] Determined from the HRTEM images. [2] Determined from the CO chemisorption. The dispersion was shown in the parentheses. [3] Determined from the CO_2-TPD measurement. [4] The maximum rate of ammonia synthesis was determined from Figures 6 and 7. The corresponded temperature with a unit of Celsius degree (°C) was shown in the parentheses.

3. Materials and Methods

3.1. Synthesis of Mesoporous Carbon Material-Supported Cs-Ru Catalysts

Mesoporous carbon materials (a series of commercial CNovel®P(3)010 products denoted as MPC-xx) used in this study were prepared with a hard-template method at Toyo Tanso Co., Ltd., Osaka, Japan and were used as received [55]. Here, "xx" represents the annealing temperature. In other words, the notations MPC-15, MPC-18 and MPC-21 imply that their annealing temperatures were 1500, 1800 and 2100 °C, respectively. In a typical synthesis process, MPC-xx supports (1 g) were dispersed in 70 mL of an ethanol solution (50%, v/v) containing 0.31 g of nitrosylruthenium(III) nitrate (Ru(NO)(NO$_3$)$_3$), Ru assay = 31.4 wt %, Mitsuwa Chemicals Co., Ltd., Osaka, Japan) and slowly heated to around 70–80 °C until the solvent evaporated completely. The resulting solids were calcined at 400 °C for 3 h in N_2 at a ramp rate of 5 °C min^{-1} to produce Ru-impregnated MPC-xx samples. Caesium carbonate (Cs$_2$(CO$_3$), 0.40 g, Cs = 81.6 wt %, Alfa Aesar, Lancaster, UK) was then impregnated into the Ru-impregnated MPC-xx samples by the same procedure as described above but without calcination. Note that the Ru loading was varied at 2.5–10 wt % while the Cs loading was kept constant at 33 wt % in the MPC-xx supports. The freshly prepared Cs-Ru catalysts were labelled as yCs-zRu/MPC-xx, where y and z represent the Cs/Ru molar ratio and Ru loading, respectively. For example, the notation 2.5Cs-10Ru/MPC-18 signifies that mesoporous carbon material annealed at 1800 °C (namely MPC-18) was impregnated with a Ru loading of 10 wt % at a Cs/Ru molar ratio of 2.5, corresponding to a Cs loading of 33 wt % based on the carbon content.

3.2. Synthesis of Reference Catalysts

Microporous activated carbon (denoted as AC, product code HG15-119, Osaka Gas Chemical Co., Ltd., Japan) was used as received and after mild thermal treatment at 500 °C for 3 h in an H_2 environment. Ru and Cs species were impregnated into the AC support using the same procedures as described in the previous section. Thus, a 2.5Cs-10Ru/AC reference catalyst with a Ru loading of 10 wt % and a Cs/Ru molar ratio of 2.5 was prepared. Further, a 10Ru/MPC18 sample with a Ru loading of 10 wt % and a 2.5Cs/MPC18 sample with a Cs loading of 33 wt % were also prepared for comparison.

3.3. Characterization

The specific surface area and porosity of the prepared catalysts were analysed by N_2 physisorption on a BELSORP-max instrument (MicrotracBEL Corp., Osaka, Japan) at 77 K. The pore size distribution (PSD) was calculated using the nonlinear density function theory (NLDFT) using a slit-pore model. The crystallinity of the prepared catalysts was determined on a Rigaku MiniFlex600 diffractometer (Tokyo, Japan) with Cu Kα radiation (λ = 0.15418 nm) and operating at 40 kV and 15 mA. The Ru particle size and size distribution were statistically analysed by high-resolution transmission electron microscopy (HRTEM) on a TOPCON EM002B instrument (Tokyo, Japan) operating at 120 kV. The microstructure of the prepared Cs-Ru catalyst was captured and mapped using a FEI Tecnai Osiris instrument (Santa Clara, CA, USA) equipped with an electron-dispersive X-ray spectroscopy (EDS) (Oregon, USA). High-angle annular dark-field scanning transmission electron microscopy-scanning transmission electron microscopy (HAADF-STEM) images were captured and analysed using the Bruker Esprit software (Massachusetts, USA). Temperature-programmed reduction (TPR) measurements of the prepared Cs-Ru catalysts were recorded on a BELCATII instrument equipped with a thermal conductive detector (TCD) and a BELMass mass spectrometer (MS) (MicrotracBEL Corp., Osaka, Japan). Freshly prepared samples were finely packed in a quartz tube and connected to the BELCATII instrument; purging was carried out with a standard gas of 5% H_2 in Ar at a flow rate of 30 mL min^{-1} until the TCD signal was stable. The TPR-TCD and TPR-MS profiles were recorded without using a molecular sieve at the downstream in the temperature range of 50 to 800 °C at a ramp rate of 5 °C min^{-1}. The temperature-programmed desorption of carbon dioxide (CO_2-TPD) of the prepared Cs-Ru catalysts were also measured by a BELCATII instrument. Before the CO_2-TPD measurement, freshly prepared samples were reduced at 450 °C for 2 h, followed by purging with an Ar flow (50 mL min^{-1}) until the temperature was decreased to 50 °C. The reduced samples were then treated by a mixed gas of 10%CO_2 in Ar (50 mL min^{-1}) at 50 °C for 30 min, followed by purging with an Ar flow (50 mL min^{-1}) until the TCD signal was stable. The uptakes of CO_2 over the reduced samples were calculated by the CO_2-TPD profiles recorded in the range of 50-800 °C at a ramp rate of 5 °C min^{-1}. The pulse chemisorption of carbon monoxide (CO) was determined by an Ohkura Riken R6015 instrument (Saitama, Japan). For the pre-treatment, freshly prepared samples were reduced at 450 °C for 2 h, followed by purging with an Ar flow (50 mL min^{-1}) until the TCD signal was stable. After the pre-treatment, a sequential pulse using a standard gas of 10%CO in He was injected to the reduced samples at 50 °C until no CO was adsorbed. Carbon, hydrogen and nitrogen (CHN) elemental analysis was performed on a PerkinElmer 2400II instrument (Massachusetts, USA). X-ray fluorescence (XRF) analysis was conducted on a Rigaku EDXL300 instrument (Tokyo, Japan) to monitor the Ru and Cs contents in the prepared and used Cs-Ru catalysts.

3.4. Mild Ammonia Synthesis

Typically, ammonia synthesis over the prepared Cs-Ru catalysts was carried out on a fixed-bed reactor at mild reaction conditions (280–450 °C, <1 MPa). It is specially noted that high pressure gas safety act of Japan has defined that "high pressure gas" is the pressure of the compressed gas equal to or higher than 1 MPa at 35 °C [56]. In this study, we specifically carried out mild ammonia synthesis at the reaction pressure of lower than 1 MPa using G1 grade N_2 and H_2 standard gases as feedstocks. The H_2/N_2 ratio in the feed gas was kept at 3. Typically, for ammonia synthesis, the prepared Cs-Ru catalysts sandwiched in between quartz woods were finely packed in a quartz inlet and inserted into a stainless-steel cylindrical reactor controlled by an automatic reaction test system (Taiyo system Corp., Kanagawa, Japan). Prior to the reaction, the prepared Cs-Ru catalysts were reduced on-line at 450 °C for 2 h using a H_2 flow (SV = 10,000 h^{-1}). To start ammonia synthesis, a mixed gas of hydrogen and nitrogen (H_2/N_2 ratio = 3) was fed and the downstream was quantitatively analysed with an online Shimadzu gas chromatograph (GC-2014) equipped with a TCD detector and a column of Thermon-3000 + KOH (2 + 2)% Sunpak-N 60/100 mesh (2.1 m length and 3.2 mm internal diameter, Shinwa Chemical Industries Ltd., Kyoto, Japan). For intermittently variable ammonia synthesis, the influence of SV (3000–20,000 h^{-1}) was studied. Further, cycling tests of a temperature jump (ca. 60 °C) were conducted to monitor the activities of 2.5Cs-10Ru/MPC-18 and 2.5Cs-10Ru/AC for mild ammonia synthesis. The procedures for pre-reduction treatment and downstream analysis were the same as described earlier. The SV value was kept constant at 9000 h^{-1} during these processes. The heating and cooling rates were maintained at 5 °C min^{-1}.

4. Conclusions

Nanostructured Cs-Ru catalysts supported on mesoporous carbon materials with different porosities and crystallinities of the graphite structure were prepared by a wet impregnation method and thermal treatment in an inert atmosphere. The studies of CO chemisorption and HRTEM-HAADF images showed that the prepared Cs-Ru/MPC catalysts contained nanosized Ru particles (around 2–3 nm) close to the Cs species, which were homogeneously impregnated inside mesoporous carbon materials of different degrees of crystallinity, which in turn were influenced by the annealing temperature. TPR studies showed that nanosized RuO_2 particles were formed in the mesoporous pores of the prepared Cs-Ru/MPC catalysts and they could be reduced to metallic Ru particles at around 100–200 °C—this reduction temperature was higher than the pure RuO_2 particles due to the strong interaction between Cs and Ru species. Moreover, gaseous CO_2, CH_4 and CO were observed at 200–500 °C, corresponding to the activation temperature region for ammonia synthesis, due to the conversion of the Cs precursor to form CsOH species and methanation of surface oxygenated species to form clean carbon surface. As a result, the metallic Ru and CsOH species close to each other as the catalytically active sites for ammonia synthesis could be confined firmly inside the mesoporous carbon framework. At higher temperatures (>600 °C), CO gas was formed due to the thermal decomposition of carbon materials; however, it could be reduced by increasing the annealing temperature of the carbon materials. For ammonia synthesis, the prepared Cs-Ru/MPC catalysts exhibited high activity under mild reaction conditions; in particular, the 2.5Cs-10Ru/MPC-18 catalyst with a proper size of metallic Ru nanoparticles (2.4 nm), which were co-impregnated with CsOH species inside the mesoporous carbon framework and surface Ru and Cs concentrations of ca. 3–4 µmol m^{-2} corresponding to a surface Cs/Ru ratio of ca. 1 exhibited excellent activity at lower temperatures (7.3–10.2 mmol g^{-1} h^{-1} at 360–370 °C). Ru particle size and size distribution in the fresh and used 2.5Cs-10Ru/MPC-18 catalysts were similar, whereas those of the microporous catalyst (2.5Cs-10Ru/AC) changed significantly, resulting in a low activity and stability for ammonia synthesis due to serious deactivation by Ru-particle sintering. Moreover, the 2.5Cs-10Ru/MPC-18 catalyst with its mesoporous carbon framework and small Ru size and narrow size distribution showed high responsibility and durability in intermittently variable ammonia synthesis in a wide SV region and in cycling tests with a large temperature variation. Therefore, we are demonstrating, for the first time, that sustainable ammonia synthesis can be carried out by the nanostructured Cs-Ru catalysts under mild conditions using CO$_2$-free hydrogen derived from renewable energy with intermittent operation in Fukushima Renewable Energy Institute (FREA) of AIST, Japan and the results will be reported in the near future.

Supplementary Materials: The following are available online at http://www.mdpi.com/2073-4344/9/5/406/s1, The TPR analysis, electronic microscopy and supplementary of catalytic tests of prepared catalysts, in comparison to those of reference catalysts. Table S1: TPR-MS data of the prepared Cs-Ru catalysts, Figure S1: HRTEM images of carbon supports (a) AC, (b) MPC-15, (c) MPC-18 and (d) MPC-21, Figure S2: HAADF-STEM images of 2.5Cs-10Ru/MPC-15 catalysts. (a) Fresh and (b) used samples, Figure S3: HAADF-STEM images of 2.5Cs-10Ru/MPC-18 catalysts. (a) Fresh and (b) used samples, Figure S4: HAADF-STEM images of 2.5Cs-10Ru/MPC-21 catalysts. (a) Fresh and (b) used samples, Figure S5: HAADF-STEM images of 2.5Cs-10Ru/AC catalysts. (a) Fresh and (b) used samples, Figure S6: TPR-TCD and TPR-MS profiles of RuO$_2$, Figure S7: TPR-TCD and TPR-MS profiles of MPC-18, Figure S8: TPR-TCD and TPR-MS profiles of 10Ru/MPC-18 obtained by the dispersion of MPC-18 (1 g) in 70 mL of ethanol (50%, v/v) containing 0.31 g of nitrosylruthenium(III) nitrate (Ru(NO)(NO$_3$)$_3$) and slowly heating to around 70–80 °C until the solvent completely evaporated. This was followed by calcination at 400 °C for 3 h in N$_2$ at a ramp rate of 5 °C min^{-1}, Figure S9: TPR-TCD and TPR-MS profiles of 2.5Cs/MPC-18 obtained by the dispersion of MPC-18 (1 g) in 70 mL of ethanol (50%, v/v) containing a 0.40 g of caesium carbonate (Cs$_2$(CO$_3$)) and slowly heating to around 70–80 °C until the solvent completely evaporated, Figure S10: TPR-MS profiles of freshly prepared catalysts (a) 2.5Cs-10Ru/AC, (b) 2.5Cs-10Ru/MPC-15, (c) 2.5Cs-10Ru/MPC-18 and (d) 2.5Cs-10Ru/MPC-21, Figure S11: TPR-MS profiles (m/z = 18) of freshly prepared Cs-Ru catalysts, Figure S12: HRTEM images and Ru particle size distributions of the used catalysts. (a) 2.5Cs-10Ru/MPC-AC, (b) 2.5Cs-10Ru/MPC-15, (c) 2.5Cs-10Ru/MPC-18 and (d) 2.5Cs-10Ru/MPC-21, Figure S13: Wide-angle XRD patterns of used catalysts. (a) 2.5Cs-10Ru/AC, (b) 2.5Cs-10Ru/MPC-15, 2.5Cs-10Ru/MPC-18 and (d) 2.5Cs-10Ru/MPC-21, Figure S14: Rate of ammonia synthesis as a function of reaction temperature over the 10Ru/MPC-18 and 2.5Cs/MPC-18 catalysts at an SV value of 9000 h^{-1}.

Author Contributions: M.N. designed and performed the experiments including the preparation and characterization of the catalysts and their catalyst activity tests and wrote the original paper; S.-Y.C. conceived of the characterization and catalytic tests of the prepared catalysts as well as reviewed and edited the paper; H.T. proposed and supervised the project. All the authors discussed and commented on the paper.

Funding: This research was funded by Japan Science and Technology Agency (JST), the Council for Science, Technology and Innovation (CSTI), the Cross-ministerial Strategic Innovation Promotion Program (SIP) and the Energy Carriers program.

Acknowledgments: The authors acknowledge financial support from the Council for Science, Technology and Innovation (CSTI), the Cross-ministerial Strategic Innovation Promotion Program (SIP) and the Energy Carriers program funded by Japan Science and Technology Agency (JST). Furthermore, the authors would like to express their gratitude to Mr. Akira Takatsuki of RIEF, AIST, for assisting with the HRTEM and HAADF-STEM measurements, Dr. Koji Kuramoto of RIEF, AIST, for assistance with XRD measurements, Dr. Takehisa Mochizuki of RIEF, AIST, for his help constructing the CO chemisorption instrument and Mr. Kiyoaki Imoto of RIEF, AIST, for his help conducting mild ammonia synthesis. Special thanks to Editage (https://www.editage.jp/) for English language editing.

Conflicts of Interest: The authors declare no conflict of interest.

References

1. U.S. Energy Information Administration (EIA). International Energy Outlook 2017. Available online: https://www.eia.gov/outlooks/ieo/pdf/0484(2017).pdf (accessed on 11 July 2018).
2. United Nations Framework Convention on Climate Change (UNFCCC). Paris agreement. Available online: https://unfccc.int/sites/default/files/english_paris_agreement.pdf (accessed on 8 April 2019).
3. Ministry of the Environment, Outline of Long-term Low-carbon Vision. Available online: http://www.env.go.jp/press/103822/713.pdf (accessed on 11 July 2018).
4. Ministry of Economy, Trade and Industry, Basic Hydrogen Strategy. Available online: http://www.meti.go.jp/english/press/2017/pdf/1226_003a.pdf (accessed on 6 August 2018).
5. Gandía, L.M.; Oroz, R.; Ursúa, A.; Sanchis, P.; Diéguez, P.M. Renewable hydrogen production: performance of an alkaline water electrolyzer working under emulated wind conditions. *Energy Fuels* **2007**, *21*, 1699–1706. [CrossRef]
6. Felice, L.D.; Courson, C.; Jand, N.; Gallucci, K.; Foscolo, P.U.; Kiennemann, A. Catalytic biomass gasification: Simultaneous hydrocarbons steam reforming and CO_2 capture in a fluidised bed reactor. *Chem. Eng. J.* **2009**, *154*, 375–383. [CrossRef]
7. Mukherjee, S.; Devaguptapu, S.V.; Sviripa, A.; Lund, C.R.F.; Wu, G. Low-temperature ammonia decomposition catalysts for hydrogen generation. *Appl. Catal. B Environ.* **2018**, *226*, 162–181. [CrossRef]
8. Ju, X.; Liu, L.; Yu, P.; Guo, J.; Zhang, X.; He, T.; Wu, G.; Chen, P. Mesoporous Ru/MgO prepared by a deposition-precipitation method as highly active catalyst for producing CO_x-free hydrogen from ammonia decomposition. *Appl. Catal. B Environ.* **2017**, *211*, 167–175. [CrossRef]
9. Yin, S.F.; Xu, B.Q.; Ng, C.F.; Au, C.T. Nano Ru/CNTs: a highly active and stable catalyst for the generation of CO_x-free hydrogen in ammonia decomposition. *Appl. Catal. B Environ.* **2004**, *48*, 237–241. [CrossRef]
10. Wang, S.J.; Yin, S.F.; Li, L.; Xu, B.Q.; Ng, C.F.; Au, C.T. Investigation on modification of Ru/CNTs catalyst for the generation of CO_x-free hydrogen from ammonia. *Appl. Catal. B Environ.* **2004**, *52*, 287–299. [CrossRef]
11. US Geological Survey, Nitrogen (Fixed)–Ammonia. Available online: https://minerals.usgs.gov/minerals/pubs/commodity/nitrogen/mcs-2019-nitro.pdf (accessed on 1 February 2019).
12. Smil, V. Detonator of the population explosion. *Nature* **1999**, *400*, 415. [CrossRef]
13. Schrock, R.R. Reduction of dinitrogen. *Proc. Natl. Acad. Sci. USA* **2006**, *103*, 17087. [CrossRef] [PubMed]
14. Farla, J.C.M.; Hendriks, C.A.; Blok, K. Carbon dioxide recovery from industrial processes. *Energy Convers. Manag.* **1995**, *36*, 827–830. [CrossRef]
15. Ozaki, A.; Aika, K.; Hori, H. A new catalyst system for ammonia synthesis. *Bull. Chem. Soc. Jpn.* **1971**, *44*, 3216. [CrossRef]
16. Aika, K.; Hori, H.; Ozaki, A. Activation of nitrogen by alkali metal promoted transition metal I. Ammonia synthesis over ruthenium promoted by alkali metal. *J. Catal.* **1972**, *27*, 424–431. [CrossRef]
17. Truszkiewicz, E.; Raróg-Pilecka, W.; Schmidt-Szałowski, K.; Jodzis, S.; Wilczkowska, E.; Łomot, D.; Kaszkur, Z.; Karpiński, Z.; Kowalczyk, Z. Barium-promoted Ru/carbon catalyst for ammonia synthesis: State of the system when operating. *J. Catal.* **2009**, *286*, 181–190. [CrossRef]
18. Rossetti, I.; Mangiarini, F.; Forni, L. Promoters state and catalyst activation during ammonia synthesis over Ru/C. *Appl. Catal. A Gen.* **2007**, *323*, 219–225. [CrossRef]
19. Lin, B.; Qi, Y.; Guo, Y.; Lin, J.; Ni, J. Effect of potassium precursors on the thermal stability of K-promoted Ru/carbon catalysts for ammonia synthesis. *Catal. Sci. Technol.* **2015**, *5*, 2829–2838. [CrossRef]
20. Fernández, C.; Sassoye, C.; Debecker, D.P.; Sanchez, C.; Ruiz, P. Effect of the size and distribution of supported Ru nanoparticles on their activity in ammonia synthesis under mild reaction conditions. *Appl. Catal. A Gen.* **2014**, *474*, 194–202.
21. Hansen, T.W.; Hansen, P.L.; Dahl, S.; Jacobsen, C.J.H. Support effect and active sites on promoted ruthenium catalysts for ammonia synthesis. *Catal. Lett.* **2002**, *84*, 7–12. [CrossRef]
22. Brown, D.E.; Edmonds, T.; Joyner, R.W.; McCarroll, J.J.; Tennison, S.R. The genesis and development of the commercial BP doubly promoted catalyst for ammonia synthesis. *Catal. Lett.* **2014**, *144*, 545–552. [CrossRef]
23. Sato, K.; Imamura, K.; Kawano, Y.; Miyahara, S.; Yamamoto, T.; Matsumura, S.; Nagaoka, K. A low-crystalline ruthenium nano-layer supported on praseodymium oxide as an active catalyst for ammonia synthesis. *Chem. Sci.* **2017**, *8*, 674–679. [CrossRef]

24. Ogura, Y.; Sato, K.; Miyahara, S.; Kawano, Y.; Toriyama, T.; Yamamoto, T.; Matsumura, S.; Hosokawa, S.; Nagaoka, K. Efficient ammonia synthesis over a Ru/La$_{0.5}$Ce$_{0.5}$O$_{1.75}$ catalyst pre-reduced at high temperature. *Chem. Sci.* **2018**, *9*, 2230–2237. [CrossRef]
25. Kitano, M.; Kanbara, S.; Inoue, Y.; Kuganathan, N.; Sushko, P.V.; Yokoyama, T.; Hara, M.; Hosono, H. Electride support boosts nitrogen dissociation over ruthenium catalyst and shifts the bottleneck in ammonia synthesis. *Nat. Commun.* **2015**, *6*, 6731. [CrossRef] [PubMed]
26. Kitano, M.; Inoue, Y.; Sasase, M.; Kishida, K.; Kobayashi, Y.; Nishiyama, K.; Tada, T.; Kawamura, S.; Yokoyama, T.; Hara, M.; et al. Self-organized ruthenium–barium core–shell nanoparticles on a mesoporous calcium amide matrix for efficient low-temperature ammonia synthesis. *Angew. Chem. Int. Ed.* **2018**, *57*, 2648–2652. [CrossRef]
27. Mao, C.; Yu, L.; Li, J.; Zhao, J.; Zhang, L. Energy-confined solar thermal ammonia synthesis with K/Ru/TiO$_{2-x}$H$_x$. *Appl. Catal. B Environ.* **2018**, *224*, 612–620. [CrossRef]
28. Raróg-Pilecka, W.; Miśkiewicz, E.; Szmigiel, D.; Kowalczyk, Z. Structure sensitivity of ammonia synthesis over promoted ruthenium catalysts supported on graphitised carbon. *J. Catal.* **2005**, *231*, 11–19. [CrossRef]
29. Raróg-Pilecka, W.; Miśkiewicz, E.; Jodzis, S.; Petryk, J.; Łomot, D.; Kaszkur, Z.; Karpiński, Z.; Kowalczyk, Z. Carbon-supported ruthenium catalysts for NH$_3$ synthesis doped with caesium nitrate: Activation process, working state of Cs–Ru/C. *J. Catal.* **2006**, *239*, 313–325. [CrossRef]
30. Kowalczyk, Z.; Jodzis, S.; Raróg, W.; Zielinski, J.; Pielaszek, J.; Presz, A. Carbon-supported ruthenium catalyst for the synthesis of ammonia. The effect of the carbon support and barium promoter on the performance. *Appl. Catal. A Gen.* **1999**, *184*, 95–102. [CrossRef]
31. Rossetti, I.; Pernicone, N.; Forni, L. Graphitised carbon as support for Ru/C ammonia synthesis catalyst. *Catal. Today* **2005**, *102–103*, 219–224. [CrossRef]
32. Zeng, H.S.; Inazu, K.; Aika, K. The working state of the barium promoter in ammonia synthesis over an active-carbon-supported ruthenium catalyst using barium nitrate as the promoter precursor. *J. Catal.* **2002**, *211*, 33–41. [CrossRef]
33. Nishi, M.; Chen, S.Y.; Takagi, H. A mesoporous carbon-supported and Cs-promoted Ru catalyst with enhanced activity and stability for sustainable ammonia synthesis. *ChemCatChem* **2018**, *10*, 3411–3414. [CrossRef]
34. Kresge, C.T.; Leonowicz, M.E.; Roth, W.J.; Vartuli, J.C.; Beck, J.S. Orderd mesoporous molecular sieves synthesized by a liquid-crystal template mechanism. *Nature* **1992**, *359*, 710–712. [CrossRef]
35. Beck, J.S.; Vartuli, J.C.; Roth, W.J.; Leonowicz, M.E.; Kresge, C.T.; Schmitt, K.D.; Chu, C.T.W.; Olson, D.H.; Sheppard, E.W.; McCullen, S.B.; Higgins, J.B.; Schlenker, J.L. A new family of mesoporous molecular sieves prepared with liquid crystal templates. *J. Am. Chem. Soc.* **1992**, *114*, 10834–10843. [CrossRef]
36. Zhao, D.Y.; Feng, J.L.; Huo, Q.S.; Melosh, N.; Fredrickson, G.H.; Chmelka, B.F.; Stucky, G.D. Triblock copolymer syntheses of mesoporous silica with periodic 50 to 300 angstrom pores. *Science* **1998**, *279*, 548–552. [CrossRef] [PubMed]
37. Zhao, D.Y.; Huo, Q.S.; Feng, J.L.; Chmelka, B.F.; Stucky, G.D. Nonionic triblock and star diblock copolymer and oligomeric surfactant syntheses of highly ordered, hydrothermally stable, mesoporous silica structures. *J. Am. Chem. Soc.* **1998**, *120*, 6024–6036. [CrossRef]
38. Fukuoka, A.; Kimura, J.; Oshio, T.; Sakamoto, Y.; Ichikawa, M. Preferential oxidation of carbon monoxide catalysed by platinum nanoparticles in mesoporous silica. *J. Am. Chem. Soc.* **2007**, *129*, 10120–10125. [CrossRef] [PubMed]
39. Olkhovyk, O.; Jaroniec, M. Periodic mesoporous organosilica with large heterocyclic bridging groups. *J. Am. Chem. Soc.* **2004**, *127*, 60–61. [CrossRef] [PubMed]
40. Inagaki, S.; Guan, S.; Fukushima, Y.; Ohsuna, T.; Terasaki, O. Novel mesoporous materials with a uniform distribution of organic groups and inorganic oxide in their frameworks. *J. Am. Chem. Soc.* **1999**, *121*, 9611–9614. [CrossRef]
41. Liang, C.D.; Dai, S. Synthesis of mesoporous carbon materials via enhanced hydrogen-bonding interaction. *J. Am. Chem. Soc.* **2006**, *128*, 5216–5317. [CrossRef]
42. Lee, J.S.; Joo, S.H.; Ryoo, R. Synthesis of mesoporous silicas of controlled pore wall thickness and their replication to ordered nanoporous carbons with various pore diameters. *J. Am. Chem. Soc.* **2002**, *124*, 1156–1157. [CrossRef] [PubMed]

43. Zhang, F.Q.; Meng, Y.; Gu, D.; Yan, Y.; Yu, C.Z.; Tu, B.; Zhao, D.Y. A facile aqueous route to synthesize highly ordered mesoporous polymers and carbon frameworks with Ia(3)over-bard bicontinuous cubic structure. *J. Am. Chem. Soc.* **2005**, *127*, 13508–13509. [CrossRef]
44. Kasahara, N.; Shiraishi, S.; Oya, A. Heterogeneous graphitization of thin carbon fiber derived from phenol-formaldehyde resin. *Carbon* **2003**, *41*, 1654–1656. [CrossRef]
45. Eslava, J.L.; Iglesias-Juez, A.; Agostini, G.; Fernández-García, M.; Guerrero-Ruiz, A.; Rodríguez-Ramos, I. Time-resolved XAS investigation of the local environment and evolution of oxidation states of a Fischer–Tropsch Ru–Cs/C catalyst. *ACS Catal.* **2016**, *6*, 1437–1445. [CrossRef]
46. Rossetti, I.; Forni, L. Effect of Ru loading and of Ru precursor in Ru/C catalysts for ammonia synthesis. *Appl. Catal. A Gen.* **2005**, *282*, 315–320. [CrossRef]
47. Hill, A.K.; Torrente-Murciano, L. Low temperature H_2 production from ammonia using ruthenium-based catalysts: Synergetic effect of promoter and support. *Appl. Catal. B Environ.* **2015**, *172–173*, 129–135. [CrossRef]
48. Lin, B.; Guo, Y.; Lin, J.; Ni, J.; Lin, J.; Jiang, L.; Wang, Y. Deactivation study of carbon-supported ruthenium catalyst with potassium promoter. *Appl. Catal. A Gen.* **2017**, *541*, 1–7. [CrossRef]
49. Addoun, A.; Dentzer, J.; Ehrburger, P. Porosity of carbons obtained by chemical activation: effect of the nature of the alkaline carbonates. *Carbon* **2002**, *40*, 1140–1143. [CrossRef]
50. Li, C.; Shao, Z.; Pang, M.; Williams, C.T.; Zhang, X.; Liang, C. Carbon nanotubes supported mono- and bimetallic Pt and Ru catalysts for selective hydrogenation of phenylacetylene. *Ind. Eng. Chem. Res.* **2012**, *51*, 4934–4941. [CrossRef]
51. Li, Z.; Liang, C.; Feng, Z.; Ying, P.; Wang, D.; Li, C. Ammonia synthesis on graphitic-nanofilament supported Ru catalysts. *J. Mol. Catal. A Chem.* **2004**, *211*, 103–109. [CrossRef]
52. Jacobsen, C.J.H.; Dahl, S.; Hansen, P.L.; Törnqvist, E.; Jensen, L.; Topsøe, H.; Prip, D.V.; Møenshaug, P.B.; Chorkendorff, I. Structure sensitivity of supported ruthenium catalysts for ammonia synthesis. *J. Mol. Catal. A Chem.* **2000**, *163*, 19–26. [CrossRef]
53. Aika, K. Role of alkali promoter in ammonia synthesis over ruthenium catalysts—Effect on reaction mechanism. *Catal. Today* **2017**, *286*, 14–20. [CrossRef]
54. Liang, C.; Wei, Z.; Xin, Q.; Li, C. Ammonia synthesis over Ru/C catalysts with different carbon supports promoted by barium and potassium compounds. *Appl. Catal. A Gen.* **2001**, *208*, 193–201. [CrossRef]
55. Morishita, T.; Tsumura, T.; Toyoda, M.; Przepiorski, J.; Morawski, A.W.; Konno, H.; Inagaki, M. A review of the control of pore structure in MgO-templated nanoporous carbons. *Carbon* **2010**, *48*, 2707. [CrossRef]
56. Overview of the High Pressure Gas Safety Act. Available online: https://www.khk.or.jp/Portals/0/resources/english/dl/overview_hpg_act.pdf (accessed on 1 October 2016).

© 2019 by the authors. Licensee MDPI, Basel, Switzerland. This article is an open access article distributed under the terms and conditions of the Creative Commons Attribution (CC BY) license (http://creativecommons.org/licenses/by/4.0/).

Article

Differences in the Catalytic Behavior of Au-Metalized TiO₂ Systems During Phenol Photo-Degradation and CO Oxidation

Oscar H. Laguna [1,*], Julie J. Murcia [2], Hugo Rojas [2], Cesar Jaramillo-Paez [3], Jose A. Navío [1] and Maria C. Hidalgo [1]

[1] Instituto de Ciencia de Materiales de Sevilla, Centro Mixto Universidad de Sevilla-CSIC, Avenida Américo Vespucio 49, 41092 Seville, Spain; navio@us.es (J.A.N.); mchidalgo@icmse.csic.es (M.C.H.)
[2] Grupo de Catálisis, Escuela de Ciencias Químicas, Universidad Pedagógica y Tecnológica de Colombia UPTC, Avenida Central del Norte 39-115, Tunja, Boyacá, Colombia; julie.murcia@uptc.edu.co (J.J.M.); hugo.rojas@uptc.edu.co (H.R.)
[3] Departamento de Química, Universidad del Tolima, Barrio Santa Elena, Ibagué 730006299, Colombia; cajaramillopa@ut.edu.co
* Correspondence: olaguna@us.es

Received: 7 March 2019; Accepted: 28 March 2019; Published: 3 April 2019

Abstract: For this present work, a series of Au-metallized TiO₂ catalysts were synthesized and characterized in order to compare their performance in two different catalytic environments: the phenol degradation that occurs during the liquid phase and in the CO oxidation phase, which proceeds the gas phase. The obtained materials were analyzed by different techniques such as XRF, SBET, XRD, TEM, XPS, and UV-Vis DRS. Although the metallization was not totally efficient in all cases, the amount of noble metal loaded depended strongly on the deposition time. Furthermore, the differences in the amount of loaded gold were important factors influencing the physicochemical properties of the catalysts, and consequently, their performances in the studied reactors. The addition of gold represented a considerable increase in the phenol conversion when compared with that of the TiO₂, despite the small amount of noble metal loaded. However, this was not the case in the CO oxidation reaction. Beyond the differences in the phase where the reaction occurred, the loss of catalytic activity during the CO oxidation reaction was directly related to the sintering of the gold nanoparticles.

Keywords: phenol photo-degradation; CO oxidation; Au–TiO₂; gold catalysts; titania

1. Introduction

During the last decade, different chemical reactions which have focused on environmental pollution remediation have been extensively studied, with the aim of removing pollutants from the atmosphere and water. Regarding gaseous pollutants whose elimination has been studied to a greater extent, CO is one of the most relevant. In fact, in 2002 Haruta and coworkers [1] suggested that CO oxidation could be the most extensively studied reaction in the history of heterogeneous catalysis, and this fact has been confirmed in recent years due to the fundamental role of such a reaction in the cleaning of the air and the control of emissions produced by the automotive sector [2–5].

Up until now, several systems have been analyzed and different approaches for the design of feasible catalysts for CO abatement have been proposed [6]. Particularly, the results of the noble metal-based catalysts have been attractive due to their tolerance to the presence of water [7], and although high temperatures are often required for the catalysts to achieve optimal performance, the CO oxidation induced by noble metal nanoparticles is a widely studied reaction, taking into account that it is a sensitive surface structure reaction. Therefore, the main lines of research on this topic have been focused on the development of optimized catalysts for obtaining the maximum CO conversion using low temperatures and short periods of time [1–3]. Additionally, the use of different noble metals [8], such as Au and Pt, which have demonstrated high activity and selectivity in oxidation reactions. Nevertheless, their catalytic performance, especially in the case of gold, largely depends on the particle size, and this is a parameter that should be controlled during synthesis by selecting the most suitable noble metal precursor, procedures of deposition, and temperatures of calcination in order to avoid sintering and loss of the electronic properties of the metal nanoparticles [9–11].

Another aspect of obtaining active catalysts for oxidation reactions based on noble metals is the selection of a suitable support, and some ceramic materials such as Al_2O_3 and SiO_2 have been widely used. However, TiO_2, which has attracted increasing attention in the electronics industry due to its high dielectric and semi-conducting properties, is a very interesting system as a support for catalytic reactions due to its electronic properties [12]. In fact, there is evidence that TiO_2, as a catalyst support, enforces an electric interaction between the d electrons of Ti^{3+} cations with those of noble metals supported over its surface, and this results, for instance, in a decrease in the adsorption energy of CO intermediates [13–15].

The TiO_2 has also been widely studied in photo-catalytic oxidation processes, and the addition of noble metals over this oxide has demonstrated an enhancement in its photo-catalytic performance. This is thanks to the decrease in the recombination rate of photo-generated charges in the systems since the noble metals act as electron collectors. Within the different photo-degradation processes [10], phenol oxidation is more widely studied since this pollutant may contaminate water and generate other compounds even more contaminant than the bare phenol. Therefore, its mineralization is the main goal. In fact, in a previous study, we examined the photo-catalytic phenol oxidation in Au/TiO_2 catalysts prepared from sulphated TiO_2 by means of a photo-deposition of gold. We observed that the conditions during the preparation of the materials (light intensity and photo-deposition time) were determinant for controlling the properties of the gold deposits over the surface of the TiO_2.

It is clear that the differences during the preparation of the gold photo-catalysts resulted in alterations of their surface and structural properties that control their catalytic performance in an aqueous medium. However, the interaction between gold and TiO_2 has been demonstrated to be also closely related to the presence of structural defects such as the oxygen vacancies that also modify the electronic properties of Au/TiO_2 and which have been studied in CO abatement reactions [2,16]. Subsequently, the following question arises: could these properties generated in photo-catalysts be applied in a different catalytic environment such as a gas phase oxidation process where there is no light irradiation? Therefore, the present work proposes the exploration of a series of different gold catalysts prepared by the photo-deposition method over TiO_2 in both the phenol photo-oxidation process and the CO oxidation reaction. Special attention is paid to the study of the electronic properties of the prepared materials and their effect on the catalytic performance in the cited reactions.

2. Results and Discussion

2.1. Physicochemical Characterization of the Obtained Materials

The real content of noble metal measured by XRF is presented in Table 1, as well as other physico-chemical features of the materials. It has to be noted that the amounts of deposited gold were notably lower than the theoretical values in all cases. However, regardless of the poor loading of gold, it was found that the real content of deposited noble metal increased with the deposition time in both cases, with 2 and 5 wt.%. Regarding the efficiency of the deposition of noble metal, in a previous work, it was observed that the loading of gold was strongly connected with the light intensity and the irradiation time during the photo-deposition [17].

Table 1. Miscellaneous physico-chemical properties of the prepared materials including: noble metal content, specific area BET, band gap, and particle sizes of the metal nanoparticles.

Sample	Noble Metal Content (Au wt.%)	Area BET (m^2/g)	Direct Band Gap (eV)	* Crystallite Size of the Metallic Species (nm)	$^\Delta$ Average Particle Size of the Metallic Species (nm)
S–TiO_2	-	58	3.20	-	-
2Au-15	0.28	53	3.56		4.8
2Au-120	0.77	53	3.53	not detected	6.4
5Au-15	0.43	50	3.51		5.6
5Au-120	0.73	48	3.51		6.2
2Au-15C	0.28	32	3.16	29	22
2Au-120C	0.77	35	3.02	25	27
5Au-15C	0.43	33	3.03	17	16
5Au-120C	0.73	34	3.04	30	22

* Crystallite size of Au species calculated by XRD results and using the Scherrer equation for the calcined solids.
$^\Delta$ Average particle size of Au species calculated from the analysis carried out over the TEM micrographs of the non-calcined and calcined solids.

As indicated in the experimental section, after synthesis, a certain portion of all the catalysts was calcined at 300 °C, and in order to evaluate the effect of temperature on the surface area of the obtained catalysts, N_2 adsorption–desorption measurements were carried out before and after calcination. Prior to the calcination, the metalized solids presented similar specific surface areas around 53 m^2/g and no large differences can be appreciated after the metallization process in all cases (Table 1). The BET area values were in agreement with those reported in the literature for similar sulfated titania [18]. However, after the calcination at 300 °C, all the materials exhibited a decrease in the specific area around 20 m^2/g, which may possibly be due to the powder's particles' agglomeration during the calcination process, thus leading to a decrease in the S_{BET} value. The similarity in the decrease of the specific area values for all the materials confirmed once again that no large differences concerning the textural properties were generated during the metallization process regardless of the nature of the noble metal or its amount.

The XRD profiles of the support and the catalysts, prior to and after the calcination at 300 °C, are presented in Figure 1. Before calcination (Figure 1A), all the materials presented had a similar profile where the principal reflections coincided with those of the anatase phase (JCPDS no. 00-021-1272). This result agrees with the fact that the sulphation of TiO_2 stabilizes the anatase phase, avoiding the Rutile formation during the support synthesis (S–TiO_2) [18]. Moreover, no clear signals of the presence of gold prior to the calcination were observed. This could indicate a high dispersion of the noble metal, although considering the poor loading of this, the main gold peak corresponding to the crystal plane (111) located at 2θ = 38.18° could be overlapped with the reflections (004) and (112) of the anatase, therefore, it cannot be clearly observed.

After the calcination, the XRD pattern of the support (S–TiO_2) exhibited the evolution of the reflections due to the presence of the rutile phase (JCPSD no. 00-021-1276) along with those of the anatase (Figure 1B). However, this new phase of the TiO_2 was not observed in the XRD patterns of the metallized catalysts that preserved the main reflections of the Anatase phase.

The presence of gold was noticeable in the XRD patterns after calcination through the evolution of the reflections associated to this metal, thus indicating the increase in the crystalline domain. An evolution of the crystal plane (111) of the gold was observed; however, it is still not clear, due to the overlap with the crystalline planes (004) and (112) of the anatase. However, in the materials 2Au-120C and 2Au-15C, a low-intensity peak was observed at a 2θ~44°, which could correspond to the crystalline plane (200) of gold (JCPDS No. 00-004-0784).

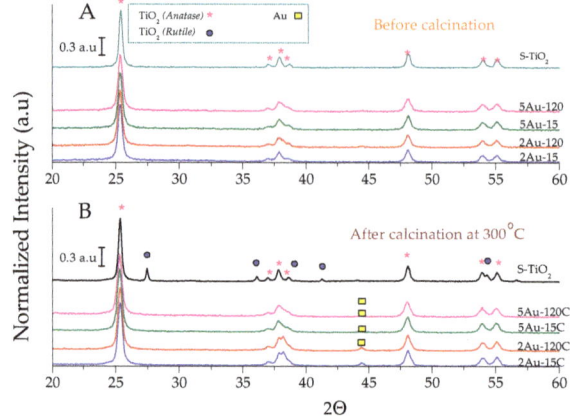

Figure 1. XRD patterns of the prepared materials: (**A**) before calcination at 300 °C; (**B**) after calcination at 300 °C.

The crystallite sizes of the metal particles in the calcined catalysts are presented in Table 1 and the systems loaded with 2 wt.% showed practically the same gold crystallite size, while for the catalysts with 5 wt.%, a superior size of the nanoparticles was achieved for the solid submitted to the longer photo-deposition procedure (120 min). Calcined and non-calcined catalysts were also analyzed by means of TEM, and in this case, a particle size distribution was obtained for both families of solids. Thus, representative micrographs of the studied materials are presented in Figure 2.

Regarding the non-calcined materials, these had a similar morphology and the clusters of gold (confirmed by EDX analyses and highlighted by yellow circles) presented sizes below ~10 nm. Although the gold average particle size was similar for all the catalysts, the 2Au-120 solid exhibited a higher population of gold nanoparticles below 3 nm.

The calcined materials presented a similar morphology of small irregular agglomerates. In all cases, the majority of the black spots with higher contrast corresponded to clusters of gold (confirmed by EDX analyses), and for establishing the particle size distributions, several particles were measured, including those of micrographs not presented in the manuscript. In this sense, the agreement between the crystallite sizes obtained by the Scherrer equation and the average particle sizes established from the TEM micrographs needs to be noted (See Table 1). A broader particle size distribution was produced with the longer photo-reduction treatment that promoted the growth of the gold nanoparticles.

The surface elemental composition of the materials was analyzed by XPS (Figure 3). It can be observed that O (1s) peaks in all materials (Figure 3a,d,g,j), can be formed by the contribution of two peaks, where the smallest contribution varies between 12% and 15%. It is notable that these higher percentages were produced in materials with a longer photo-deposition time of Au. The contribution of the peak at 530 eV can be assigned to the lattice oxygen of the TiO_2, while the peak to 531 eV can be assigned to hydroxyl groups or chemisorbed O species [19].

On the other hand, all the materials show Ti $2p_{1/2}$ and Ti $2p_{3/2}$ peaks, located at 465.5 eV and 458.8 eV, respectively; the distance between the Ti 2p peaks was 5.7 eV, which was induced by the coupling spin-orbital, indicating the presence of Ti^{4+} on the surface [19]. In all samples (Figure 3c,f,i,l), peaks were found centered around 87.6 eV and 84.0 eV, which can be assigned to Au $4f_{5/2}$ and Au $4f_{7/2}$, respectively, the distance between the peaks was about 3.6 eV, which confirms the presence of metallic Au particles deposited on the surface of TiO_2 [20].

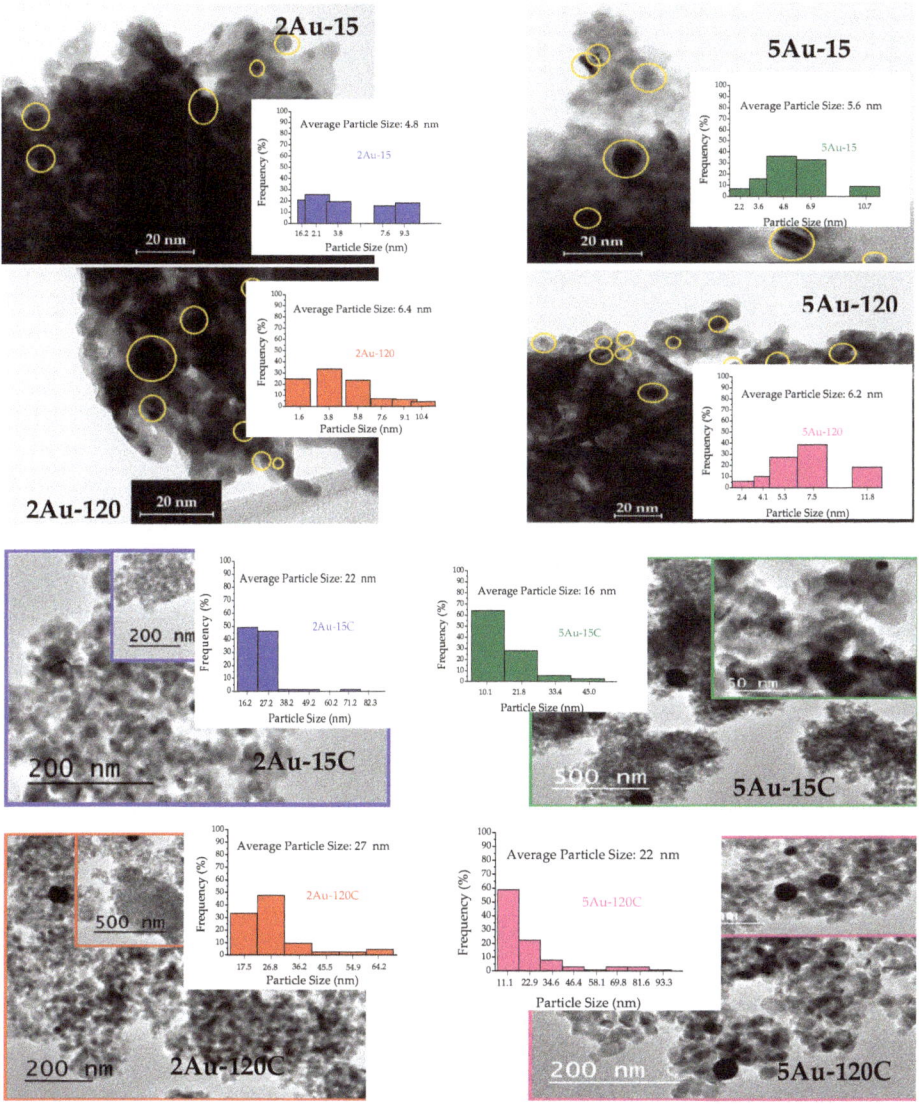

Figure 2. TEM micrographs of the non-calcined and calcined catalysts including the corresponding particle size distributions of the metal nanoparticles.

Although metallic Au was present in all the materials, it was remarkable to observe that for 2 wt.% nominal percentage, the amounts of deposited metal showed around 0.12%, while ~0.08% for materials with 5 wt.% nominal percentage, possibly due to an inverse relationship between the concentration of photo-deposited metal and the size of the Au particle on the TiO_2 surface [21,22].

The XPS spectra also confirmed the presence of S (2p) located at 168.9 eV, which corresponds to S^{6+}, which could indicate a substitution of Ti^{4+} cations by S^{6+} on the surface, due to the large size of the S^{6+}. It is a viable candidate to replace the Ti site and would be able to compensate the Ti^{4+} deficiencies [19]. The atomic percentage of all samples was 0.5% (data not shown).

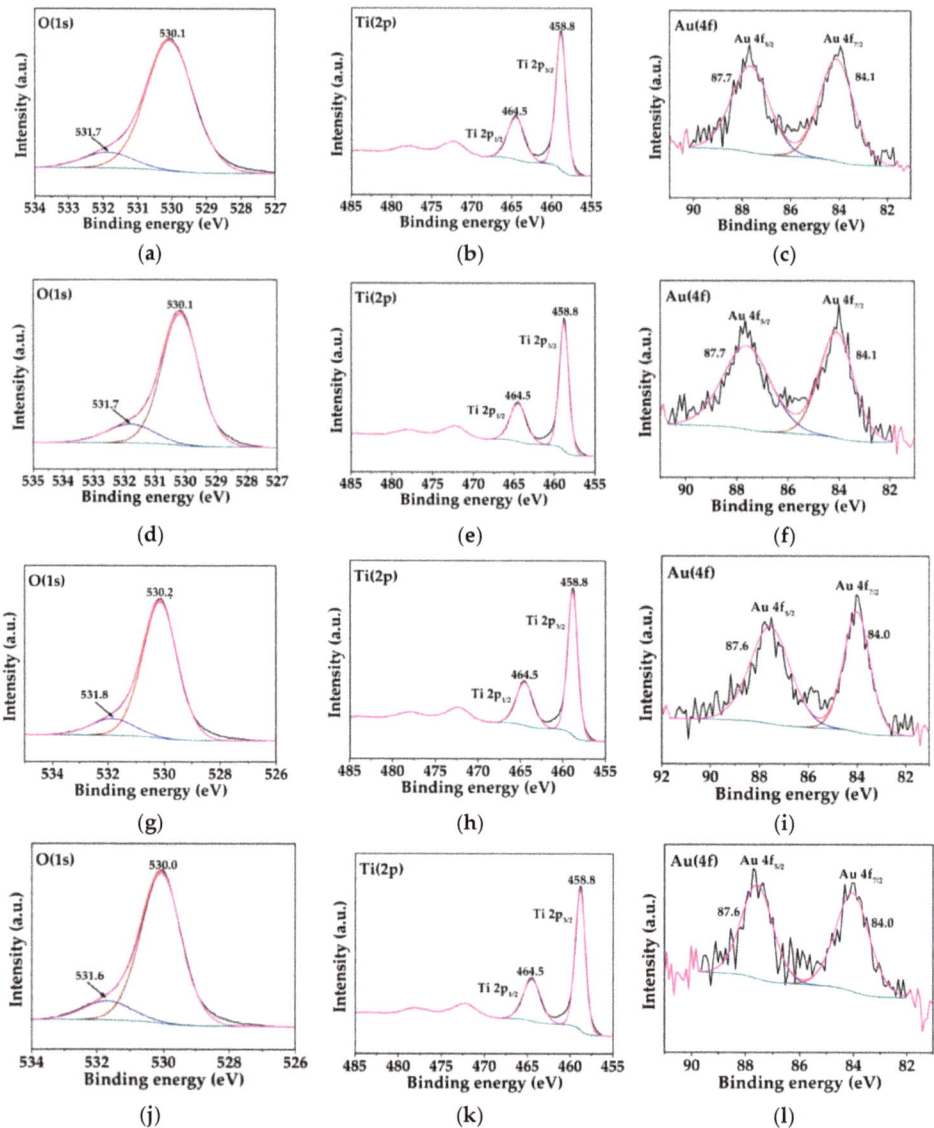

Figure 3. (a–c) 5Au-120; (d–f) 5Au-15; (g–i) 2Au-120; (j–l) 2Au-15.

By means of XPS analysis, it was also possible to calculate the O/Ti atomic ratio (Table 2). For the S–TiO$_2$ support, the O/Ti ratio was 1.70, thus the O/Ti atomic ratio of these species was below the stoichiometric one. Therefore, a certain number of oxygen vacancies on the surface of the material should be expected. This is because at the calcination temperature of 650 °C used in the S–TiO$_2$ preparation, the elimination of sulfate groups promotes the creation of several oxygen vacancies, which have been reported as preferential sites for noble metal adsorption [23,24], this being one of the main reason for the application of sulphation treatment on the titania surface in this study. Furthermore, it is important to note that after gold photo-deposition, the O/Ti ratio increased to values around 1.83–1.95, suggesting that oxygen vacancies were partially annihilated during the metal deposition process.

Table 2. Binding energies of the Ti ($2p_{3/2}$) and O (1s) peaks, and the O/Ti atomic ratio for the prepared catalysts.

Catalyst	Binding Energy (eV)		O/Ti Atomic Ratio
	Ti ($2p_{3/2}$)	O (1s)	
S–TiO$_2$	458.5	529.8	1.70
2Au-15	458.6	530.0	1.83
2Au-120	458.5	529.8	1.85
5Au-15	458.9	530.2	1.86
5Au-120	458.5	529.9	1.95

The UV-Vis DR spectra of the catalysts prior to and after calcination are presented in Figure 4. In all cases, the characteristic sharp absorption threshold of TiO$_2$ around 350 nm was observed [25,26]. Furthermore, a broad signal around 546 nm was observed for the catalysts with gold. This absorption in the visible region agrees with the purple color of the solids, whose hue depends on the gold content being more intense for the solids with the highest loadings of gold. The modifications in the hue of the colors of the powders, which is a qualitative feature, was confirmed by the variable intensity of the band in the visible region, this being more intense for the 5Au-15 and 5Au-120 solids.

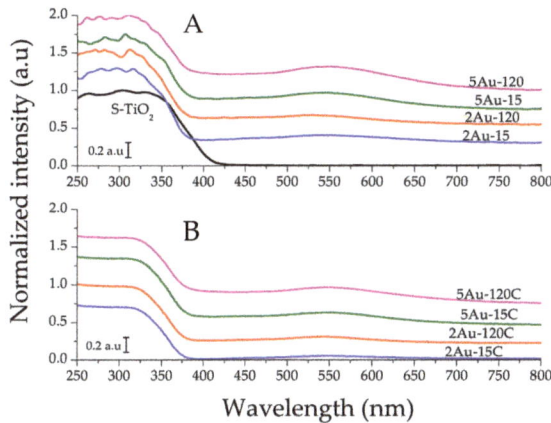

Figure 4. DR/UV-Vis spectra of the prepared catalysts: (**A**) before calcination at 300 °C; (**B**) after calcination at 300 °C.

The absorption band around 550 nm was associated with the localized surface plasmon of resonance of gold nanoparticles supported on TiO$_2$ [25–27], and this signal intensified as the loading of gold increased. Different authors have reported a close relationship between the size of the clusters of gold and the position, shape, and intensity of the absorption band [28]. Furthermore, other aspects such as the dielectric constant of the support, and the surrounding medium may affect the position and shape of the surface plasmon resonance in the spectrum [27]. The alteration of the electronic environment of TiO$_2$ produced by the modification with the noble metals may be observed in the change of the band gap values presented in Table 1. Before calcination, all the systems which contained gold presented a superior band gap to the bare support. Nevertheless, the calcined catalysts demonstrated a decrease in their band gap values. The surroundings of the TiO$_2$ that were in contact with the metal depended on the amount of this and the size of the obtained nanoparticles. In this sense, if the calcination produces alterations in the size of the metal nanoparticles, it is to be expected that their interaction with the support will also be modified. Therefore, the catalytic performance, which in this sort of materials is strongly related to their electronic properties, will depend on whether or not the materials were calcined.

2.2. Catalytic Activity Measurements

-Catalytic activity during the phenol photo-catalytic oxidation: The photo-catalytic activity was evaluated with the degradation of a contaminating substrate such as phenol using different percentages of photo-deposited metallic Au on TiO_2. Figure 5 shows that all materials evaluated had a phenol conversion greater than 95% after two hours of treatment. These results are in agreement with those reported in previous studies [17,29,30].

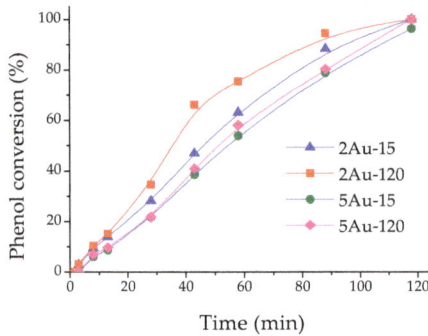

Figure 5. Catalytic activity of the prepared catalysts during the phenol photo-catalytic oxidation.

However, at short illumination times (40 min), it can be observed that the material with the best photo-catalytic behavior was 2Au-120, reaching a conversion higher than 60%, followed by the material 2Au-15, with a percentage lower than 50%. With respect to the materials 5Au-15 and 5Au-120, their behavior was lower still than the two previously described, with values lower than 40% and both very similar. The superior activity of the 2Au-120 material may be related to the size of the gold nanoparticles, since this material presented a superior population of smaller particles (below 3 nm) as was observed by means of TEM.

It is possible that the metallized materials with Au had an improved photo-catalytic activity, since the photo-deposited nanoparticles on the surface of the TiO_2 act as sinks for the photo-generated electrons after the excitation with UV radiation. Generally, the Fermi levels of the photo-deposited noble metals were lower than those in the TiO_2 conduction band [29], causing electrons to be efficiently transferred from the gold nanoparticles to the TiO_2 conduction band, thus preventing the recombination of the charge carriers and generating a more efficient photo-catalytic process [31]. However, as can be observed at higher percentages of Au, it is possible that a greater recombination process of the h^+/e^- pairs is present, as suggested by K. Sornalingam et al. [32] at higher percentages of photo-deposited noble metals which can generate recombination processes by decreasing the photo-catalytic activity of the material [31,33]. In addition, at a high particle size, the active centers of TiO_2 can be blocked, thus decreasing the photo-catalytic activity. Additionally, the blocked active centers may turn into recombination centers, which also results in the decrease in the catalytic activity [17].

Gold has a high affinity to suffer photo-deposition, where a direct relationship between the size of the nanoparticles and the nominal amount of the metal is observed. In addition, the photo-reduction of this metal is very fast, especially in the TiO_2's vacancies that stabilize gold deposits due to the high-adhesion energy that is generated between the Au and TiO_2 [17,34]. This is why the leaching of gold was not observed in the prepared materials, according to the XRF analysis of the post-reaction catalysts (not presented data) that exhibited the same amount of noble metal than that of the catalysts prior to the catalytic activity measurements.

-*Catalytic activity during the CO oxidation:* The performance of the prepared catalysts during the CO oxidation reaction is presented in Figure 6.

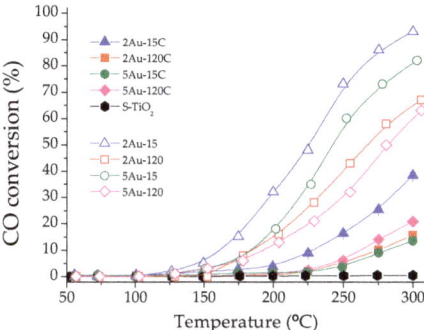

Figure 6. Catalytic activity of the prepared catalysts during the CO oxidation.

The bare support did not show activity in the CO oxidation reaction, while the CO conversion was noticeable after the deposition of noble metals. Furthermore, for all the catalysts, the CO conversion increased along with the temperature, although the performance of the calcined materials was lower than that of the non-calcined ones. For this last fact, the first explanation that arises and agrees with that reported in the literature [35] is that the additional thermal treatment applied over the calcined catalysts resulted in the agglomeration of the gold nanoparticles, even though it has been proposed that the our calcination temperature (300 °C) was within a temperature range where strong sintering was not expected [35]. In this sense, the activation process itself may alter the size of the gold nanoparticles making them probably different to those presented in Table 1. This is a fact that has to be confirmed since it may allow establishing a clearer relationship between the size of the gold nanoparticles and the catalytic performance.

Another relevant aspect that has to be considered about the activation process is the reductant character of the applied atmosphere, since the use of H_2 may result firstly in the total reduction of all gold species. Therefore, despite the possible negative effect of the thermal treatment, in the present work, the total reduction of the gold species was aimed at, since it has been demonstrated that Au^0 species are more active in the activation of the reactants involved in CO oxidation [36–38].

The effects of the reductant conditions of the activation atmosphere may also modify the support, since TiO_2 is a reducible oxide and the generation of Ti^{3+} species may result in the generation of oxygen vacancies. These punctual defects, generally stabilized in the surface of the oxide, are closely related with the role of this material in the activation of oxygen molecules during the oxidation of CO [39]. In addition, it has been demonstrated that these may enhance also the interaction between gold and the surface of the oxide [2]. Consequently, the reduction of the catalyst could counteract the sintering produced by the increase in temperature, taking also into account the possible dynamic re-dispersion of deposited gold nanoparticles in different supports depending on the atmosphere reported in literature [2,40,41]. However, this has to be confirmed and deeply analyzed through further experiments revealing more insights about the effects of the activation process in the studied catalysts.

Moreover, an approach to understanding the different performance of the catalysts, regardless of whether this performance is measured prior to or after the calcination, could be based on the different amounts of loaded noble metal. In this sense, the systems with higher amounts of gold (2Au-120 and 5Au-120) should be the most active systems, but this was not the case. Therefore, the particle size of the metal clusters, and consequently their interaction with the electronic properties of the support, have to play a more determinant role than the total amount of gold loaded. Different examples of studies combining DFT calculations have demonstrated alterations in the performance of the Au and Pt catalysts during the CO oxidation depending on the electronic environment of the metallic clusters [42,43]. Therefore, the different catalytic behaviors may be produced not only by the different loadings of noble metals but also by the interaction between these and the support. In fact, such interaction has been confirmed by means of the different characterization techniques discussed above. For instance, the strong interaction noble metal-support seemed to inhibit the evolution of the rutile phase (see Figure 1) of the prepared materials. In this sense, the surface O/Ti atomic ratio was also altered by such interaction (see Table 2), and consequently the electronic environment, resulting in the modification of the band gap values (see Table 1).

Bamwenda et al. [14] studied the effect of the preparation method of Pt and Au catalysts for the CO oxidation, supported on the TiO_2. Furthermore, the deposition–precipitation and photo-deposition approaches were used by them and the former resulted in the more active systems. In fact, the gold catalysts prepared by the deposition–precipitation method presented a superior activity than the platinum catalysts. This agrees not only with the relatively low activity of the solids studied in the present work prepared by photo-deposition, but also with the fact that the synthesis procedure may determine the interaction between the noble metal and the surface of TiO_2.

Regarding the calcined catalysts, the systems 2Au-120C and 5Au-120 that presented a comparable gold content (see Table 1), exhibited differences in their performance. In this case, it has to be remarked that the particle sizes obtained by means of the Scherrer equation and with the TEM micrographs (see Table 1) were very similar for these two solids. Therefore, the dispersion of gold in these two solids should be comparable, demonstrating that the electronic environment may be being influenced by other aspects such as the synthesis procedure that resulted in modifications of the support's structure. This has also been remarked on by Bamwenda et al. [14] and other authors [44,45] that highlighted the structure sensitivity of Au/TiO_2 catalysts due to the contribution from the perimeter interface between Au nanoparticles and the surface of the support.

3. Materials and methods

3.1. Synthesis of the Different Materials

The synthesis of the catalysts has previously been reported by Hidalgo et al. [17]. Firstly, the TiO_2 was prepared through the hydrolysis of titanium tetraisopropoxide (Aldrich, 97%) dissolved in isopropanol solution (1.6 M), by means of the slow addition of distilled water (volume ratio isopropanol/water 1:1). The generated precipitate was filtered afterward and dried at 110 °C overnight. Then, the TiO_2 powder was sulfated by immersion in H_2SO_4 (aq.) 1 M for 1 hour, and finally calcined at 650 °C for 2 hours (S–TiO_2).

Concerning the synthesis of the Au-modified solids, the photo-deposition method was applied following the procedure previously described [46] and using Gold (III) chloride trihydrate ($HAuCl_4 \cdot 3H_2O$, Aldrich 99.9%) as metal precursor. Therefore, under an inert atmosphere (N_2), a suspension of 5 g/L S–TiO_2 in distilled water containing 0.3 M isopropanol (Merck 99.8%) which acts as sacrificial donor was prepared. Then, the appropriate amount of metallic precursor was added to obtain the desired loading of gold. Two loadings of noble metal (2 and 5 wt.%) were proposed. The photo-chemical deposition in every system was performed by illuminating the suspension with an Osram® Ultra-Vitalux lamp (300 W), with a sun-like radiation spectrum and the main emission line in the UVA range at 365 nm. Two different photo-deposition times were applied (15 and 120 min). Furthermore, the light intensity on the TiO_2 surface was 0.15 W/m² for Au photo-chemical deposition. After the noble metal depositions, the powders were recovered by filtration and dried at 110 °C overnight [17]. In all cases, a portion of the catalysts was calcined at 300 °C for 2 h in air. The materials have been labelled as follows: indicating firstly, the intended wt.% of noble metal it was indicated; secondly, the noble metal, and finally the deposition time in minutes (2Au-15, 2Au-120, 5Au-15, 5Au-120). Additionally, regarding the calcined materials, the letter C was added: 2Au-15C, 2Au-120C, 5Au-15C, 5Au-120C.

3.2. Characterization of the Obtained Materials

The chemical composition and the total noble metals content in the samples was determined by X-ray fluorescence spectrometry (XRF) in a Panalytical® AXIOS sequential spectrophotometer (Malvern Panalytical, Malvern, United Kingdom) equipped with a rhodium tube as the source of radiation. The XRF measurements were performed onto pressed pellets (sample included in 10 wt.% of wax).

The specific surface area BET measurements were carried out using low-temperature N_2 adsorption in a ASAP 2010 instrument (Micromeritics, Norcross, GA, USA). Degasification of the samples was performed at 150 °C for two hours under vacuum.

The XRD analyses were carried out on a Siemens® D500 diffractometer (Siemens, Munich, Germany) using the Cu Kα radiation (40 mA, 40 kV). The patterns were recorded with a 0.05° step size and 300 s of step time. Furthermore, the crystallite sizes of the metallic particles were calculated by means of the Scherrer equation.

Transmission electron microscopy (TEM) was performed with a Philips CM200 instrument (Philips, Amsterdam, Netherlands) and the samples were dispersed in ethanol using an ultrasonicator and dropped on a carbon grid prior to the analysis. With regards to the analysis of the micrographs, the different particle sizes of the metallic clusters were measured with the program ImageJ 1.51g, and afterwards, the corresponding histograms were made, following the Sturges' rule for the establishment of the categories [47].

X-ray photoelectron spectroscopy (XPS) studies were carried out on a Leybold–Heraeus LHS-10 spectrometer (Leybold, Cologne, Germany), working with constant pass energy of 50 eV. The spectrometer main chamber, working at a pressure <2×10^{-9} Torr, was equipped with an EA-200MCD hemispherical electron analyzer with a dual X-ray source working with Al Kα (hv = 1486.6 eV) at 120 W and 30 mA. C 1s signal (284.6 eV) was used as internal energy reference in all the experiments. Samples were outgassed in the pre-chamber of the instrument at 150 °C up to a pressure <2×10^{-8} Torr to remove chemisorbed water.

The light absorption properties of the samples were studied by means of the diffuse reflectance UV–Vis spectrophotometry (DR/UV-Vis) using a Varian® spectrophotometer model Cary 100 (Varian, Palo Alto, CA, USA), equipped with an integrating sphere and using $BaSO_4$ as reference. All the spectra were recorded in diffuse reflectance mode and transformed into a magnitude proportional to the extinction coefficient through the Kubelka–Munk function, (Fα). Concerning the direct band gap, this may be estimated from the adsorption edge wavelength of the inter-band transition using the Tauc function. For insulators and semiconductors, the square root of the Kubelka–Munk function multiplied by the photon energy is depicted versus the photon energy and extrapolating the linear zone of the rising curve to zero [48].

3.3. Catalytic Activity Measurements

3.3.1. Phenol Photo-Catalytic Oxidation

The phenol photo-catalytic oxidation was carried out following the procedure previously reported by Hidalgo et al. [17]. Briefly, an aqueous suspension that contains 50 ppm of phenol and the photo-catalyst (1 g/L) was magnetically stirred in a 400-mL Pyrex batch reactor foiled with aluminum in the presence of a continuous oxygen flow for 20 minutes in the dark, in order to favor the adsorption–desorption equilibrium, since in this type of liquid phase systems it is important to reach equilibrium in order to obtain a high interaction between the adsorbate and the surface of the photocatalyst [34], and thus obtaining a better performance during the photocatalytic process, avoiding errors while the decrease of Phenol concentration was followed.

The light source used was an Osram Ultra-Vitalux lamp (300 W). The intensity of the incident UVA light on the solution was 140 W/m^2, using a UV-transparent Plexiglas® top window with the threshold absorption at 250 nm. This parameter was determined with a PMA 2200 UVA photometer (Solar Light Co.).

The HPLC technique was used for monitoring the phenol concentration during the photo-catalytic activity, using an Agilent Technologies 1200 device, equipped with an Elipse XDB-C18 column (5 µm, 4.6 mm × 150 mm), with the water/methanol (65:35) mobile phase, and a flow rate of 0.8 mL/min.

3.3.2. CO Oxidation Reaction

The catalytic activity measurements were carried out on a fixed-bed cylindrical stainless-steel reactor with an internal diameter of 0.9 mm, coupled to a Microactivity Reference Unit (PID Eng&Tech®), which allowed for the temperature to be controlled, as well as the composition of the different feed-streams passed through the samples. For every experiment, 100 mg of catalyst (particle size 100 < ø < 200 µm) was diluted with SiC VWR Prolabo® (particle size 0.125 µm) to achieve a bed of about 5 mm in height.

Firstly, the catalysts were activated under a 100 mL/min total flow of H_2 (50 vol.%) and N_2 (50 vol.%) at 300 °C for 2 h. Subsequently, the catalysts were cooled and the temperature was stabilized at 50 °C, and then the feed-stream of activation was replaced by the mixture of the reaction (100 mL/min: CO 3 vol.%; O_2 15 vol.%; N_2 82 vol.%). The reactants and products were analyzed and quantified (from 50 to 300 °C every 25 °C) by gas chromatography on a micro GC (Varian® CP-4900), equipped with a Porapak® Q, a Molecular Sieve 5A, and two TCD detectors. The carrier gas for the chromatographic analyses was He. The CO conversion was calculated according to Equation (1), where $F_{in,CO}$ and $F_{out,CO}$ refer to the molar flow rates at the reactor inlet and outlet, respectively.

$$CO\ conversion = 100 \times \frac{F_{in,CO} - F_{out,CO}}{F_{in,CO}} \qquad (1)$$

4. Conclusions

A series of Au–TiO$_2$ catalysts were successfully prepared by means of the photo-deposition method. Although the loading of noble metal in all cases was low, compared with the intended value, a strong interaction between the deposited metallic species and TiO$_2$ was noticed. One of the remarkable consequences of such interaction was that in all cases, the presence of noble metal inhibited the transition from anatase–rutile, preserving the anatase as the main phase in the calcined catalysts. Moreover, the spectroscopic characterization of the studied materials showed alterations of the electronic properties appreciable in the modification of the UV-Vis spectra that also resulted in different band gap values depending on the loading and particle size of the deposited nanoparticles. In all cases, the calcination of the materials generated considerable changes in their electronic properties due to the sintering of the noble metal clusters.

Regarding the catalytic activity studies, on the one hand, the uncalcined materials showed considerable activity during the photo-catalytic degradation of phenol, with the 2Au-120 systems being the most active one. On the other hand, during the CO oxidation reaction, although the uncalcined materials also exhibited catalytic activity, this seemed to be below that of other gold catalysts reported in literature. Furthermore, an important reduction of the catalytic activity was obtained after the calcination of the catalyst principally due to the sintering of the metal clusters.

Therefore, the studied materials present structural and electronic properties that seem to be more determinant in photo-oxidation reactions rather than in the CO oxidation. Despite this, it is clear that the cited properties also determine the catalytic performance during the CO oxidation reaction. Therefore, the challenge is to increase the number of active sites without sacrificing the dispersion of the metallic phase, optimizing the synthesis parameters to also improve the noble metal load, while aiming to obtain more active systems for heterogeneous solid–gas reaction environment.

Author Contributions: O.H.L. and J.J.M. conceived and designed the experiments. O.H.L., J.J.M. and C.J.-P. performed the experiments. O.H.L., J.J.M. and C.J.-P. analyzed the data and wrote the paper. H.R., J.A.N. and

M.C.H., contributed reagents, materials and analysis tools. O.H.L. and J.J.M. looked for funding for the mobility of researchers involved in this paper.

Funding: This research received no external funding.

Acknowledgments: The authors thank COLCIENCIAS and Universidad Pedagógica y Tecnológica de Colombia for the economic support through the aid: "Complementos para proyectos con financiación internacional para la comunidad COLCIENCIAS en el exterior—2014". O.H. Laguna thanks the Spanish Ministry of Economy and Competitiveness for the support through the project ENE2015-66975-C3-2-R.C. Jaramillo-Paez thanks the Universidad del Tolima for the financial support through the Studies Commission.

Conflicts of Interest: The authors declare no conflict of interest.

References

1. Xie, X.; Li, Y.; Liu, Z.-Q.; Haruta, M.; Shen, W. Low-temperature oxidation of CO catalysed by Co_3O_4 nanorods. *Nature* **2009**, *458*, 746–749. [CrossRef]
2. Sarria, F.R.; Plata, J.J.; Laguna, O.H.; Márquez, A.M.; Centeno, M.A.; Sanz, J.F.; Odriozola, J.A. Surface oxygen vacancies in gold based catalysts for CO oxidation. *RSC Adv.* **2014**, *4*, 13145–13152. [CrossRef]
3. Choudhary, T. CO-free fuel processing for fuel cell applications. *Catal. Today* **2002**, *77*, 65–78. [CrossRef]
4. Costello, C.K.; Yang, J.H.; Law, H.Y.; Wang, Y.; Lin, J.N.; Marks, L.D.; Kung, M.C.; Kung, H.H. On the potential role of hydroxyl groups in CO oxidation over Au/Al_2O_3. *Appl. Catal. A Gen.* **2003**, *243*, 15–24. [CrossRef]
5. Avgouropoulos, G.; Ioannides, T. Selective CO oxidation over $CuO-CeO_2$ catalysts prepared via the urea–nitrate combustion method. *Appl. Catal. A Gen.* **2003**, *244*, 155–167. [CrossRef]
6. Laguna, O.H.; Bobadilla, L.F.; Hernández, W.Y.; Centeno, M.A. Chapter 20: Low-Temperature CO oxidation. In *Perovskites and Related Mixed Oxides: Concepts and Applications*; Granger, P., Parvulescu, V.I., Kaliaguine, S., Prellier, W., Eds.; Wiley-VCH: Weinheim, Germany, 2016; pp. 453–475.
7. Corma, A.; García, H. Supported Gold Nanoparticles as Oxidation Catalysts. In *Nanoparticles and Catalysis*; Astruc, D., Ed.; Willey-VCH Verlag GmbH & Co. KGaA: Weinheim, Germany, 2008; pp. 389–429.
8. Santos, V.P.; Carabineiro, S.A.C.; Tavares, P.B.; Pereira, M.F.R.; Órfão, J.J.M.; Figueiredo, J.L. Oxidation of CO, ethanol and toluene over TiO_2 supported noble metal catalysts. *Appl. Catal. B Environ.* **2010**, *99*, 198–205. [CrossRef]
9. Bond, G.C.; Louis, C.; Thompson, D.T. *Catalysis by Gold*; Hutchings, G.J., Ed.; Catalytic Science Series; Imperial College Press: London, UK, 2006; pp. 180–182.
10. Ide, Y.; Nakamura, N.; Hattori, H.; Ogino, R.; Ogawa, M.; Sadakane, M.; Sano, T. Sunlight-induced efficient and selective photocatalytic benzene oxidation on TiO_2-supported gold nanoparticles under CO_2 atmosphere. *Chem. Commun.* **2011**, *47*, 11531–11533. [CrossRef] [PubMed]
11. Doustkhah, E.; Rostamnia, S.; Tsunoji, N.; Henzie, J.; Takei, T.; Yamauchi, Y.; Ide, Y. Templated synthesis of atomically-thin Ag nanocrystal catalysts in the interstitial space of a layered silicate. *Chem. Commun.* **2018**, *54*, 4402–4405. [CrossRef] [PubMed]
12. Galusek, D.; Ghillányová, K. Ceramic Oxides. In *Ceramics Science and Technology*; Riedel, R., Chen, I.-W., Eds.; Wiley-VCH Verlag GmbH & Co.: Weinheim, Germany, 2010.
13. Bagheri, S.; Julkapli, N.M.; Hamid, S.B.A. Titanium Dioxide as a Catalyst Support in Heterogeneous Catalysis. *Sci. J.* **2014**, *2014*, 1–21. [CrossRef]
14. Bamwenda, G.; Tsubota, S.; Nakamura, T.; Haruta, M. The influence of the preparation methods on the catalytic activity of platinum and gold supported on TiO_2 for CO oxidation. *Catal. Lett.* **1997**, *44*, 83–87. [CrossRef]
15. Sui, X.-L.; Wang, Z.-B.; Yang, M.; Huo, L.; Gu, D.-M.; Yin, G.-P. Investigation on $C-TiO_2$ nanotubes composite as Pt catalyst support for methanol electrooxidation. *J. Sources* **2014**, *255*, 43–51. [CrossRef]
16. Plata, J.J.; Romero-Sarria, F.; Amaya-Suarez, J.; Márquez, A.M.; Laguna, O.H.; Odriozola, J.A.; Sanz, J.F.; Ramos, J.J.P.; Sanz, J.F. Improving the activity of gold nanoparticles for the water-gas shift reaction using $TiO_2-Y_2O_3$: An example of catalyst design. *Phys. Chem. Chem. Phys.* **2018**, *20*, 22076–22083. [CrossRef] [PubMed]
17. Hidalgo, M.C.; Murcia, J.; Navío, J.A.; Colón, G.; Mesa, J.J.M. Photodeposition of gold on titanium dioxide for photocatalytic phenol oxidation. *Appl. Catal. A Gen.* **2011**, *397*, 112–120. [CrossRef]

18. Colón, G.; Hidalgo, M.C.; Navío, J.A. Photocatalytic behaviour of sulphated TiO_2 for phenol degradation. *Appl. Catal. B Environ.* **2003**, *45*, 39–50. [CrossRef]
19. Chen, X.; Sun, H.; Zhang, J.; Guo, Y.; Kuo, D.-H. Cationic S-doped TiO_2/SiO_2 visible-light photocatalyst synthesized by co-hydrolysis method and its application for organic degradation. *J. Mol. Liq.* **2019**, *273*, 50–57. [CrossRef]
20. Murcia, J.J.; Navio, J.A.; Hidalgo, M.C. Insights towards the influence of Pt features on the photocatalytic activity improvement of TiO_2 by platinisation. *Appl. Catal. B Environ.* **2012**, *126*, 76–85. [CrossRef]
21. Chenakin, S.; Kruse, N. Combining XPS and ToF-SIMS for assessing the CO oxidation activity of Au/TiO_2 catalysts. *J. Catal.* **2018**, *358*, 224–236. [CrossRef]
22. Chenakin, S.P.; Kruse, N. Au 4f spin–orbit coupling effects in supported gold nanoparticles. *Phys. Chem. Chem. Phys.* **2016**, *18*, 22778–22782. [CrossRef]
23. Okazaki, K.; Morikawa, Y.; Tanaka, S.; Tanaka, K.; Kohyama, M. Electronic structures of Au onTiO_2 (110) by first-principles calculations. *Phys. Rev. B* **2004**, *69*, 235404. [CrossRef]
24. Vittadini, A.; Selloni, A. Small gold clusters on stoichiometric and defected TiO_2 anatase (101) and their interaction with CO: A density functional study. *J. Chem. Phys.* **2002**, *117*, 353–361. [CrossRef]
25. Li, B.; Hao, Y.; Shao, X.; Tang, H.; Wang, T.; Zhu, J.; Yan, S. Synthesis of hierarchically porous metal oxides and Au/TiO_2 nanohybrids for photodegradation of organic dye and catalytic reduction of 4-nitrophenol. *J. Catal.* **2015**, *329*, 368–378. [CrossRef]
26. Sanchez, V.M.; Martínez, E.D.; Ricci, M.L.M.; Troiani, H.; Soler-Illia, G.J.A.A. Optical Properties of Au Nanoparticles Included in Mesoporous TiO_2 Thin Films: A Dual Experimental and Modeling Study. *J. Phys. Chem. C* **2013**, *117*, 7246–7259. [CrossRef]
27. Chen, H.; Shao, L.; Li, Q.; Wang, J. Gold nanorods and their plasmonic properties. *Chem. Soc. Rev.* **2013**, *42*, 2679–2724. [CrossRef] [PubMed]
28. Link, S.; El-Sayed, M.A. Size and Temperature Dependence of the Plasmon Absorption of Colloidal Gold Nanoparticles. *J. Phys. Chem. B* **1999**, *103*, 4212–4217. [CrossRef]
29. Ayati, A.; Ahmadpour, A.; Bamoharram, F.F.; Tanhaei, B.; Mänttäri, M.; Sillanpää, M. A review on catalytic applications of Au/TiO_2 nanoparticles in the removal of water pollutant. *Chemosphere* **2014**, *107*, 163–174. [CrossRef] [PubMed]
30. Maicu, M.; Hidalgo, M.C.; Colón, G.; Navío, J.A. Comparative study of the photodeposition of Pt, Au and Pd on pre-sulphated TiO_2 for the photocatalytic decomposition of phenol. *J. Photochem. Photobiol. A Chem.* **2011**, *217*, 275–283. [CrossRef]
31. Gołąbiewska, A.; Malankowska, A.; Jarek, M.; Lisowski, W.; Nowaczyk, G.; Jurga, S.; Zaleska-Medynska, A. The effect of gold shape and size on the properties and visible light-induced photoactivity of $Au-TiO_2$. *Appl. Catal. B Environ.* **2016**, *196*, 27–40. [CrossRef]
32. Sornalingam, K.; McDonagh, A.; Zhou, J.L.; Johir, M.A.H.; Ahmed, M.B. Photocatalysis of estrone in water and wastewater: Comparison between $Au-TiO_2$ nanocomposite and TiO_2, and degradation by-products. *Sci. Total Environ.* **2018**, *610*, 521–530. [CrossRef] [PubMed]
33. Kaur, R.; Pal, B. Size and shape dependent attachments of Au nanostructures to TiO_2 for optimum reactivity of $Au–TiO_2$ photocatalysis. *J. Mol. Catal. A Chem.* **2012**, *355*, 39–43. [CrossRef]
34. Panayotov, D.A.; Morris, J.R. Surface chemistry of Au/TiO_2: Thermally and photolytically activated reactions. *Surf. Sci. Rep.* **2016**, *71*, 77–271. [CrossRef]
35. Maciejewski, M.; Fabrizioli, P.; Grunwaldt, J.-D.; Becker, O.S.; Baiker, A. Supported gold catalysts for CO oxidation: Effect of calcination on structure, adsorption and catalytic behaviour. *Phys. Chem. Chem. Phys.* **2001**, *3*, 3846–3855. [CrossRef]
36. Yang, J.H.; Henao, J.D.; Raphulu, M.C.; Wang, Y.; Caputo, T.; Groszek, A.J.; Kung, M.C.; Scurrell, M.S.; Miller, J.T.; Kung, H.H. Activation of Au/TiO_2 Catalyst for CO Oxidation. *J. Phys. Chem. B* **2005**, *109*, 10319–10326. [CrossRef]
37. Weiher, N.; Beesley, A.M.; Tsapatsaris, N.; Delannoy, L.; Louis, C.; Van Bokhoven, J.A.; Schroeder, S.L.M. Activation of Oxygen by Metallic Gold in Au/TiO_2 Catalysts. *J. Am. Chem. Soc.* **2007**, *129*, 2240–2241. [CrossRef]
38. Wei, S.; Fu, X.-P.; Wang, W.-W.; Jin, Z.; Song, Q.-S.; Jia, C. Au/TiO_2 Catalysts for CO Oxidation: Effect of Gold State to Reactivity. *J. Phys. Chem. C* **2018**, *122*, 4928–4936. [CrossRef]

39. Laguna, O.H.; Domínguez, M.I.; Romero-Sarria, F.; Odriozola, J.A.; Centeno, M.A. *Role of Oxygen Vacancies in Gold Oxidation Catalysis*; Royal Society of Chemistry (RSC): London, UK, 2014; Chapter 13; pp. 489–511.
40. Sarria, F.R.; Martínez T, L.M.; Centeno, M.A.; Odriozola, J.A. Surface Dynamics of Au/CeO$_2$ Catalysts during CO Oxidation. *J. Phys. Chem. C* **2007**, *111*, 14469–14475. [CrossRef]
41. Kamiuchi, N.; Sun, K.; Aso, R.; Tane, M.; Tamaoka, T.; Yoshida, H.; Takeda, S. Self-activated surface dynamics in gold catalysts under reaction environments. *Nat. Commun.* **2018**, *9*, 2060. [CrossRef] [PubMed]
42. An, T.; Selloni, A.; Wang, H. Effect of reducible oxide–metal cluster charge transfer on the structure and reactivity of adsorbed Au and Pt atoms and clusters on anatase TiO$_2$. *J. Chem. Phys.* **2017**, *146*, 184703.
43. Kandoi, S.; Gokhale, A.; Grabow, L.; Dumesic, J.; Mavrikakis, M. Why Au and Cu Are More Selective Than Pt for Preferential Oxidation of CO at Low Temperature. *Catal. Lett.* **2004**, *93*, 93–100. [CrossRef]
44. Haruta, M.; Tsubota, S.; Kobayashi, T.; Kageyama, H.; Genet, M.; Delmon, B. Low-Temperature Oxidation of CO over Gold Supported on TiO$_2$, α-Fe$_2$O$_3$, and Co$_3$O$_4$. *J. Catal.* **1993**, *144*, 175–192. [CrossRef]
45. Cunningham, D.; Tsubota, S.; Kamijo, N.; Haruta, M. Preparation and catalytic behaviour of subnanometer gold deposited on TiO$_2$ by vaccum calcination. *Res. Chem. Intermed.* **1993**, *19*, 1–13. [CrossRef]
46. Murcia, J.J.; Ávila-Martínez, E.G.; Rojas, H.; Navío, J.A.; Hidalgo, M.C. Study of the E. coli elimination from urban wastewater over photocatalysts based on metallized TiO$_2$. *Appl. Catal. B Environ.* **2017**, *200*, 469–476. [CrossRef]
47. Sturges, H.A. The choice of a class interval Case I Computations involving a single. *J. Am. Stat. Assoc.* **1926**, *21*, 65–66. [CrossRef]
48. Channei, D.; Inceesungvorn, B.; Wetchakun, N.; Ukritnukun, S.; Nattestad, A.; Chen, J.; Phanichphant, S. Photocatalytic Degradation of Methyl Orange by CeO$_2$ and Fe–doped CeO$_2$ Films under Visible Light Irradiation. *Sci. Rep.* **2014**, *4*, 5757. [CrossRef] [PubMed]

© 2019 by the authors. Licensee MDPI, Basel, Switzerland. This article is an open access article distributed under the terms and conditions of the Creative Commons Attribution (CC BY) license (http://creativecommons.org/licenses/by/4.0/).

Article

Immobilization of Stabilized Gold Nanoparticles on Various Ceria-Based Oxides: Influence of the Protecting Agent on the Glucose Oxidation Reaction

Meriem Chenouf [1,2], Cristina Megías-Sayago [1], Fatima Ammari [2], Svetlana Ivanova [1], Miguel Angel Centeno [1,*] and José Antonio Odriozola [1]

1. Departamento de Química Inorgánica e Instituto de Ciencia de Materiales de Sevilla, Universidad de Sevilla-CSIC, Américo Vespucio 49, 41092 Sevilla, Spain; meriemc94@gmail.com (M.C.); cristina.megias@icmse.csic.es (C.M.-S.); svetlana@icmse.csic.es (S.I.); odrio@us.es (J.A.O.)
2. LGPC, Department of chemical process engineering, Ferhat-Abbas Sétif-1 University, Sétif 19000, Algeria; ammarifatima@yahoo.fr
* Correspondence: centeno@icmse.csic.es

Received: 31 December 2018; Accepted: 17 January 2019; Published: 31 January 2019

Abstract: The influence of the protecting agent's nature on gold particle size and dispersion was studied in this work over a series of gold-based catalysts. CO and glucose oxidation were chosen as catalytic reactions to determine the catalyst's structure–activity relationship. The nature of the support appeared to be the predominant factor for the increase in activity, as the oxygen mobility was decisive for the CO oxidation in the same way that the Lewis acidity was decisive for the glucose oxidation. For the same catalyst composition, the use of montmorillonite as the stabilizing agent resulted in better catalytic performance.

Keywords: gold nanoparticles; clay; PVA; stabilizing agent; glucose oxidation

1. Introduction

The use of renewable feedstocks to produce platform chemicals as an alternative to the classical petrochemical route is attracting more and more scientific attention [1,2]. The number of possibilities to convert lignocellulosic biomass into value-added chemicals is continuously increasing, focusing the research interest on this topic. New technologies have arisen, but the majority of these processes are still under investigation. One example of such a process is the production of gluconic acid, industrially available as an enzymatically catalyzed process. It is a process with great potential to become a heterogeneously catalyzed process in the future [3–5]. Gluconic acid and its salts are extensively used as detergents, food and beverage additives, and intermediates for the pharmaceutical and cosmetic industries [6,7]. Glucose to gluconic acid conversion goes through selective oxidation of the aldehyde function in anomeric position (C1), where all other alcoholic groups (C2-C6) must remain unaltered. Therefore the choice of a robust and highly selective catalyst is crucial to avoid side products [8].

Gold catalysts are often reported to be good candidates for oxidation reactions due to their remarkable oxidation ability and resistance to oxygen poisoning [9]. In a liquid phase, gold catalysts are especially useful, making it possible to use cheap and non-corrosive oxidants (air instead of $KMnO_4$ or H_2O_2) [10,11]. Gold catalysts are, in fact, reported to be promising for the selective production of gluconic acid [12–15].

Although they are very stable and effective in the presence of a base [16], gold catalysts deactivate rapidly in base-free conditions due principally to metal leaching and sintering [12]. The base-free conditions present some advantages and are highly desirable from a technological point of view. They are based on green chemistry principles and present some economic advantages due to the absence of time and money spent on reaction steps, such as gluconate salt separation and conversion to acid. This is why the most important requisite for using gold catalysts is to improve their stability in base-free conditions. For this purpose, the use of nanoparticle stabilizing agents, or a highly specific surface support could be useful to prevent both gold leaching and sintering.

On the other hand, nanoparticle homogeneity is of great importance to establish a correlation between catalytic activities, the nature of the active species, support and their interactions. Although several methods for the preparation of gold catalysts have been reported, the gold colloidal route seems to be one of the most appropriate to obtain homogeneous and reproducible gold nanoparticle size distribution. This route includes the utilization of stabilizing agents influencing the nucleation/growth mechanisms during the nanoparticle formation process. The stabilization of gold nanoparticles (AuNPs) can be achieved using surfactants [17], polymers [18,19] and metal oxides [20–22].

Recently, clay minerals have been reported as stabilizing agents of nanoparticles [23,24]. Clay minerals are natural materials that are environmentally benign, costless and abundant. They are largely used as sorbents [25,26] or in catalysis, for glycerol dehydration [27] and aromatic hydrocarbon production [28]. Their use as support for gold nanoparticles is also reported [28–32].

In this study, two different agents were used to stabilize gold nanoparticles prepared by the colloidal route, i.e. clay (montmorillonite (Mt)) and polymer (polyvinyl alcohol (PVA)). The catalytic performance in CO and glucose oxidation reactions of those nanoparticles immobilized over a series of supports were used to determine the influence of the stabilizing agent on gold particle size, dispersion and activity.

2. Results and Discussion

The specific surface areas, particle size and gold contents are reported in Table 1. A total of 25 to 50% of gold loss is observed within the series and it is hardly related to the nature of the supports and/or the specific surface area. Nevertheless, it could be explained by the stability of the gold colloids, depending on the nature of the protecting agent. The higher the stability of the colloids, the lower the quantity of gold deposited during the immobilization. It appears that PVA as a "liquid state" stabilizing agent (polymer completely dissolved within the media) helps particle adsorption and results in a higher final gold loading. As for the Mt stabilizer, the higher gold loss could be assigned either to a greater colloid stabilization or to a decrease of the metal adsorption rate caused by the changes produced in the isoelectric point of the supports with the addition of the clay material. The use of clays as the stabilizers of gold particles is also prone to higher experimental errors due to the delicate execution of the procedure.

Table 1. Catalysts metal loadings and structural properties.

Samples	Au Loading, wt.%	BET (m^2/g)	Pore Size (nm)	Pore Volume (cm^3/g)	Mean Particle Diameter dp(nm) TEM ± 1.2
CeO_2	-	137	6.5	0.23	-
Au-Mt/Ce	1.5	128	6	0.21	4 (6.8 *)
Au-PVA/Ce	1.6	137	5.5	0.21	3.9 (7 *)
50Ce50Zr	-	55	11	0.18	-
Au-Mt/50CeZr	1.2	50	10.4	0.16	5.1
Au-PVA/50CeZr	2.2	47	10.6	0.15	5.7
20CeAl	-	186	6.9	0.42	-
Au-Mt/20CeAl	1	168	7.7	0.36	5.6
Au-PVA/20CeAl	1.7	154	8.1	0.37	5.7

* particle size of the spent catalysts

The The Brunauer-Emmett-Teller (BET)-specific surface areas of the catalysts were similar to those of the corresponding supports. The use of different stabilizing agents did not significantly influence this parameter. All samples were mesoporous solids with a pore size within the 5–10 nm range and pore total volume increasing in order 50 CeZr < Ce < 20 CeAl.

XRD analysis of all samples, as seen in Figure 1, showed the presence of one dominating phase, the cubic fluorite CeO_2 structure (JCPDS #00-034-0394). With the addition of Zr (50CeZr sample), the fluorite diffractions shifted to higher 2θ angles, an effect caused by the formation of a solid solution. The shift is also indicative of the ceria lattice contraction due to the replacement of Ce^{4+} (0.098 nm ionic radii) with smaller Zr^{4+} cations (0.084 nm). Thus, the observed diffractions for this sample corresponded to the cubic Ce-Zr solid solution (JCPDS #00-028-0271). For the 20CeAl sample, the presence of g- alumina (JCPDS #00-048-0367) was also detected.

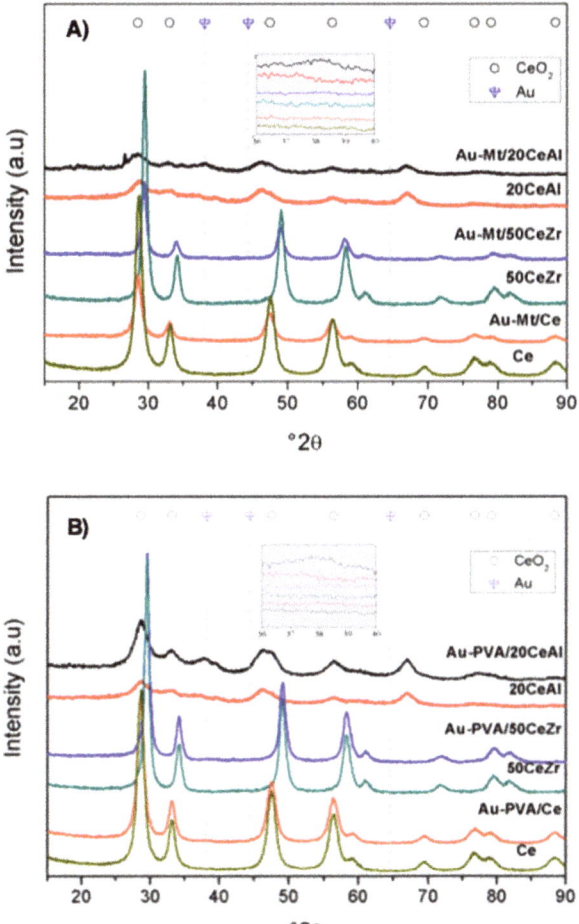

Figure 1. X-Ray diffractograms of supports and catalysts: (**A**) Mt- and (**B**) Polyvinylalcohol (PVA)-protected solids, insets zoom on Au (111) diffraction.

For all clay-stabilized samples, the presence of Mt was not evident, probably due to the high intensity of the ceria phase diffractions. Nevertheless, for the 20CeAl support two additional diffractions at 19 and 26 ° 2Q appeared, which could be attributed to the Mt phase but also to g-alumina.

For the catalysts, diffractions corresponding to gold phases were not detected, suggesting an average gold particle size inferior to 4 nm. The in-zone XRD analysis (36–40 2θ range) also suggested important gold dispersion. Only for the Au/20CeAl catalyst did a diffraction situated at the typical angle of Au(111) plane appear, but it could also be ascribed to the overlapping g-alumina phase.

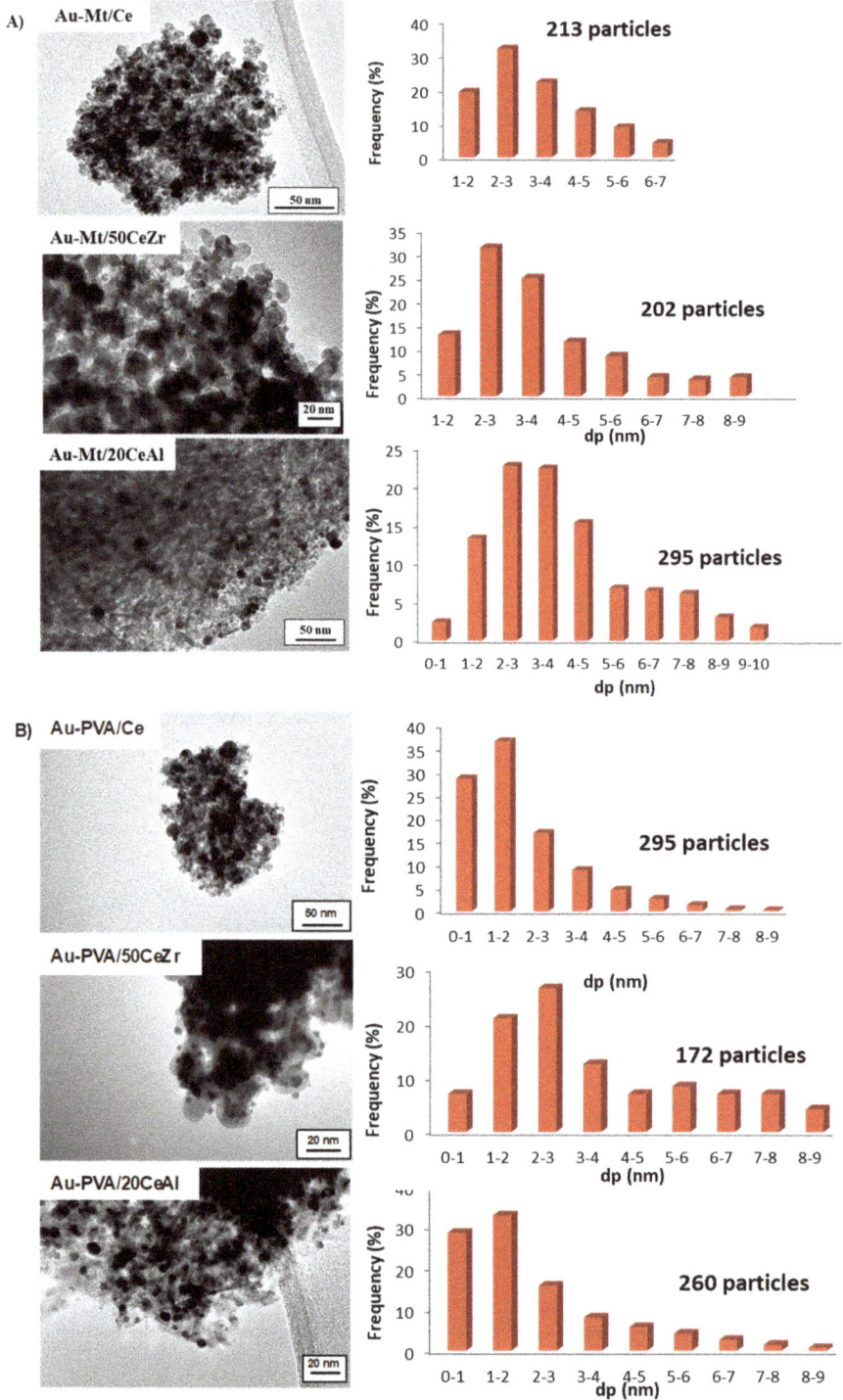

Figure 2. TEM micrographs and particle size distributions of (**A**) Mt- and (**B**) PVA-solids.

The TEM analysis, seen in Figure 2, showed that the calculated average particle size, seen in Table 1, was not affected by the stabilizing agent but by the support. The gold particles were smaller on the bare ceria support than on its mixed analogues.

The reaction of CO oxidation is often described as very sensitive to the gold metal state (loadings and dispersion) and to the nature of the support [33,34]. The CO oxidation activity in terms of the light off conversion curves is presented in Figure 3.

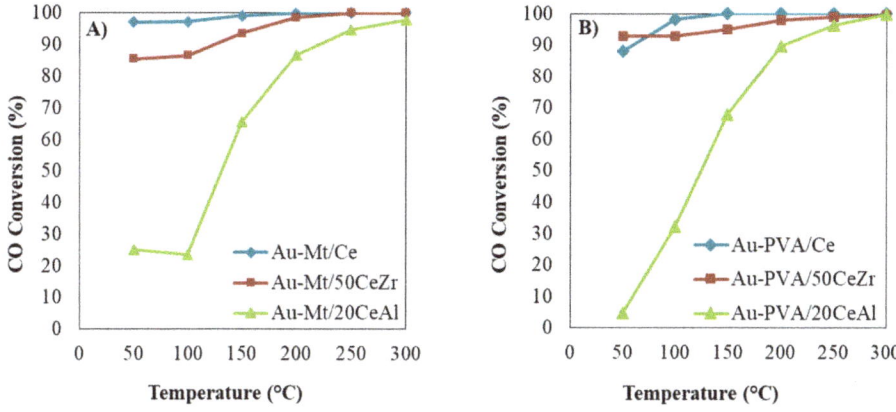

Figure 3. CO conversion (%) over (**A**) Mt-solids and (**B**) PVA-solids.

It was observed that the activity was especially influenced by the support composition. Whereas the Ce and 50CeZr-supported samples attained almost full conversion at the lowest temperature of measurement, the 20CeAl-supported catalyst was active at a higher temperature, regardless of the stabilizing agent. Taking into account the different gold loadings and dispersion for every catalyst, a normalization of the catalytic activity was made for comparison. The specific activity was expressed in millimoles of CO converted per gram of catalyst and second, as well as the turnover frequency (TOF), which was defined as the number of moles of CO converted per mole of surface gold atom and second, were calculated for all samples at 50 °C. The results are summarized in Table 2. The required dispersion was calculated using the mathematical model for cuboctahedral particles [35,36], which is also presented in Table 2.

Table 2. Catalyst' specific activity and gold particles dispersion.

Sample	Specific Reaction Rate, mmolesCO$_{conv}$·g$_{cat}^{-1}$·s^{-1}	TOF(CO), s^{-1} ×10^2	TOF(Glucose) ×10^3, s^{-1}	Dispersion
Au-Mt/Ce	0.69	41.3	4.7	0.32
Au-PVA/Ce	0.58	34.4	3.8	0.33
Au-Mt/50CeZr	0.77	58.5	6.3	0.26
Au-PVA/50CeZr	0.45	37.9	3.9	0.23
Au-Mt/20CeAl	0.13	10.5	8.5	0.23
Au-PVA/20CeAl	0.031	2.6	6.5	0.23
Au/Ce [13]	-	-	1.7	0.33
Au/20CeAl [13]	-	-	2.6	0.38
Au/50CeZr [13]	-	-	1.6	0.38

The normalization of the activity by the active phase loading (specific activity) and dispersion (TOF) presents the same main trend. It is clear that the catalyst activity depends on the nature of both the support and the stabilizing agent. However, the predominant factor is the nature of the support, over catalytic activity. The CO oxidation over ceria-based catalysts proceeds via the Mars van Krevelen mechanism, where the support participates in the reaction, supplying active oxygen species.

The increase in oxygen mobility, reported to occur with Zr addition [37,38], results in a better catalyst (Au/Ce50Zr > Au/Ce > Au20CeAl). As long as the ceria surface oxygen is accessible and abundant, the catalyst activity is higher.

As for the nature of the stabilizing agent, the use of Mt is more important than the use of polymer (PVA). A few probable causes for this, is the existence of some carbonaceous residues after the calcination procedure, a change of the gold–support interaction in the presence of organic agents and/or the electronic effect of metal. A more detailed study is needed to distinguish between the causes. Contrary to a gas phase reaction (CO oxidation), where the oxidant is continuously renewed, the liquid phase oxidation (glucose oxidation) depends on the concentration of dissolved oxygen in the media, which is more affected by the pressure than by the temperature. Higher partial pressures and lower temperatures promote a full glucose conversion. In order to observe and evaluate the real effect of the stabilizing agent and support nature, we decided to decrease the conversion by using higher temperatures and atmospheric pressure. The temperature did not affect the glucose conversion over the Au-Mt/Ce catalyst (Figure 4), but produced interesting changes in selectivity. An increase in the temperature promoted lactic acid formation, disfavoring the glucose to fructose reaction. Both products affected the gluconic acid selectivity, with the best activity/selectivity ratio observed at 100 °C.

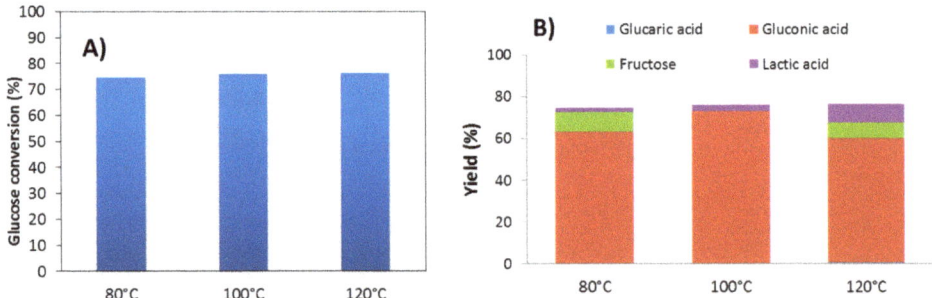

Figure 4. (**A**) Glucose conversion (%) as a function of the temperature over Au-Mt/Ce (600 rpm, 18 h); (**B**) yield (%).

As the major changes in selectivity occurred at 120 °C, this temperature was chosen for all other experiments. The glucose conversion and calculated TOF at the final time of the reaction (18 h) are summarized in Figure 5 and Table 2, respectively.

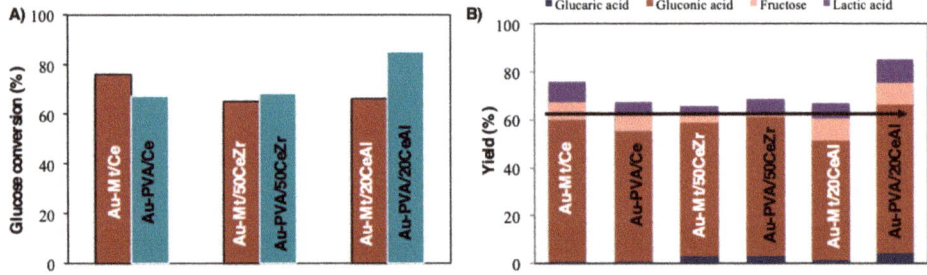

Figure 5. Glucose conversion (**A**) and product yields (**B**) as a function of support.

The glucose conversion did not follow a clear trend. However, the TOF values (calculated as mole of converted glucose per mole of exposed Au sites (moles Au*dispersion) and per hour) showed that the PVA-stabilized samples were less active than the clay-stabilized materials. The observed CO-TOF values were slightly inferior to those reported for other gold catalysts [34]. As for the glucoseTOF values, the use of gold-stabilized colloids resulted in higher specific activity in comparison to very similar catalysts over the same supports, as presented in Table 2. It appears that this method of metal deposition was more effective, especially for the liquid phase reactions. However, the Au-Mt/Ce and Au-PVA/Ce catalysts were less active than the Au/C systems prepared by the same colloidal immobilization method used in this study [39].

This was true for both reactions, CO and glucose oxidation, which means that they were influenced in the same manner. Taking into account that the variable here was the stabilizing agent and that all the samples had similar particle size, we propose two more reasons explaining this variation: (i) electronic state stabilization, and/or (ii) the presence of carbonaceous leftovers covering the active sites.

As expected, the type of support also played an important role, and the order of activity was 20 CeAl > 50 CeZr > Ce. The activity trend did not correlate with the surface area. However, the larger pore size appeared to be beneficial to obtain higher conversion. On the other hand, a similar activity trend, previously reported for similar Au catalysts, was related to the decrease in the strength of the Lewis acid sites in the same order as the activity (20 CeAl > 50 CeZr > Ce) [13]. The product yield distribution was similar, suggesting that the selectivity was a function of temperature and not catalyst state, as seen in Figure 5B.

The addition of base, as seen in Figure 6, promotes glucose conversion and changes the selectivity balance whatever the nature of catalyst. An important effect of the base addition was the complete absence of gold metal leaching, observed for the spent samples, which on first sight appeared satisfactory. However, this fact and even the slight increase in glucose conversion in the base conditions did not justify their use since a decrease in the gluconic acid yield in favor of fructose formation was observed [40].

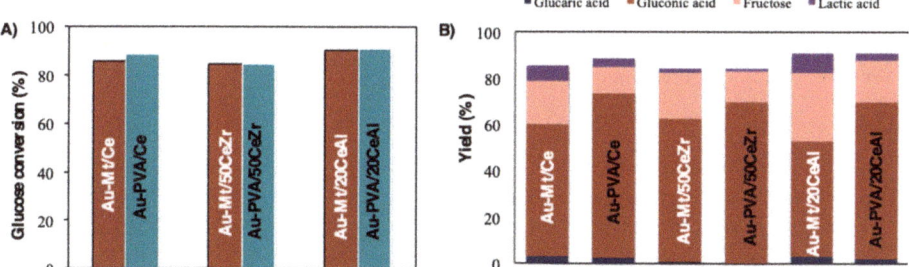

Figure 6. (**A**) Glucose conversion (%) in NaOH presence, glucose: NaOH molar ratio 1:2 (600 rpm, 18 h); (**B**) yield (%).

Four cycles of utilization were carried out over the Au-Mt/Ce catalyst with a constant glucose-to-catalyst ratio, as seen in Figure 7. This sample was selected to representing all of the samples, as a similar trend was observed in all cases. Between the cycles, the catalysts were recovered by centrifugation, washed in water and dried overnight prior to their re-utilization. A decrease in the initial activity was observed after the first cycle and remained constant afterwards. Our previous study of gold catalyst recycling in base-free reactions proposed three reasons for catalyst deactivation: (i) gold particle size, (ii) gold leaching, and (iii) catalyst surface covered with reaction intermediates [12]. The third reason was quickly discarded, and only the change of gold metal state (loadings and dispersion) was considered as the major reason for catalyst deactivation. In the case of the Au-Mt/Ce catalyst, both factors should be considered. The size of the gold particles increased to 6.8 nm after the first cycle, and the same was observed for the Au-PVA/Ce sample (7 nm). However, the excessive metal leaching detected after the first cycle (around 20% occurring between the 1st and the 2nd cycle and not changing after this) suggests that the probable main reason for this activity change was gold metal loss, which was also observed for the gold catalysts supported over mineral supports [13].

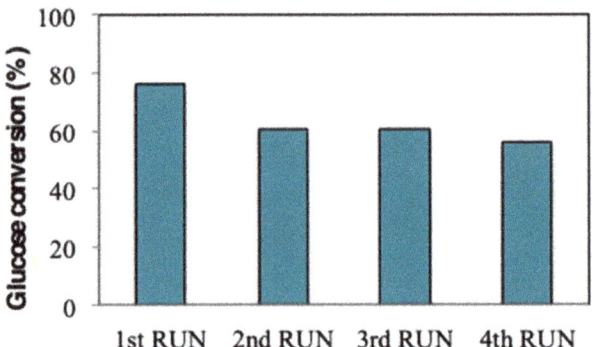

Figure 7. Reusability study over the Au-Mt/Ce catalyst.

3. Experimental

3.1. Stabilizing Agents

Polyvinylalcohol (PVA) (Sigma Aldrich, Mw 31000, San Luis, MO, USA) and natural bentonite, obtained from the Roussel deposit of Maghnia (Algeria), were the stabilizing agents.

Prior to use, the natural bentonite underwent purification by sedimentation to remove all impurities (mostly sand, feldspar and calcite). In a typical preparation, 10 g of bentonite were vigorously stirred in 1 L of distilled water for 3 h and left to sediment for 24 h. A total of 2/3 of the uppermost supernatant portion was then separated and dried overnight at 80 °C. The purified bentonite, mainly montmorillonite, was subsequently transformed to its homoionic Na-exchanged form by treatment with 1 M sodium chloride solution. As the stabilizer, only the <2 mm fraction of Na-montmorillonite was retained. The sample label was Mt.

3.2. Gold Catalysts

$HAuCl_4$ (Johnson Matthey, 49.99%, London, UK) and $NaBH_4$ (Sigma Aldrich, 98%, San Luis, MO, USA) were used as gold precursor and reducing agent, respectively. Three commercial ceria-based solids were used as supports for the gold nanoparticle immobilization: CeO_2(20%wt)/Al_2O_3 from Sasol (denoted as 20CeAl) and pure CeO_2 and CeO_2(50%wt)/ZrO_2 sample from Daiichi Kigenso Kagaku Kogio Co., Ltd. (Osaka, Japan) (denoted as Ce and 50CeZr, respectively).

3.2.1. Mt as Stabilizing Agent

Catalysts with a nominal gold loading of 2 wt%, were prepared via the colloidal route. A total of 50 mL of a 10^{-3} M $HAuCl_4$ solution was reduced with 10 ml $NaBH_4$ aqueous solution (1 mmol) in the presence of Mt montmorillonite (0.5 g) as the stabilizer. Finally, the obtained gold colloids were immobilized over the corresponding supports at a pH lower than its isoelectric point (4.8), with the pH adjusted by the addition of 0.1 M HCl. After centrifugation, washing, and drying (overnight at 80 °C) the catalysts labeled Au-Mt/Ce, Au-Mt/50CeZr and Au-Mt/20CeAl were used directly in the reactions.

3.2.2. PVA as Stabilizing Agent

A similar colloidal route was used for the samples using PVA (Sigma Aldrich, Mw 31000, San Luis, MO USA) as the stabilizing agent (2 wt% Au, nominal value). A total of 10^{-3} M $HAuCl_4$ solution was reduced with $NaBH_4$ in presence of PVA in Au:PVA, mass ratio of 1.5:1 and the obtained colloid was immobilized over the supports. The solids were then filtered, washed and dried at 80 °C for 24 h and calcined in air at 350 °C for 4 h with 3 °C/min^{-1} heating rate. The catalysts received the labels Au-PVA/Ce, Au-PVA/50CeZr, and Au-PVA/20CeAl, respectively.

3.3. Catalytic Test

3.3.1. CO Oxidation

The activity measurements were carried out at atmospheric pressure in a cylindrical stainless steel fixed bed reactor (7.5 mm inner diameter). The catalysts (100 mg) were activated in 50 mL min^{-1} air flow at 300 °C, for 1 h (heating rate of 10 °C.min^{-1}). Then, the system was cooled down to 50 °C and the reactive mixture, composed of 3.4% CO (Air Liquide, 99.997%), and 21% O_2 (Air Liquide, 99.999%) balanced in nitrogen, was introduced into the reactor. The total gas flow employed in the catalytic test was 42 mL/min^{-1} corresponding to a weight hourly space velocity (WHSV) of 31.5 L $g_{(cat)}^{-1}$ h^{-1}. The catalytic activity was measured at each temperature increase of 50 °C up to 300 °C. The inlet and outlet of the reactor gas concentrations were quantified by the online gas chromatograph Agilent® 6890 equipped with HPLOT Q and HP-5 columns and a TCD detector. The CO conversion was calculated according to Equation (1).

$$CO\,conversion = \frac{CO_{in} - CO_{out}}{CO_{in}} \times 100 \qquad (1)$$

where CO_{in} is the inlet CO concentration and CO_{out} is that of the outlet.

3.3.2. Glucose Oxidation

D-(+)-Glucose (anhydrous, 99%) purchased from Alfa Aesar was used as received for the catalytic test performed in a glass batch reactor (50 mL) equipped with Young valve and magnetic stirrer. The reactor containing 5 mL 0.2M glucose solution and catalyst in Glucose/Au molar ratio of 100 was contacted and saturated with oxygen at atmospheric pressure (approximate P(O_2) of 0.1 MPa). The reaction occurred upon constant stirring (600 rpm) at different temperatures (80 °C–120 °C range) in base free conditions as a function of time. After reaction, a 500 µL aliquot from the final mixture, diluted in 500 µL of MilliQ water, was immediately analyzed by HPLC, in a Varian 360-LC instrument equipped with a refractory index detector and Hi- Plex H column preheated at 40 °C using water as the mobile phase with a 0.4 mL.min^{-1} flow rate.

Conversion, selectivity and yield calculations were based on the HPLC measurements. The reported conversions were obtained after comparing the glucose concentration before and after the reaction, as shown in Equation (2). On the other hand the selectivity was calculated on the basis of the analyzed carbon moles, as described in Equation (3). Finally, the yields were calculated by Equation (4).

$$conversion = \frac{[glucose]_i - [glucose]_f}{[glucose]_i} \times 100 \qquad (2)$$

$$selectivity = \frac{n_i}{n_T} \times 100 \qquad (3)$$

$$Yield = \frac{conversion}{100} \times selectivity \qquad (4)$$

where $[glucose]_i$ and $[glucose]_f$ are the initial and final number of glucose moles and n_i and n_T are the C moles of the *i* products and the total number of analyzed C moles.

3.4. Characterization

BET-specific surface areas and pore diameter, calculated by the Barrett-Joyner-Halenda (BJH) method, were obtained using the Micromeritics Tristar II equipment. The samples were outgassed before analysis at 250 °C in vacuum 2 h prior to N_2 adsorption.

The XRD analyses were performed on an X'Pert Pro PANanalytical instrument (Almelo, The Netherlands) using CuKa radiation in the 5–80° 2θ range with a 0.01° step size and 300 s step time.

X-ray Fluorescence (XRF) using Panalytical AXIOS spectrometer (Almelo, The Netherlands) with an Rh tube as the source of radiation determined the actual gold loadings.

The transmission electron microscopy (TEM) analysis was carried out on the TOPCON-002B apparatus (Tokyo, Japan). Prior to analysis, all samples were dispersed in ethanol and deposited on a holey carbon copper grid. The particle size distribution was estimated to be over 30 micrographs and the mean gold particle diameter was considered based on its homogeneity over an important number of particles (>100 counted particles).

4. Conclusions

Gold nanoparticles were successfully prepared and stabilized using montmorillonite and PVA as the protecting agents. A similar particle size and distribution was observed in both cases. The catalytic activity of the samples depends strongly on the support nature. The presence of mobile oxygen species was very important for the gas phase CO oxidation. Meanwhile, the Lewis sites strength and the rate of oxygen dissolution were the predominant parameters for the glucose liquid phase oxidation. In both reactions, the use of montmorillonite as the stabilizing agent resulted in more active catalysts tentatively ascribed to a different gold electronic state and/or to the presence of carbonaceous leftovers influencing the gold/support interaction and subsequent CO/glucose adsorption step.

Author Contributions: Conceptualization, S.I.; methodology, F.A.; validation, S.I., F.A. and M.A.C.; investigation, M.C. and C.M.-S.; resources, J.A.O. and M.A.C.; writing—original draft preparation, S.I.; writing—review and editing, M.A.C.; visualization, S.I. and M.A.C.; supervision, J.A.O.; funding acquisition, J.A.O.

Funding: This research received no external funding.

Acknowledgments: Meriem Chenouf acknowledges the Algerian Ministry of Higher Education and Scientific Research (MESRS) and the University of Seville, Spain (Program for international students mobility, 3er Plan Propio de Docencia).

Conflicts of Interest: The authors declare no conflict of interest.

References

1. Chheda, J.N.; Huber, G.W.; Dumesic, J.A. Liquid-phase catalytic processing of biomass-derived oxygenated hydrocarbons to fuels and chemicals. *Angew. Chem. Int. Ed.* **2007**, *46*, 7164–7183. [CrossRef] [PubMed]
2. Serrano-Ruiz, J.C.; Luque, R.; Sepúlveda-Escribano, A. Transformations of biomass-derived platform molecules: From high added-value chemicals to fuels via aqueous-phase processing. *Chem. Soc. Rev.* **2011**, *40*, 5266–5281. [CrossRef] [PubMed]
3. Ishida, T.; Kinoshita, N.; Okatsu, H.; Akita, T.; Takei, T.; Haruta, M. Influence of the support and the size of gold clusters on catalytic activity for glucose oxidation. *Angew. Chem. Int. Ed.* **2008**, *47*, 9265–9268. [CrossRef] [PubMed]
4. Mirescu, A.; Prüße, U. Selective glucose oxidation on gold colloids. *Catal. Commun.* **2006**, *7*, 11–17. [CrossRef]
5. Rautiainen, S.; Lehtinen, P.; Vehkamäki, M.; Niemelä, K.; Kemell, M.; Heikkilä, M.; Repo, T. Microwave-assisted base-free oxidation of glucose on gold nanoparticle catalysts. *Catal. Commun.* **2016**, *74*, 115–118. [CrossRef]
6. Climent, M.J.; Corma, A.; Iborra, S. Converting carbohydrates to bulk chemicals and fine chemicals over heterogeneous catalysts. *Green Chem.* **2011**, *13*, 520–540. [CrossRef]
7. Hustede, E.S.H.; Haberstroh, H.J.; Schinzig, E. *Ullmann's Encyclopedia of Industrial Chemistry*; Wiley-VCH Verlag GmbH & Co. KGaA: Weinheim, Germany, 2000.
8. Wojcieszak, R.; Cuccovia, I.M.; Silva, M.A.; Rossi, L.M. Selective oxidation of glucose to glucuronic acid by cesium-promoted gold nanoparticle catalyst. *J. Mol. Catal. A Chem.* **2016**, *422*, 35–42. [CrossRef]
9. Biella, S.; Prati, L.; Rossi, M. Selective oxidation of D-glucose on gold catalyst. *J. Catal.* **2002**, *206*, 242–247. [CrossRef]
10. Amarasekara, A.S.; Green, D.; McMillan, E. Efficient oxidation of 5-hydroxymethylfurfural to 2,5-diformylfuran using Mn (III)-salen catalysts. *Catal. Commun.* **2008**, *9*, 286–288. [CrossRef]
11. Saliger, R.; Decker, N.; Prüße, U. D-Glucose oxidation with H_2O_2 on an Au/Al2O3 catalyst. *Appl. Catal. B Environ.* **2011**, *102*, 584–589. [CrossRef]
12. Megías-Sayago, C.; Bobadilla, L.F.; Ivanova, S.; Penkova, A.; Centeno, M.A.; Odriozola, J.A. Gold catalyst recycling study in base-free glucose oxidation reaction. *Catal. Today* **2018**, *301*, 72–77. [CrossRef]

13. Megías-Sayago, C.; Ivanova, S.; López-Cartes, C.; Centeno, M.A.; Odriozola, J.A. Gold catalysts screening in base-free aerobic oxidation of glucose to gluconic acid. *Catal. Today* **2017**, *279*, 148–154. [CrossRef]
14. Miedziak, P.J.; Alshammari, H.; Kondrat, S.A.; Clarke, T.J.; Davies, T.E.; Morad, M.; Morgan, D.J.; Willock, D.J.; Knight, D.W.; Taylor, S.H.; et al. Base-free glucose oxidation using air with supported gold catalysts. *Green Chem.* **2014**, *16*, 3132–3141. [CrossRef]
15. Qi, P.; Chen, S.; Chen, J.; Zheng, J.; Zheng, X.; Yuan, Y. Catalysis and reactivation of ordered mesoporous carbon-supported gold nanoparticles for the base-free oxidation of glucose to gluconic acid. *ACS Catal.* **2015**, *5*, 2659–2670. [CrossRef]
16. Prüße, U.; Herrmann, M.; Baatz, C.; Decker, N. Gold-catalyzed selective glucose oxidation at high glucose concentrations and oxygen partial pressures. *Appl. Catal. A Gen.* **2011**, *406*, 89–93. [CrossRef]
17. Schrekker, H.S.; Gelesky, M.A.; Stracke, M.P.; Schrekker, C.M.L.; Machado, G.; Teixeira, S.R.; Rubim, J.C.; Dupont, J. Disclosure of the imidazolium cation coordination and stabilization mode in ionic liquid stabilized gold (0) nanoparticles. *J. Colloid Interface Sci.* **2007**, *316*, 189–195. [CrossRef] [PubMed]
18. Shang, K.; Geng, Y.; Xu, X.; Wang, C.; Lee, Y.I.; Hao, J.; Liu, H.G. Unique self-assembly behavior of a triblock copolymer and fabrication of catalytically active gold nanoparticle/polymer thin films at the liquid/liquid interface. *Mater. Chem. Phys.* **2014**, *146*, 88–98. [CrossRef]
19. Zhang, W.; Liu, B.; Zhang, B.; Bian, G.; Qi, Y.; Yang, X.; Li, C. Synthesis of monodisperse magnetic sandwiched gold nanoparticle as an easily recyclable catalyst with a protective polymer shell. *Colloids Surf. A Physicochem. Eng. Asp.* **2015**, *466*, 210–218. [CrossRef]
20. Boronat, M.; Corma, A.; Illas, F.; Radilla, J.; Ródenas, T.; Sabater, M.J. Mechanism of selective alcohol oxidation to aldehydes on gold catalysts: Influence of surface roughness on reactivity. *J. Catal.* **2011**, *278*, 50–58. [CrossRef]
21. Ivanova, S.; Pitchon, V.; Zimmermann, Y.; Petit, C. Preparation of alumina supported gold catalysts: Influence of washing procedures, mechanism of particles size growth. *Appl. Catal. A Gen.* **2006**, *298*, 57–64. [CrossRef]
22. Zanella, R.; Giorgio, S.; Shin, C.H.; Henry, C.R.; Louis, C. Characterization and reactivity in CO oxidation of gold nanoparticles supported on TiO_2 prepared by deposition-precipitation with NaOH and urea. *J. Catal.* **2004**, *222*, 357–367. [CrossRef]
23. Agarwal, S.; Ganguli, J.N. Selective hydrogenation of monoterpenes on rhodium (0) nanoparticles stabilized in Montmorillonite K-10 clay. *J. Mol. Catal. A Chem.* **2013**, *372*, 44–50. [CrossRef]
24. Sarmah, P.P.; Dutta, D.K. Stabilized Rh0-nanoparticles-Montmorillonite clay composite: Synthesis and catalytic transfer hydrogenation reaction. *Appl. Catal. A Gen.* **2014**, *470*, 355–360. [CrossRef]
25. Eren, E.; Afsin, B. An investigation of Cu (II) adsorption by raw and acid-activated bentonite: A combined potentiometric, thermodynamic, XRD, IR, DTA study. *J. Hazard. Mater.* **2008**, *151*, 682–691. [CrossRef]
26. Sarma, G.K.; Sen Gupta, S.; Bhattacharyya, K.G. Adsorption of crystal violet on raw and acid-treated montmorillonite, K10, in aqueous suspension. *J. Environ. Manag.* **2016**, *171*, 1–10. [CrossRef] [PubMed]
27. Zhao, H.; Zhou, C.H.; Wu, L.M.; Lou, J.Y.; Li, N.; Yang, H.M.; Tong, D.S.; Yu, W.H. Catalytic dehydration of glycerol to acrolein over sulfuric acid-activated montmorillonite catalysts. *Appl. Clay Sci.* **2013**, *74*, 154–162. [CrossRef]
28. Letaief, S.; Grant, S.; Detellier, C. Phenol acetylation under mild conditions catalyzed by gold nanoparticles supported on functional pre-acidified sepiolite. *Appl. Clay Sci.* **2011**, *53*, 236–243. [CrossRef]
29. Álvarez, A.; Moreno, S.; Molina, R.; Ivanova, S.; Centeno, M.A.; Odriozola, J.A. Gold supported on pillared clays for CO oxidation reaction: Effect of the clay aggregate size. *Appl. Clay Sci.* **2012**, *69*, 22–29. [CrossRef]
30. Carriazo, J.G.; Martínez, L.M.; Odriozola, J.A.; Moreno, S.; Molina, R.; Centeno, M.A. Gold supported on Fe, Ce, and Al pillared bentonites for CO oxidation reaction. *Appl. Catal. B Environ.* **2007**, *72*, 157–165. [CrossRef]
31. Martínez, L.M.T.; Domínguez, M.I.; Sanabria, N.; Hernández, W.Y.; Moreno, S.; Molina, J.A.; Odriozola, J.A.; Centeno, M.A. Deposition of Al-Fe pillared bentonites and gold supported Al-Fe pillared bentonites on metallic monoliths for catalytic oxidation reactions. *Appl. Catal. A Gen.* **2009**, *364*, 166–173. [CrossRef]
32. Zhu, L.; Letaief, S.; Liu, Y.; Gervais, F.; Detellier, C. Clay mineral-supported gold nanoparticles. *Appl. Clay Sci.* **2009**, *43*, 439–446. [CrossRef]
33. Carabineiro, S.A.C.; Silva, A.M.T.; Draić, G.; Tavares, P.B.; Figueiredo, J.L. Gold nanoparticles on ceria supports for the oxidation of carbon monoxide. *Catal. Today* **2010**, *154*, 21–30. [CrossRef]
34. Grunwaldt, J.-D.; Kiener, C.; Wögerbauer, C.; Baiker, A. Preparation of supported gold catalysts for low-temperature CO oxidation via "size-controlled" gold colloids. *J. Catal.* **1999**, *181*, 223–232. [CrossRef]

35. Ivanova, S.; Pitchon, V.; Petit, C.; Caps, V. Support effects in the gold-catalyzed preferential oxidation of CO. *ChemCatChem* **2010**, *2*, 556–563. [CrossRef]
36. Ivanova, S.; Pitchon, V.; Petit, C. Application of the direct exchange method in the preparation of gold catalysts supported on different oxide materials. *J. Mol. Catal. A Chem.* **2006**, *256*, 278–283. [CrossRef]
37. Madier, Y.; Descorme, C.; Le Govi, A.M.; Duprez, D. Oxygen mobility in CeO_2 and $Ce_x Zr_{(1-x)} O_2$ compounds: Study by CO transient oxidation and 18O/16O isotopic exchange. *J. Phys. Chem. B* **1999**, *103*, 10999–11006. [CrossRef]
38. Comotti, M.; Pina, C.; Della Rossi, M. Mono- and bimetallic catalysts for glucose oxidation. *J. Mol. Catal. A Chem.* **2006**, *251*, 89–92. [CrossRef]
39. Megías-Sayago, C.; Santos, J.L.; Ammari, F.; Chenouf, M.; Ivanova, S.; Centeno, M.A.; Odriozola, J.A. Influence of gold particle size in Au/C catalysts for base-free oxidation of glucose. *Cat. Today* **2018**, *306*, 183–190. [CrossRef]
40. Kooyman, C.; Vellenga, K.; De Wilt, H.G.J. The isomerization of D-glucose into D-fructose in aqueous alkaline solutions. *Carbohydr. Res.* **1977**, *54*, 33–44. [CrossRef]

© 2019 by the authors. Licensee MDPI, Basel, Switzerland. This article is an open access article distributed under the terms and conditions of the Creative Commons Attribution (CC BY) license (http://creativecommons.org/licenses/by/4.0/).

Article

Properties of Carbon-supported Precious Metals Catalysts under Reductive Treatment and Their Influence in the Hydrodechlorination of Dichloromethane

Alejandra Arevalo-Bastante, Maria Martin-Martinez *, M. Ariadna Álvarez-Montero, Juan J. Rodriguez and Luisa M. Gómez-Sainero

Departamento de Ingeniería Química, Facultad de Ciencias, Universidad Autónoma de Madrid, Cantoblanco, 28049 Madrid, Spain; alejandrarevalo.b@gmail.com (A.A.-B.); ariadna.alvarez@uam.es (M.A.Á.-M.); juanjo.rodriguez@uam.es (J.J.R.); luisa.gomez@uam.es (L.M.G.-S.)
* Correspondence: maria.martin.martinez@uam.es; Tel.: +34-91-497-5725

Received: 27 November 2018; Accepted: 14 December 2018; Published: 18 December 2018

Abstract: This study analyzes the effect of the reduction temperature on the properties of Rh, Pt and Pd catalysts supported on activated carbon and their performance in the hydrodechlorination (HDC) of dichloromethane (DCM). The reduction temperature plays an important role in the oxidation state, size and dispersion of the metallic phase. Pd is more prone to sintering, followed by Pt, while Rh is more resistant. The ratio of zero-valent to electro-deficient metal increases with the reduction temperature, with that effect being more remarkable for Pd and Pt. The higher resistance to sintering of Rh and the higher stability of electro-deficient species under thermal reductive treatment can be attributed to a stronger interaction with surface oxygen functionalities. Dechlorination activity and a TOF increase with reduction temperature (250–450 °C) occurred in the case of Pt/C catalyst, while a great decrease of both was observed for Pd/C, and no significant effect was found for Rh/C. Pt^0 represents the main active species for HDC reaction in Pt/C. Therefore, increasing the relative amount of these species increased the TOF value, compensating for the loss of dispersion. In contrast, Pd^{n+} appears as the main active species in Pd/C and their relatively decreasing occurrence together with the significant decrease of metallic area reduces the HDC activity. Rh/C catalyst suffered only small changes in dispersion and metal oxidation state with the reduction temperature and thus this variable barely affected its HDC activity.

Keywords: precious metals; reduction temperature; hydrodechlorination; XPS; dispersion; turnover frequency

1. Introduction

Catalysis plays a crucial role on the path of building a more sustainable industrial chemistry and diminishes the impact of the industrial processes in the environment. Synthesizing novel, more stable, active and selective catalysts, or effectively improving the existing ones, is a fundamental task.

Chloromethanes are chlorinated volatile organic compounds with serious environmental impact, due to their carcinogenic and highly toxic character, in addition to the production of photochemical smog, the depletion of ozone and global warming [1–4]. Regardless of their noxious effects, some of these compounds, like dichloromethane (DCM) and chloroform (TCM), remain irreplaceable in some of their applications because of their singular chemical and physical properties (high stability, volatility and solvent capacity, and low flammability). Hence, they are still extensively used in industry, and large amounts are discharged into the environment through gaseous and liquid streams.

Catalytic hydrodechlorination (HDC) is a promising technology for the treatment of these contaminated streams. It allows the conversion of organochlorinated species like chloromethanes into harmless chlorine-free compounds under relatively mild conditions. It is effective within a wide range of concentrations, and thus presents environmental and economic advantages over other techniques [5–7]. Furthermore, it may be applied 'in situ', in combination with other physical separation processes like adsorption, for the end-of-pipe treatment of residual streams generated in different industries [8].

A wide diversity of catalysts has been used in the HDC of several chlorocarbons and chlorofluorocarbons. Those based on noble metals supported by a porous material like alumina, silica or activated carbon, are the most common [6,9,10]. Among the publications related to the HDC with activated carbon-based catalysts, metals like Pd, Pt, Rh, Ru or Ni are the most frequent active phases found, with Pd being by far the most preferred one, due to its high capacity for the hydrogenolysis of C-Cl bonds [11–13]. This metal, as well as Pt and Rh, have demonstrated high activity and dechlorination capacity [12,14–23]. Several authors have reported relationships between the catalyst performance and different properties, like metal particles structure, catalyst porosity or surface chemistry. However, there is no consensus on the trends observed, being highly dependent on the reactants and the catalytic systems used. Several authors relate better HDC performances to small metal particle sizes [24–29]. On the contrary, higher turnover frequency (TOF) values were obtained when increasing metal particle size in other HDC studies [30–33]. The optimum particle size seems to depend on the particular reaction. Diaz et al. [25] associated Pd and Rh particles of sizes within 3–4 nm with higher activities in the HDC of 4-chlorophenol. The same reaction was studied by Baeza et al. [28] using size-controlled Rh nanoparticles within the 1.9–4.9 nm range. They reported the highest activity with the smaller nanoparticles, in agreement with the optimum size found by Ren et al. [27], who claimed that small and uniform Rh particles of 1.7 nm induced higher activity. Dantas Ramos et al. [30], found a decrease of TOF with increasing dispersion in the HDC of CFC-12 and TCM, and Bedia et al. [31] observed a TOF decrease in the HDC of DCM when decreasing metal particle size (from 2.26 to 1.86 nm). The metal oxidation state also affects the catalyst performance. A higher proportion of zero-valent Rh^0 has been associated with higher activity and selectivity towards cyclohexanol in the HDC of 4-chlorophenol, while a higher proportion of electro-deficient Rh^{n+} induced a higher selectivity to phenol [28]. Cobo et al. [34] also found that a higher Rh^0/Rh^{n+} ratio favors the HDC of trichloroethane with catalysts supported on CeO_2. But with reduced graphene oxide, Rh^{n+} better interacts with the surface functional groups of the support, favoring the HDC [27]. In previous studies of our group [12,19,35], four different activated carbon-supported catalysts (Ru/C, Rh/C, Pd/C and Pt/C) were compared for the gas phase HDC of chloromethanes, finding attractive results in terms of catalytic activity. All the catalysts were considerably active, showing Rh/C, Pd/C and Pt/C have the best dechlorination capacity, although with significant differences in terms of activity, selectivity and stability. Pt/C led mainly to methane formation, while Rh/C and Pd/C were more selective to higher hydrocarbons (C2 and C3). Besides, with Rh/C higher amounts of olefins were obtained [12,14,16,19]. On the other hand, while Pt/C showed itself to be highly stable (demonstrated in long-term experiments up to 26 days on stream), the other two metallic catalysts experienced a progressive loss of activity [16,19,36]. This exceptional stability was also found for a catalyst supported on sulfated zirconia with a bimetallic Pd-Pt active phase during the HDC of dichloromethane (DCM) [37]. In all of the cases, the characterization of the catalysts revealed important differences in their physico-chemical properties. In particular, the oxidation state and particle size of the metallic active phase were revealed as determinant properties on the selectivity, activity and stability of the catalysts [19,36,37]. Higher proportions of electro-deficient metallic (M^{n+}) species combined with lower metal dispersions favored the selectivity to olefins and alkanes other than methane. On the contrary, higher proportions of zero-valent species (M^0) and high metal dispersions promoted better stability. Understanding how these properties are conferred to the catalysts and their optimization to achieve a better performance in the HDC reaction is of crucial interest. These properties can be, to some extent, modulated varying the reduction conditions during the preparation of the catalysts, which may be used as an easy strategy for improving the catalytic activity or tuning the reaction selectivity.

Hence, in the current study, the effect of the reduction temperature (Tred) of Rh/C, Pd/C and Pt/C catalysts on their properties and behavior in the HDC of cloromethanes is analyzed, using DCM as the target compound.

2. Results and Discussion

2.1. Characterization of the Catalysts

The reducibility of the metallic phase in each catalyst was determined in a previous work by temperature-programmed reduction (TPR) [14]. In that study, it was shown to be a single reduction peak for Pd and Pt corresponding to the reduction of M^{2+} to M^0, centered at 233 and 244 °C, respectively. Rh/C displayed three peaks, centered at 86, 125 and 222 °C. From these results, 250 °C was selected as the minimum reduction temperature to be tested and the range of 250–450 °C was covered.

The metal content of the non-reduced and reduced catalysts, as determined by mass spectrometry with inductive coupling plasma (ICP-MS), was always close to the nominal value (1.0% w/w) and no significant loss of metal occurred at any of the reduction temperatures. The reduction of the catalyst even at the highest temperature did not alter significantly the amount of metallic phase.

As can be seen in Table 1, all the catalysts showed an important BET surface area (S_{BET}), above 1200 $m^2\ g^{-1}$, and no considerable differences were observed after reduction at any of the temperatures tested.

Table 1. BET surface area ($m^2 \cdot g^{-1}$) of the catalysts reduced at different temperatures.

Tred (°C)	Pt/C	Pd/C	Rh/C
250	1236	1200	1210
350	1266	1246	1253
450	1247	1251	1209

The most significant effects associated with the reduction temperature were observed in the dispersion and the oxidation state of the metal on the surface of the catalyst (Table 2). The metal dispersion values, as determined by CO chemisorption, prove that Rh was well dispersed with only some slight dispersion decrease at the highest reduction temperature. Pt/C and Pd/C yielded significantly lower dispersion values, although were still quite good at the lowest reduction temperatures. Increasing reduction temperature caused a gradual dispersion decrease, more marked in the case of Pd/C. The lowest reduction temperature led in the three cases to the highest dispersion. This temperature allows the reduction of most of the metal particles, as confirmed by the TPR profiles [14].

Table 2 also presents the relative distribution of the surface metallic species (zero-valent and electro-deficient) at the different reduction temperatures tested. These results were achieved from the deconvolution of the Pt 4f, Pd 3d, and Rh 3d X-ray photoelectron spectroscopy (XPS) profiles (see Figures S1–S3 in the supplementary material). Both metallic species exist in all of the reduction temperatures tested, even at the highest one, although, as expected, the relative occurrence of M^0 increased with the reduction temperature. However, this effect is of a significantly different intensity depending on the catalyst, being much more noticeable in the case of Pt/C and Pd/C, and showing lower significance for Rh/C. Moreover, in the latter, the electro-deficient species always remains the most abundant. This can be attributed to a different interaction of the metal precursor with the support, as pointed out by the TPR results, where three peaks of H_2 consumption were observed with Rh/C, in comparison with the single peak of Pd/C and Pt/C. The higher amount of M^{n+} species points to a solid interaction of rhodium particles with the activated carbon, allowing a higher stability of M^{n+} species even at the highest reduction temperature. This would also explain the very low change of dispersion upon reduction. It is noteworthy that in Pt/C, most of the metal appears as zero-valent species, even at the lowest reduction temperature, in good agreement with other carbon-supported platinum catalysts reported in previous works, where platinum was reduced at 250 and 300 °C [14,16,19,38].

Table 2. Dispersion values, surface metal (M_{XPS}) and oxygen (O_{XPS}) atomic concentrations, relative distribution of the surface metallic species (M^0/M^{n+}) and O_{COOH}/O_{total} ratios of the catalysts reduced at different temperatures.

Catalyst	Tred	Dispersion	M_{XPS}	M^0/M^{n+}	O_{XPS}	O_{COOH}/O_{total}
Pt/C	250	28	0.06	2.8	5.05	0.20
	300	26	0.07	2.9	6.70	0.21
	350	23	0.05	3.2	4.11	0.20
	400	19	0.06	5.1	4.11	0.22
	450	12	0.07	7.5	4.35	0.23
Pd/C	250	23	0.08	1.0	5.65	0.28
	300	22	0.09	1.3	4.96	0.29
	350	16	0.13	1.5	4.96	0.27
	400	13	0.09	2.4	4.89	0.34
	450	10	0.08	2.4	3.93	0.36
Rh/C	250	51	1.5	0.7	6.86	0.12
	300	48	1.5	0.8	6.64	0.11
	350	49	1.4	0.8	6.86	0.12
	400	48	1.4	0.9	7.00	0.12
	450	45	1.3	0.9	6.46	0.13

The values of oxygen atomic concentrations on the surface of the catalysts are also included in Table 2. A higher amount of surface oxygen can be observed for Rh/C, which also remains more stable under reduction, while some significant decrease is observed in Pd/C and Pt/C at increasing reduction temperatures. O 1s orbital was also deconvoluted according to literature [39] as shown in Figure 1. The following groups were assessed: C=O (531.1 eV), C-OH (532.3 eV), C-O-C (533.3 eV), C-OOH (534.2 eV) and H_2O (535.9 eV) [39]. All the XPS profiles are included in Figures S4–S6 of supplementary material. The O_{COOH}/O_{total} ratio (Table 2) represents the proportion of carboxylic groups (C-OOH) with respect to the total concentration of surface oxygenated groups. It provides information of the surface acidity of the catalysts. The amount of surface groups of the catalysts and their acid-base character may have some important effects on the changes occurring during the H_2 treatment. A higher amount of surface oxygen groups favors the interaction between metal and support [40,41], avoiding the sintering of particles. On the other hand, the low ratios of carboxylic groups (Table 2) indicate that all the catalysts have a basic character on the surface, suggesting the prominent role of basic groups in the formation of electro-deficient species. Furthermore, Rh/C catalyst, which show the higher distribution of electro-deficient species, presents the higher amount of surface oxygen (O_{XPS}) and the lowest O_{COOH}/O_{total} ratio, which does not vary with reduction temperature, suggesting a stronger interaction of Rh precursor with the support that results in a higher resistance of Rh particles to sintering, in accordance with the results obtained. In contrast, Pt/C and Pd/C show a lower amount of surface groups and a lower basic character, observing some variations with reduction temperature.

Some representative transmission electron microscopy (TEM) images and the corresponding metal particle size distributions of the catalysts reduced at 450 °C, are shown in Figure 2. The images reveal smaller metallic particles in Rh/C, showing a mean particle diameter of 1.9 nm and a narrower distribution. These results agrees with the dispersion values calculated from CO chemisorption included in Table 2. The mean metal particle size obtained for Pd/C (6.0 nm) after reduction at 450 °C was significantly higher than the one previously reported [38] after reduction at 250 °C (1.9 nm), consistent with the loss of dispersion at increasing reduction temperatures (Table 2). Other authors have reported similar behavior in carbon-supported Pd catalysts [30]. With regard to Pt/C, the mean particle size after reduction at the highest temperature is 2.5 nm, also higher than the 1.6 and 1.7 nm previously obtained for the catalyst reduced at 250 °C [23,38]. Therefore, it can be concluded that sintering of metallic particles takes place during reduction at increasing temperatures, although to different extents, in the cases of Pt and Pd catalysts.

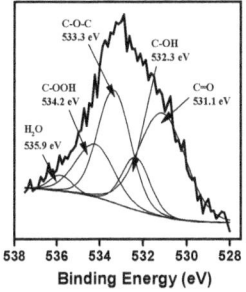

Figure 1. Deconvolution of O 1s spectra of Pd/C catalyst reduced at 300 °C.

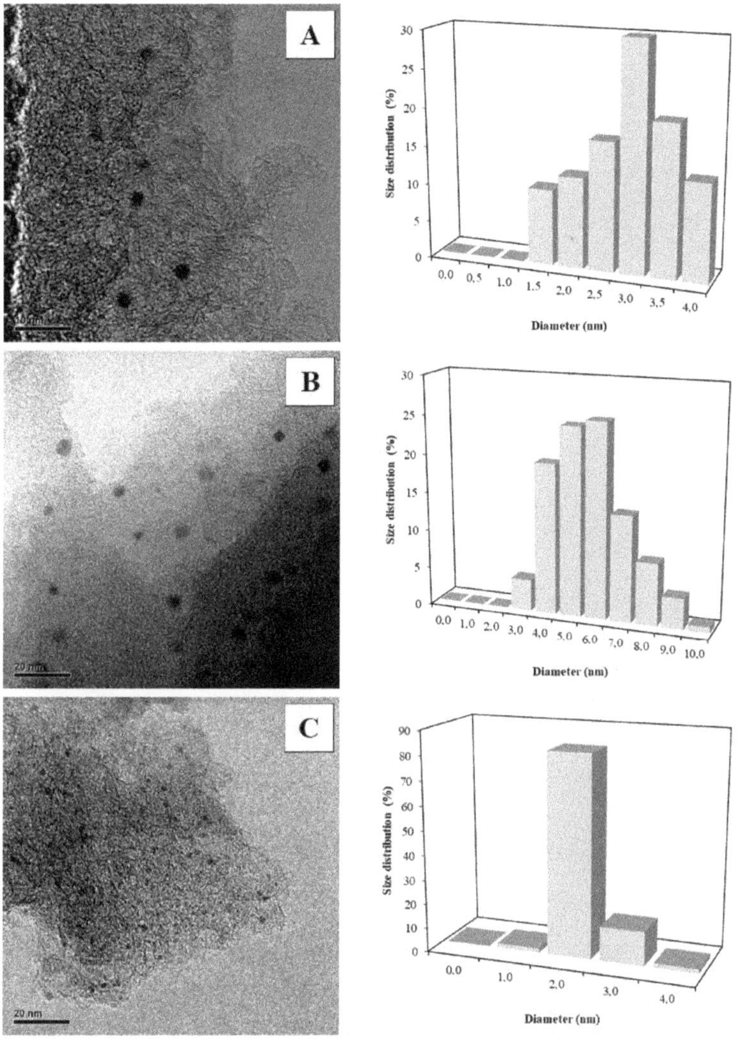

Figure 2. TEM images (left) and metal particles size distributions (right) of catalysts reduced at 450 °C: (**A**) Pt/C; (**B**) Pd/C; (**C**) Rh/C.

The X-ray diffraction (XRD) profiles of the catalysts reduced at different temperatures (Figure 3) show the characteristic peaks of the carbon support (dash lines) at 2θ values nearly 26° (planes 002), 43° (planes 100, 101 and 102, indistinguishable) and 80° (planes 110) [42–44]. As can be seen, XRD profiles were similar regardless of the reduction temperature. The peaks associated with zero-valent metal, which should be centered at 39.9°, 40.1° and 41.1°, respectively [12,14,44,45], were not observed in these patterns, supporting the small particle size of these metals for all the reduction temperatures investigated, as observed by TEM (Figure 2). In addition, in all cases, one peak appeared at 35° (*) at the lowest reduction temperature (250 °C), associated with the presence of metal chlorides (PtCl$_2$, PdCl$_2$, and RhCl$_3$) from the precursors used in the preparation of the catalysts. Furthermore, these compounds seem to disappear at higher reduction temperatures, since this favors the desorption of Cl from the surface of the activated carbon.

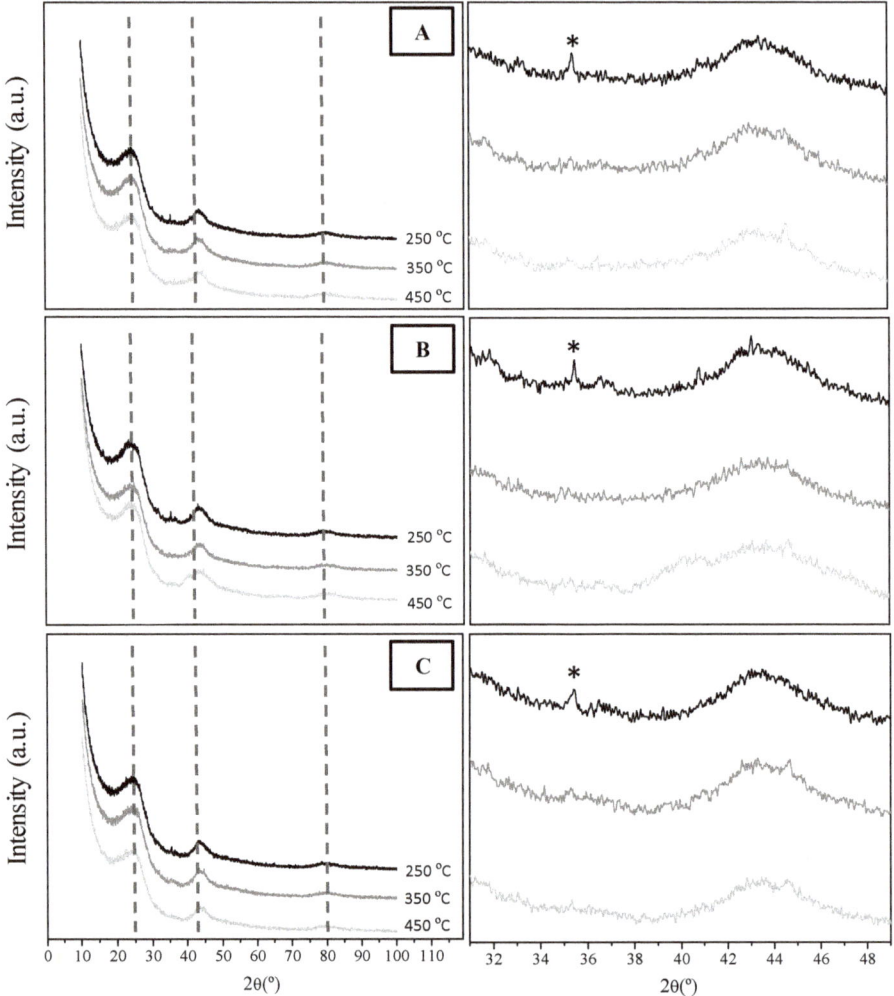

Figure 3. XRD diffractograms of: (**A**) Pt/C; (**B**) Pd/C; (**C**) Rh/C.

2.2. Hydrodechlorination Results

Figure 4 shows the results obtained in the HDC of DCM with the Pt/C catalyst reduced at different temperatures, within the range of 150–250 °C reaction temperature. As can be seen, increasing the reduction temperature improves moderately overall dechlorination, in spite of the decrease of Pt dispersion. Therefore, some other effect must be compensating for the loss of dispersion. As shown in Table 2, the relative amount of zero-valent Pt (Pt^0) rises significantly with the reduction temperature. These Pt^0 species have proven to be the most active on the HDC of DCM with Pt/C catalyst in previous studies [16,23]. Figure 4 includes the evolution of TOF with the reduction temperature of the catalyst. The TOF values are compared at a 150 °C reaction temperature, where conversion throughout the reactor becomes low so that initial reaction rate values can be taken. A significant increase in TOF can be observed with the reduction temperature, the higher proportion of Pt^0 then compensating the activity loss due to the dispersion downturn. This effect is clearly observed in Figure 4, which shows the TOF, metal dispersion and Pt^0/Pt^{n+} ratio values of Pt/C reduced at different temperatures.

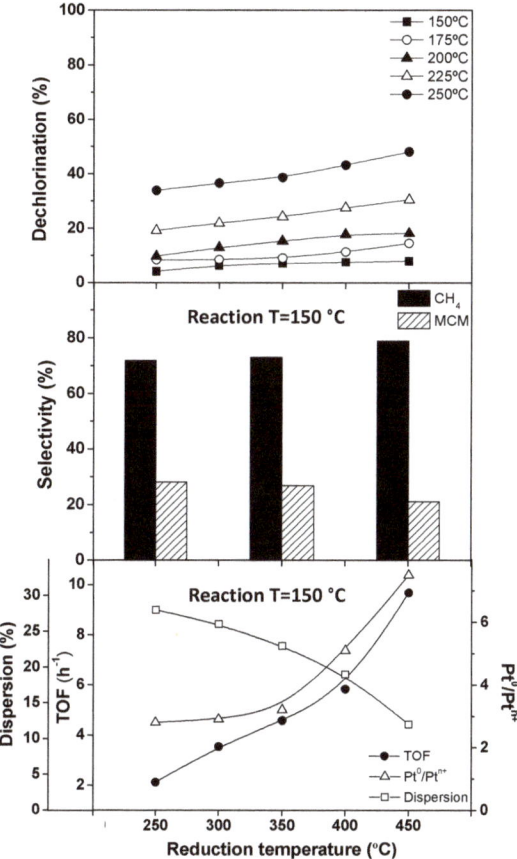

Figure 4. Results on the HDC of DCM within 150–250 °C reaction temperatures, at 0.6 $kg_{cat}·h·mol^{-1}$ space-time, with the Pt/C catalyst reduced at different temperatures.

As in previous works with platinum catalysts [14,16,19], the HDC of DCM with Pt/C only yielded methane (selectivity > 85%) and methyl chloride (MCM) (see also Table S1 of supplementary material), independently of the reduction temperature used, due to the high zero-valent proportion of platinum.

No significant variations of selectivity with the reduction temperature were found other than some slight increase towards methane, in accordance with the higher contribution of the Pt^0 species in the dissociation of hydrogen during HDC.

Figure 5 shows the results with the Pd/C catalyst. Substantially different trends are observed with respect to Pt/C. As can be seen, in contrast with Pt/C, overall dechlorination decreased dramatically at an increasing reduction temperature, with the effect being more pronounced at higher reaction temperatures. This can be attributed to the combination of two negative effects (see Table 2): (i) the decrease of the concentration of electro-deficient species, which are known to be the most active species in the HDC of DCM with palladium catalysts [12,19,38], and (ii) the decrease of palladium dispersion. Among the catalysts tested in this study, Pd/C has been proven to be the most prone to sintering. Then, raising the reduction temperature of Pd/C results in an important loss of accessible metallic surface, and the subsequent loss of activity observed. This results in a decrease in DCM conversion (Table S2 in Supplementary Material). In previous studies, it was stablished the optimum Pd^0/Pd^{n+} ratio around 1 for HDC reactions with Pd/C catalysts [19,36]. This is the value found in this study for the catalyst reduced at the lowest temperature. Despite the important loss of metal dispersion, no change in TOF is detected at reduction temperatures higher than 400 °C, where Pd^0/Pd^{n+} ratio also remains constant.

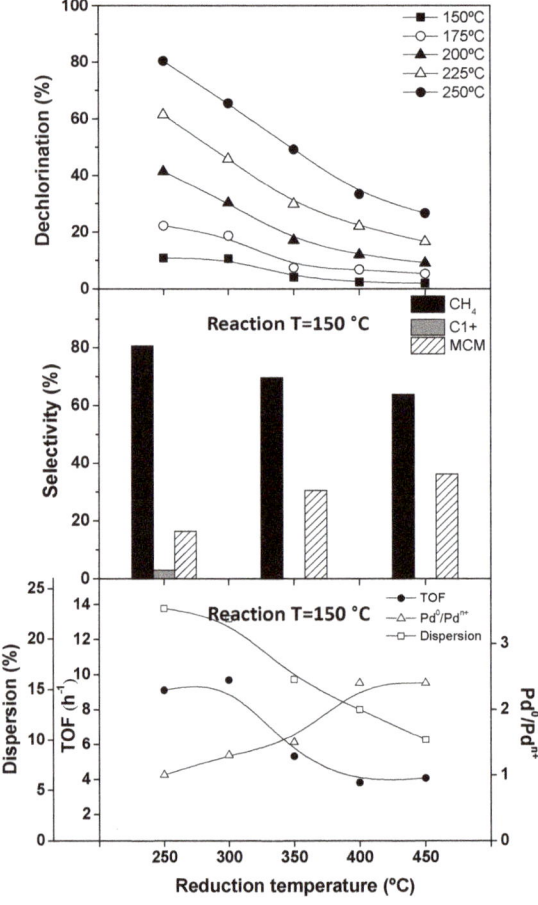

Figure 5. Results on the HDC of DCM within 150–250 °C reaction temperatures, at 0.6 $kg_{cat}\cdot h\cdot mol^{-1}$ space-time, with the Pd/C catalyst reduced at different temperatures.

With Pd/C, the HDC of DCM yielded mainly methane, ethane and MCM. At the highest reaction temperatures, small amounts of propane were also formed (Table S2 of supplementary material). The catalyst reduced at the lowest temperature yielded small amounts of hydrocarbons higher than methane (C1+), as can be observed in Figure 5. According to previous studies, this is favored by a higher electro-deficient to zero-valent palladium proportion [14,19], due to the particular ability of electro-deficient Pd to dissociate H_2. This was evidenced by molecular simulation studies [46], where the preferred adsorption of hydrogen in electro-deficient Pd^{n+} species was reported when studying DCM and H_2 interactions with Pd clusters by DFT. This would explain the decrease in methane selectivity. Hence, the concentration of MCM increases, it negatively affected the dechlorination obtained.

The results obtained with Rh/C are depicted in Figure 6. No significant effects of the reduction temperature can be pointed out, consistently with the almost unaltered characteristics given in Table 2. This catalyst yielded by far the best performance in terms of dechlorination, being almost complete at a 250 °C reaction temperature. The selectivity towards hydrocarbons higher than methane (C1+) was much higher than the obtained with the other two catalysts. Ethane was the most abundant among those hydrocarbons, but propane and even some smaller amounts of butane were also formed (see Table S3 of supplementary material).

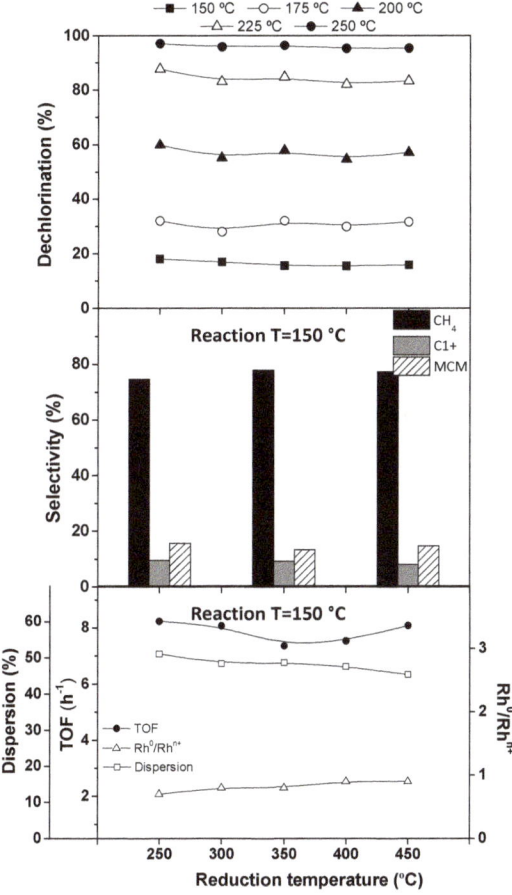

Figure 6. Results on the HDC of DCM within 150–250 °C reaction temperatures, at 0.6 $kg_{cat} \cdot h \cdot mol^{-1}$ space-time, with the Rh/C catalyst reduced at different temperatures.

3. Materials and Methods

3.1. Catalysts Preparation

The catalysts were synthesized by incipient wetness impregnation. A commercial activated carbon (*Erkimia S.A.*, S_{BET} = 1200 m^2 g^{-1}, pH slurry = 6.4 [42]), was impregnated with aqueous solutions of H$_2$PtCl$_6$, PdCl$_2$, and RhCl$_3$ (all supplied by Sigma-Aldrich, Madrid, Spain) to get 1 wt.% active phase nominal concentration. After 12 h at room temperature, the catalysts were dried at 100 °C (20 °C h^{-1}), for 2 h. They were named Pt/C, Pd/C, and Rh/C.

The catalysts were activated by heating (10 °C min^{-1}) up to the desired temperature (reduction temperatures between 250 °C and 450 °C were tested) under H$_2$ flow (50 Ncm^3 min^{-1}, delivered by Praxair, Madrid, Spain, minimum purity of 99.999%) for 2 h.

3.2. Catalysts Characterization

N$_2$ adsorption-desorption analysis were performed at −196 °C (Tristar II 3020, Micromeritics, Alcobendas, Spain) to characterized the porous texture of the catalysts. Previously, the samples were outgassed for 12 h at 150 °C (VacPrep 061, Micromeritics, Alcobendas, Spain). The BET equation was used to determine the surface areas.

The metal content (platinum, palladium and rhodium) of the synthesized catalysts was determined by ICP-MS with an *Elan 6000* (Perkin-Elmer, Tres Cantos, Spain). Previously, the samples were dissolved in HNO$_3$:3HCl, and digested for 15 min at 180 °C in a microwave oven (Milestone Ethos Plus, Madrid, Spain).

Metal dispersion was determined by CO chemisorption (PulseChemiSorb 2705, Micromeritics, Alcobendas, Spain), assuming a stoichiometry of 1:1 for the adsorption of the CO molecules on the metallic atoms [43–45,47].

The elements present on the surface of the catalysts and their concentrations were determined by XPS (5700C Multitechnique System, Physical Electronics, Madrid, Spain), scanning up to a binding energy (BE) of 1200 eV, using Mg-Kα radiation (1253.6 eV). In order to rectify the shift in BE produced by sample charging, the C 1s peak (284.6 eV) was used as an internal standard. To estimate the chemical state of Pt, Pd, Rh and O on the catalyst surface, the BE of the O 1s, Pt 4f$_{7/2}$, Pd 3d$_{5/2}$, and Rh 3d$_{5/2}$ core levels and the full width at half maximum data were used, deconvoluting the peaks by mixed Gaussian–Lorentzian functions (least-squares method) [48].

TEM was carried by a JEOL JEM-3000F microscope (300 kV, point resolution of 0.17 nm, Madrid, Spain), equipped with a high-angle annular dark field (HAADF) detector and a 1k x 1k ULTRASCAN multi sweep CCD camera. Chemical analysis was done by Energy Disperse X-ray Spectroscopy (XEDS) using an Oxford Instruments INCA Energy TEM 250 (Madrid, Spain). Previously, the catalysts were dispersed in ethanol and placed onto holey carbon-coated Cu grids (Aname, Madrid, Spain).

The XRD patterns of the catalysts at different reduction temperatures (250, 350 and 450 °C) were driven in a X'Pert PRO Panalytical Diffractometer (Madrid, Spain), scanning up to a 2θ of 100° (step size of 0.020°, 5s collection time), using CuKα monochromatic radiation (k = 0.15406 nm) and a Ge mono filter.

3.3. Catalytic Activity Experiments

The HDC experiments were performed in a Microactivity Pro (Tres Cantos, Spain) reaction system described previously [49], consisting of a quartz fixed bed micro-reactor operating under continuous flow. In order to analyze the reaction products, the system was coupled to a gas-chromatograph with a flame ionization detector (FID).

The experiments were conducted at atmospheric pressure. The total flow rate (DCM + H$_2$ + N$_2$) used was 100 Ncm^3 min^{-1}. Finally, a DCM inlet concentration of 1000 ppmv and a H$_2$/DCM molar ratio of 100 were employed. Reaction temperatures of 150–250 °C were evaluated. All the experiments were performed by triplicate, using a space-time (τ) of 0.6 kg h mol^{-1}.

The catalysts were evaluated in terms of overall dechlorination, metallic intrinsic activity or TOF, DCM conversion and selectivity to the different reaction products.

4. Conclusions

From the results obtained it follows that Pd is more prone than Pt and Rh to sintering and, therefore, increasing the reduction temperature provokes a dramatic decrease of activity, leading to poor dechlorination of DCM even at the highest reaction temperature investigated (250 °C). Sintering was of lower significance in the case of Pt and almost did not occur with Rh. On the other hand, by increasing the reduction temperature, the ratio of zero-valent to electro-deficient species increased significantly in the Pd/C catalyst and to a lower extent in Pt/C, while remaining basically unchanged in Rh/C. This led to a significant increase of TOF in Pt/C since in this catalyst Pt^0 is the main active specie for HDC, thus compensating for the loss of dispersion at a higher reduction temperature. In contrast, the decrease of Pd^{n+} (main active center for Pd/C) together with the significant decrease of dispersion caused a remarkable drop of TOF and overall dechlorination. Meanwhile, no significant changes occurred in the case of Rh/C consistently with its stable properties upon reduction at different temperatures. This higher stability of Rh/C can be ascribed to a higher concentration of surface oxygen groups, particularly of basic character, which favors the metal-support interaction, avoiding metal sintering and protecting the electro-deficient species.

It can be concluded that 250 °C is the optimum reduction temperature for Pd/C, while for Pt/C better results were obtained after reduction at 450 °C, and no significant effects were observed in the case of Rh/C.

Supplementary Materials: The following are available online at http://www.mdpi.com/2073-4344/8/12/664/s1, Figure S1. XPS deconvolution of Pt 4f on Pt/C catalyst reduced at: 250 °C (A), 300 °C (B), 350 °C (C), 400 °C (D) and 450 °C (E); Figure S2. XPS deconvolution of Pd 3d on Pd/C catalyst reduced at: 250 °C (A), 300 °C (B), 350 °C (C), 400 °C (D) and 450 °C (E); Figure S3. XPS deconvolution of Rh 3d on Rh/C catalyst reduced at: 250 °C (A), 300 °C (B), 350 °C (C), 400 °C (D) and 450 °C (E); Figure S4. XPS spectra of O 1s on Pt/C catalyst reduced at different temperatures; Figure S5. XPS spectra of O 1s on Pd/C catalyst reduced at different temperatures; Figure S6. XPS spectra of O 1s on Rh/C catalyst reduced at different temperatures; Table S1. Initial conversions and selectivities to reaction products in the HDC of DCM with Pt/C catalyst reduced at different temperatures; Table S2. Initial conversions and selectivities to reaction products in the HDC of DCM with Pd/C catalyst reduced at different temperatures; Table S3. Initial conversions and selectivities to reaction products in the HDC of DCM with Rh/C catalyst reduced at different temperatures.

Author Contributions: Conceptualization, J.J.-R. and L.M.G.-S.; investigation, A.A.-B. and M.M.-M.; methodology, A.A.-B., M.M.-M. and M.A.Á.-M.; data analysis, A.A.-B., M.M.-M, M.A.A.-M, J.J.-R. and L.M.G.-S.; supervision, M.A.Á.-M., J.J.-R. and L.M.G.-S.; writing—original draft preparation, A.A.-B. and M.M.-M.; writing—review and editing, M.A.Á.-M., J.J.-R. and L.M.G.-S.

Funding: The authors gratefully acknowledge financial support from the Spanish Ministerio de Economía y Competitividad (MINECO) through the projects CTM 2014-53008 and CTM2017-85498-R. A. Arevalo Bastante acknowledges MINECO for her research grant. M. Martín Martínez acknowledges the postdoc grant 2017-T2/AMB-5668 from Comunidad de Madrid, Programme "Atracción de talento investigador".

Conflicts of Interest: The authors declare no conflict of interest.

References

1. Hayes, W.J., Jr.; Laws, E.R., Jr. *Handbook of Pesticide Toxicology, Vol. 1: General Principles*; Academic Press: San Diego, CA, USA, 1991.
2. Lewis, N.M.; Gatchett, A.M. U.S. Environmental Protection Agency's SITE Emerging Technology Program: 1991 Update. *J. Air Waste Manag. Assoc.* **1991**, *41*, 1645–1653. [CrossRef]
3. Ciccioli, P. *Chemistry and Analysis of Volatile Organic Compounds in the Environment*; Bloemen, H.J.T., Burn, J., Eds.; Springer: Glasgow, UK, 1993; pp. 92–174.
4. Dobrzynska, E.; Posniak, M.; Szewczynska, M.; Buszewski, B. Chlorinated Volatile Organic Compounds—Old, However, Actual Analytical and Toxicological Problem. *Crit. Rev. Anal. Chem.* **2010**, *40*, 41–57. [CrossRef]
5. Noelke, C.J.; Rase, H.F. Improved Hydrodechlorination Catalysis—Chloroform Over Platinum-Alumina with Special Treatments. *Ind. Eng. Chem. Prod. Res. Dev.* **1979**, *18*, 325–328. [CrossRef]

6. Ordoñez, S.; Sastre, H.; Diez, F.V. Hydrodechlorination of aliphatic organochlorinated compounds over commercial hydrogenation catalysts. *Appl. Catal. B Environ.* **2000**, *25*, 49–58. [CrossRef]
7. Urbano, F.J.; Marinas, J.M. Hydrogenolysis of Organohalogen Compounds Over Palladium Supported Catalysts. *J. Mol. Catal. A Chem.* **2001**, *173*, 329–345. [CrossRef]
8. Elola, A.; Diaz, E.; Ordoñez, S. A New Procedure for the Treatment of Organochlorinated Off-Gases Combining Adsorption and Catalytic Hydrodechlorination. *Environ. Sci. Technol.* **2009**, *43*, 1999–2004. [CrossRef] [PubMed]
9. Prati, L.; Rossi, M. Reductive Catalytic Dehalogenation of Light Chlorocarbons. *Appl. Catal. B Environ.* **1999**, *23*, 135–142. [CrossRef]
10. Gomez-Sainero, L.M.; Cortes, A.; Seoane, X.L.; Arcoya, A. Hydrodechlorination of Carbon Tetrachloride to Chloroform in the Liquid Phase with Metal-Supported Catalysts. Effect of the Catalyst Components. *Ind. Eng. Chem. Res.* **2000**, *39*, 2849–2854. [CrossRef]
11. Gonzalez, C.A.; Bartoszek, M.; Martin, A.; de Correa, C.M. Hydrodechlorination of Light Organochlorinated Compounds and Their Mixtures over Pd/TiO$_2$-Washcoated Minimonoliths. *Ind. Eng. Chem. Res.* **2009**, *48*, 2826–2835. [CrossRef]
12. Alvarez-Montero, M.A.; Gomez-Sainero, L.M.; Martin-Martinez, M.; Heras, F.; Rodriguez, J.J. Hydrodechlorination of Chloromethanes with Pd on Activated Carbon Catalysts for the Treatment of Residual Gas Streams. *Appl. Catal. B Environ.* **2010**, *96*, 148–156. [CrossRef]
13. Keane, M.A. Supported Transition Metal Catalysts for Hydrodechlorination Reactions. *ChemCatChem* **2011**, *3*, 800–821. [CrossRef]
14. Alvarez-Montero, M.A.; Gomez-Sainero, L.M.; Juan-Juan, J.; Linares-Solano, A.; Rodriguez, J.J. Gas-Phase Hydrodechlorination of Dichloromethane With Activated Carbon-Supported Metallic Catalysts. *Chem. Eng. J.* **2010**, *162*, 599–608. [CrossRef]
15. Calvo, L.; Gilarranz, M.A.; Casas, J.A.; Mohedano, A.F.; Rodriguez, J.J. Hydrodechlorination of Diuron in Aqueous Solution with Pd, Cu and Ni on Activated Carbon Catalysts. *Chem. Eng. J.* **2010**, *163*, 212–218. [CrossRef]
16. Alvarez-Montero, M.A.; Gomez-Sainero, L.M.; Mayoral, A.; Diaz, I.; Baker, R.T.; Rodriguez, J.J. Hydrodechlorination of Chloromethanes with a Highly Stable Pt on Activated Carbon Catalyst. *J. Catal.* **2011**, *279*, 389–396. [CrossRef]
17. Shao, Y.; Xu, Z.; Wan, H.; Wan, Y.; Chen, H.; Zheng, S.; Zhu, D. Enhanced Liquid Phase Catalytic Hydrodechlorination of 2,4-Dichlorophenol Over Mesoporous Carbon Supported Pd Catalysts. *Catal. Commun.* **2011**, *12*, 1405–1409. [CrossRef]
18. Ha, J.; Kim, D.; Kim, J.; Kim, S.K.; Ahn, B.S.; Kang, J.W. Supercritical-Phase-Assisted Highly Selective and Active Catalytic Hydrodechlorination of the Ozone-Depleting Refrigerant CHClF$_2$. *Chem. Eng. J.* **2012**, *213*, 346–355. [CrossRef]
19. Martin-Martinez, M.; Gomez-Sainero, L.M.; Alvarez-Montero, M.A.; Bedia, J.; Rodriguez, J.J. Comparison of Different Precious Metals in Activated Carbon-Supported Catalysts for the Gas-Phase Hydrodechlorination of Chloromethanes. *Appl. Catal. B Environ.* **2013**, *132*, 256–265. [CrossRef]
20. Cárdenas-Lizana, F.; Hao, Y.; Crespo-Quesada, M.; Yuranov, I.; Wang, X.; Keane, M.A.; Kiwi-Minsker, L. Selective Gas Phase Hydrogenation of p-Chloronitrobenzene over Pd Catalysts: Role of the Support. *ACS Catal.* **2013**, *3*, 1386–1396. [CrossRef]
21. Díaz, E.; McCall, A.; Faba, L.; Sastre, H.; Ordóñez, S. Trichloroethylene Hydrodechlorination in Water Using Formic Acid as Hydrogen Source: Selection of Catalyst and Operation Conditions. *Environ. Prog. Sustain. Energy* **2013**, *32*, 1217–1222. [CrossRef]
22. Gregori, M.; Fornasari, G.; Marchionni, G.; Tortelli, V.; Millefanti, S.; Albonetti, S. Hydrogen-Assisted Dechlorination of CF$_3$OCFCl–CF$_2$Cl to CF$_3$OCF=CF$_2$ Over Different Metal-Supported Catalysts. *Appl. Catal. A Gen.* **2014**, *470*, 123–131. [CrossRef]
23. Arevalo-Bastante, A.; Álvarez-Montero, M.A.; Bedia, J.; Gómez-Sainero, L.M.; Rodriguez, J.J. Gas-Phase Hydrodechlorination of Mixtures of Chloromethanes with Activated Carbon-Supported Platinum Catalyst. *Appl. Catal. B Environ.* **2015**, *179*, 551–557. [CrossRef]
24. Srikanth, C.S.; Kumar, V.P.; Viswanadham, B.; Chary, K.V.R. Hydrodechlorination of 1,2,4-Trichlorbenzene Over Supported Ruthenium Catalysts on Various Supports. *Catal. Commun.* **2011**, *13*, 69–72. [CrossRef]

25. Diaz, E.; Mohedano, A.F.; Casas, J.A.; Calvo, L.; Gilarranz, M.A.; Rodriguez, J.J. Comparison of Activated Carbon-Supported Pd and Rh Catalysts for Aqueous-Phase Hydrodechlorination. *Appl. Catal. B Environ.* **2011**, *106*, 469–475. [CrossRef]
26. Baeza, J.A.; Calvo, L.; Gilarranz, M.A.; Mohedano, A.F.; Casas, J.A.; Rodriguez, J.J. Catalytic Behavior of Size-Controlled Palladium Nanoparticles in the Hydrodechlorination of 4-Chlorophenol in Aqueous Phase. *J. Catal.* **2012**, *293*, 85–93. [CrossRef]
27. Ren, Y.; Fan, G.; Wang, C. Aqueous Hydrodechlorination of 4-Chlorophenol Over an Rh/Reduced Graphene Oxide Synthesized by a Facile One-Pot Solvothermal Process Under Mild Conditions. *J. Hazard. Mater.* **2014**, *274*, 32–40. [CrossRef] [PubMed]
28. Baeza, J.A.; Calvo, L.; Gilarranz, M.A.; Rodriguez, J.J. Effect of Size and Oxidation State of Size-Controlled Rhodium Nanoparticles on the Aqueous-Phase Hydrodechlorination of 4-Chlorophenol. *Chem. Eng. J.* **2014**, *240*, 271–280. [CrossRef]
29. Baeza, J.A.; Calvo, L.; Rodriguez, J.J.; Gilarranz, M.A. Catalysts Based on Large Size-Controlled Pd Nanoparticles for Aqueous-Phase Hydrodechlorination. *Chem. Eng. J.* **2016**, *294*, 40–48. [CrossRef]
30. Dantas Ramos, A.L.; Alves, P.D.S.; Aranda, D.A.G.; Schmal, M. Characterization of Carbon Supported Palladium Catalysts: Inference of Electronic and Particle Size Effects Using Reaction Probes. *Appl. Catal. A Gen.* **2004**, *277*, 71–81. [CrossRef]
31. Bedia, J.; Arevalo-Bastante, A.; Grau, J.M.; Dosso, L.A.; Rodriguez, J.J.; Mayoral, A.; Diaz, I.; Gómez-Sainero, L.M. Effect of the Pt–Pd Molar Ratio in Bimetallic Catalysts Supported on Sulfated Zirconia on the Gas-Phase Hydrodechlorination of Chloromethanes. *J. Catal.* **2017**, *352*, 562–571. [CrossRef]
32. Malinowski, A.; Lomot, D.; Karpinski, Z. Hydrodechlorination of CH_2Cl_2 over Pd/gamma-Al_2O_3. Correlation with the Hydrodechlorination of CCl_2F_2 (CFC-12). *Appl. Catal. B Environ.* **1998**, *19*, L79–L86. [CrossRef]
33. Sánchez, C.A.G.; Patiño, C.O.M.; de Correa, C.M. Catalytic Hydrodechlorination of Dichloromethane in the Presence of Traces of Chloroform and Tetrachloroethylene. *Catal. Today* **2008**, *133–135*, 520–525. [CrossRef]
34. Cobo, M.; Becerra, J.; Castelblanco, M.; Cifuentes, B.; Conesa, J.A. Catalytic Hydrodechlorination of Trichloroethylene in a Novel NaOH/2-Propanol/Methanol/Water System on Ceria-Supported Pd and Rh Catalysts. *J. Environ. Manag.* **2015**, *158*, 1–10. [CrossRef] [PubMed]
35. Álvarez-Montero, M.A.; Martin-Martinez, M.; Gómez-Sainero, L.M.; Arevalo-Bastante, A.; Bedia, J.; Rodriguez, J.J. Kinetic Study of the Hydrodechlorination of Chloromethanes with Activated-Carbon-Supported Metallic Catalysts. *Ind. Eng. Chem. Res.* **2015**, *54*, 2023–2029. [CrossRef]
36. Martin-Martinez, M.; Alvarez-Montero, M.A.; Gomez-Sainero, L.M.; Baker, R.T.; Palomar, J.; Omar, S.; Eser, S.; Rodriguez, J.J. Deactivation Behavior of Pd/C and Pt/C Catalysts in the Gas-Phase Hydrodechlorination of Chloromethanes: Structure-Reactivity Relationship. *Appl. Catal. B Environ.* **2015**, *162*, 532–543. [CrossRef]
37. Bedia, J.; Gomez-Sainero, L.M.; Grau, J.M.; Busto, M.; Martin-Martinez, M.; Rodriguez, J.J. Hydrodechlorination of Dichloromethane With Mono- and Bimetallic Pd-Pt on Sulfated and Tungstated Zirconia Catalysts. *J. Catal.* **2012**, *294*, 207–215. [CrossRef]
38. Martin-Martinez, M.; Gómez-Sainero, L.M.; Bedia, J.; Arevalo-Bastante, A.; Rodriguez, J.J. Enhanced Activity of Carbon-Supported Pd-Pt Catalysts in the Hydrodechlorination of Dichloromethane. *Appl. Catal. B Environ.* **2016**, *184*, 55–63. [CrossRef]
39. Lesiak, B.; Jiricek, P.; Bieloshapka, I. Chemical and Structural Properties of Pd Nanoparticle-Decorated Graphene—Electron Spectroscopic Methods and QUASES. *Appl. Surf. Sci.* **2017**, *404*, 300–309. [CrossRef]
40. Castillejos, E.; García-Minguillán, A.M.; Bachiller-Baeza, B.; Rodríguez-Ramos, I.; Guerrero-Ruiz, A. When the Nature of Surface Functionalities on Modified Carbon Dominates the Dispersion of Palladium Hydrogenation Catalysts. *Catal. Today* **2018**, *301*, 248–257. [CrossRef]
41. An, N.; Zhang, M.; Zhang, Z.; Dai, Y.; Shen, Y.; Tang, C.; Yuan, X.; Zhou, W. High-Performance Palladium Catalysts for the Hydrogenation Toward Dibenzylbiotinmethylester: Effect of Carbon Support Functionalization. *J. Colloid Interface Sci.* **2018**, *510*, 181–189. [CrossRef] [PubMed]
42. Gomez-Sainero, L.M.; Seoane, X.L.; Fierro, J.L.G.; Arcoya, A. Liquid-Phase Hydrodechlorination of CCl_4 to $CHCl_3$ on Pd/Carbon Catalysts: Nature and Role of Pd Active Species. *J. Catal.* **2002**, *209*, 279–288. [CrossRef]
43. Ali, S.H.; Goodwin, J.G. SSITKA Investigation of Palladium Precursor and Support Effects on CO Hydrogenation over Supported Pd Catalysts. *J. Catal.* **1998**, *176*, 3–13. [CrossRef]

44. Mahata, N.; Vishwanathan, V. Gas phase Hydrogenation of Phenol Over Supported Palladium Catalysts. *Catal. Today* **1999**, *49*, 65–69. [CrossRef]
45. Kulkarni, P.P.; Deshmukh, S.S.; Kovalchuk, V.I.; d'Itri, J.L. Hydrodechlorination of Dichlorodifluoromethane on Carbon-Supported Group VIII Noble Metal Catalysts. *Catal. Lett.* **1999**, *61*, 161–166. [CrossRef]
46. Omar, S.; Palomar, J.; Gomez-Sainero, L.M.; Alvarez-Montero, M.A.; Martin-Martinez, M.; Rodriguez, J.J. Density Functional Theory Analysis of Dichloromethane and Hydrogen Interaction with Pd Clusters: First Step to Simulate Catalytic Hydrodechlorination. *J. Phys. Chem. C* **2011**, *115*, 14180–14192. [CrossRef]
47. Gasser-Ramirez, J.L.; Dunn, B.C.; Ramirez, D.W.; Fillerup, E.P.; Turpin, G.C.; Shi, Y.; Ernst, R.D.; Pugmire, R.J.; Eyring, E.M.; Pettigrew, K.A.; et al. A Simple Synthesis of Catalytically Active, High Surface Area Ceria Aerogels. *J. Non-Cryst. Solids* **2008**, *354*, 5509–5514. [CrossRef]
48. Wagner, C.D.; Davis, L.E.; Zeller, M.V.; Taylor, J.A.; Raymond, R.H.; Gale, L.H. Empirical Atomic Sensitivity Factors for Quantitative Analysis by Electron Spectroscopy for Chemical Analysis. *Surf. Interface Anal.* **1981**, *3*, 211–225. [CrossRef]
49. de Pedro, Z.M.; Gómez-Sainero, L.M.; González-Serrano, E.; Rodriguez, J.J. Gas-Phase Hydrochlorination of Dichloromethane at Low Concentrations with Palladium/Carbon Catalysts. *Ind. Eng. Chem. Res.* **2006**, *45*, 7760–7766. [CrossRef]

© 2018 by the authors. Licensee MDPI, Basel, Switzerland. This article is an open access article distributed under the terms and conditions of the Creative Commons Attribution (CC BY) license (http://creativecommons.org/licenses/by/4.0/).

Palladium Supported on Carbon Nanotubes as a High-Performance Catalyst for the Dehydrogenation of Dodecahydro-N-ethylcarbazole

Mengyan Zhu [1,2], Lixin Xu [1,2], Lin Du [3], Yue An [4] and Chao Wan [1,2,4,*]

1. Hexian Chemical Industrial Development Institute, School of Chemistry and Chemical Engineering, Anhui University of Technology, Ma'anshan 243002, China; AHUT.zhumy@outlook.com (M.Z.); lxxu@hotmail.com (L.X.)
2. Ahut Chemical Science & Technology Co., Ltd., Ma'anshan 243002, China
3. Anhui Haide Chemical Technology Co., Ltd., Ma'anshan 243002, China; AHUT.dulin@outlook.com
4. College of Chemical and Biological Engineering, Zhejiang University, Hangzhou 310027, China; zju.anyue@hotmail.com
* Correspondence: wanchao1219@hotmail.com; Tel.: +86-555-231-1807

Received: 20 November 2018; Accepted: 6 December 2018; Published: 8 December 2018

Abstract: Hydrogen storage in the form of liquid organic hydrides, especially N-ethylcarbazole, has been regarded as a promising technology for substituting traditional fossil fuels owing to its unique merits such as high volumetric, gravimetric hydrogen capacity and safe transportation. However, unsatisfactory dehydrogenation has impeded the widespread application of N-ethylcarbazole as ideal hydrogen storage materials in hydrogen energy. Therefore, designing catalysts with outstanding performance is of importance to address this problem. In the present work, for the first time, we have synthesized Pd nanoparticles immobilized on carbon nanotubes (Pd/CNTs) with different palladium loading through an alcohol reduction technique. A series of characterization technologies, such as X-ray diffraction (XRD), inductively coupled plasma-atomic emission spectrometer (ICP-AES), X-ray photoelectron spectroscopy (XPS) and transmission electron spectroscopy (TEM) were adopted to systematically explore the structure, composition, surface properties and morphology of the catalysts. The results reveal that the Pd NPs with a mean diameter of 2.6 ± 0.6 nm could be dispersed uniformly on the surface of CNTs. Furthermore, Pd/CNTs with different Pd contents were applied in the hydrogen release of dodecahydro-N-ethylcarbazole. Among all of the catalysts tested, 3.0 wt% Pd/CNTs exhibited excellent catalytic performance with the conversion of 99.6% producing 5.8 wt% hydrogen at 533 K, low activation energy of 43.8 ± 0.2 kJ/mol and a high recycling stability (>96.4% conversion at 5th reuse).

Keywords: palladium catalysts; CNTs; dodecahydro-N-ethylcarbazole; dehydrogenation; hydrogen storage

1. Introduction

Among numerous alternative energy, hydrogen has been deemed as one of the most important and ideal energy sources owing to its distinct merits, such as a high calorific value, non-toxic environmentally, sustainable and cost-effective [1–4]. As is known, a complete energy system that utilizes hydrogen as an energy source is composed of producing, storing, transporting and utilizing hydrogen. However, it is difficult for hydrogen to be stored and transported owing to its low density [5,6]. Therefore, the technology of hydrogen storage has been regarded as one of the bottlenecks for promoting the large-scale application of the hydrogen energy [5–8]. To search for new hydrogen storage technology satisfying the U.S. Department of Energy (DOE) requirements with minimum gravimetric of 5.5 wt% and volumetric capacity of 40 g L^{-1} remains a challenging issue for the large-scale application of hydrogen.

Among various hydrogen storage materials, such as formic acid, cyclohexane, and ammonia borane, [9–12] organic liquid hydrides have emerged as a preferred approach in existing vehicle hydrogen storage systems for its virtues like high H_2 storage density and safe transportation [12–14]. Especially, reversible hydrogen storage and release can be catalytically achieved under relatively moderate conditions [15,16]. Currently, hydrogenation reactions have been extensively studied in the previously reported literature [17–19]. Compared with the traditional organic liquid hydrides, the substitution of a heteroatom in heterocyclic aromatic molecules, such as in N-ethylcarbazole, can decrease the endothermicity of the reaction and bring down the dehydrogenation temperature [20,21]. Therefore, N-ethylcarbazole, with a gravimetric density of 5.8 wt.%, has been identified as the most prospective candidate for hydrogen storage (Scheme 1). Although there are many studies about the dehydrogenation reaction from calculations and experiments [20–28], the dehydrogenation reaction is still the key to limit its large-scale application, especially, the development of dehydrogenation catalysts with outstanding activity and stability is the hotspot of current research.

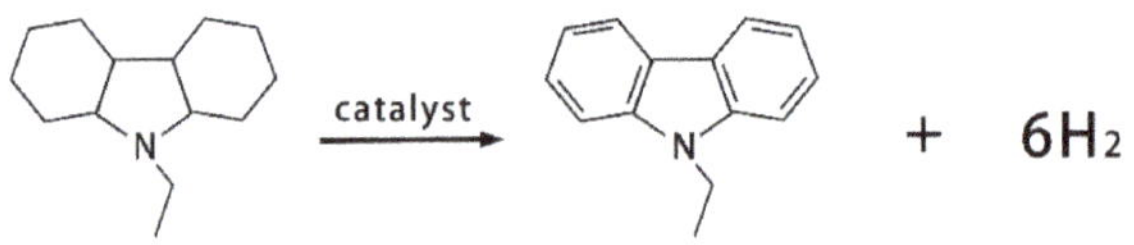

Scheme 1. Dehydrogenation pathway for dodecahydro-N-ethylcarbazole.

A large number of dehydrogenation catalysts have been extensively investigated, including homogeneous catalysts and the heterogeneous catalysts [29–35]. For the homogeneous catalysts, Wang et al. firstly reported the synthesis of homogeneous Ir-complex catalysts and explored its performance for the hydrogen release of dodecahydro-N-ethylcarbazole at 473 K but the dehydrogenation results were unsatisfactory [34]. However, the heterogeneous catalysts exhibited a notable advantage over homogeneous ones with respect to catalytic activity. For the heterogeneous catalysts, the supporting materials and NPs (Nanoparticles) are the two key factors influencing the catalytic performance. Yang et al. [36] have studied the dehydrogenation activity of perhydro-N-ethylcarbazole over a series of noble metal catalysts and the kinetics of dehydrogenation of dodecahydro-N-ethylcarbazole over a 5 wt% Pd/Al$_2$O$_3$ catalyst. The results revealed that the order is Pd > Pt > Ru > Rh according to the initial catalytic activity of the investigated noble metal catalysts in the dehydrogenation process and the rate-limiting step of the entire reaction process is the transformation from tetrahydro-N-ethylcarbazole to N-ethylcarbazole. Furthermore, Kustov et al. [37] confirmed that the catalytic activity of the catalysts can be improved under microwave activation. Although Pd based catalysts supported several supports, such as alumina, silica, TiO$_2$, MoO$_3$ and carbon [29–33,35–38], have been systematically investigated for producing hydrogen from perhydro-N-ethylcarbazole, there are no reports about CNTs as the supporting material in the hydrogen release from perhydro-N-ethylcarbazole.

In recent years, CNTs, as a one-dimensional nanomaterial, have received considerable attention as a catalyst support material due to their high specific surface area, superior electrical conductivity and outstanding chemical and thermal stability [39,40]. In addition, CNTs can endow beneficial interactions between support and metal NPs, thus improving the catalytic activity. As is well known, the synthesis method for Pd NPs, such as doping or through a supramolecular strategy, an alcohol reduction method, is another key factor for improving the dehydrogenation performance of the catalysts [35,41–44]. Fang et al. [42] have successfully synthesized Pd/rGO using ethylene glycol as a reductant for the hydrogen production of dodecahydro-N-ethylcarbazole. Constructing Pd NP catalysts using an alcohol reduction method for the dehydrogenation of dodecahydro-N-ethylcarbazole has rarely been reported.

Herein, in this work, for the first time, we have utilized an alcohol reduction method to construct CNT-supported Pd NPs (Pd/CNTs) as the catalyst for hydrogen generation from dodecahydro-N-ethylcarbazole. The catalyst has been characterized by many characterization methods, such as XRD, ICP-AES, XPS and TEM to investigate the structure, composition, surface properties and morphology of the catalysts. The dehydrogenation process of dodecahydro-N-ethylcarbazole over Pd/CNTs catalyst is also discussed.

2. Results and Discussion

The Pd/CNTs with different Pd contents were fabricated via an alcohol reduction route, as schematically shown in Scheme 2 [45,46]. Typically, PVP (Poly (N-vinyl-2-pyrrolidone))-Pd NPs were obtained by refluxing a solution containing H_2PdCl_4, ethanol, H_2O and PVP at 363 K for 3 h. Subsequently, the as-synthesized PVP-Pd NPs were put in the CNTs solution under magnetic stirring for 24 h. Then, the above-mentioned solution was evaporated, the catalyst was dried and calcined. The obtained products were denoted the X wt% Pd/CNTs catalysts.

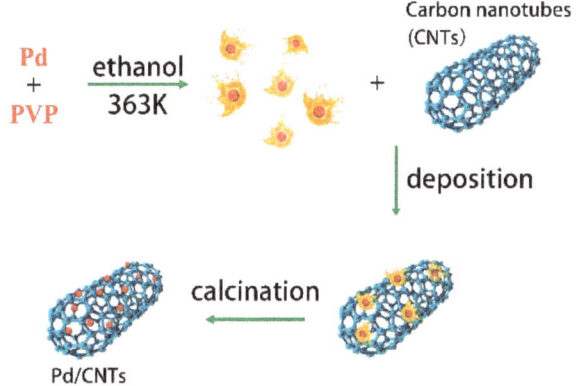

Scheme 2. A fabrication diagram for the preparation of Pd/CNTs.

Powder X-ray diffraction (XRD) patterns were collected to explore the phase and crystal structure of the acid-treated CNTs, Pd/CNTs with different Pd contents. As shown in Figure 1, a similar XRD pattern was observed for all of the samples. All of the samples showed three obvious peaks at 25.9°, 43.8°, and 54.2°, which could be ascribed to the (002), (100), and (004) reflections of graphite structure, respectively. There was only the diffraction peak of the graphite structure for the Pd/CNTs with different Pd loadings. However, no distinct characteristic diffraction corresponding to Pd NPs was detected in the XRD patterns, probably owing to the fact that the Pd loading of Pd/CNTs was too low. The diffraction peak ascribed to Pd (JCPDS (Joint Committee on Powder Diffraction Standards) no. 46-1043) could be observed for Pd/CNTs with a higher loading (20 wt%) in Figure S1 (Supporting Information). The accurate composition of Pd/CNTs was measured by an inductively coupled plasma-atomic emission spectrometer (ICP-AES), which is close to their designed content (Table S1, Supporting Information).

X-ray photoelectron spectroscopy (XPS) was performed to investigate the surface state of 3.0 wt% Pd/CNTs. As seen in Figure 2, the peaks centered at 341.0 eV ($3d^{3/2}$ state) and 335.8 eV ($3d_{5/2}$ state), lower than that of Pd/Rgo, can be ascribed to the Pd^0 species, which is consistent with the previously reported literature [39,40,42]. Furthermore, it is worth noting that the two small peaks appeared at 343.6 eV and 337.8 eV for the Pd3d spectra of 3.0 wt% Pd/CNTs can be attributed to Pd^{2+}, may relate to the sample treatment process for the XPS measurements [47]. The nitrogen adsorption-desorption isotherms and pore-size distributions for the CNTs and 3.0 wt% Pd/CNTs are displayed in Figures S2 and S3 (Supporting Information). It can be seen that the samples present similar adsorption-desorption curves (type IV isotherms) and pore-size distributions. The BET (Brunauer–Emmett–Teller) surface areas of the CNTs and 3.0 wt% Pd/CNTs were calculated to be 137 and 94 m^2 g^{-1}, respectively. Furthermore, the microstructure of 3.0 wt% Pd/CNTs was further investigated using transmission electron microscopy (TEM) measurements (Figure 3). As displayed in Figure 3, it can be observed that the Pd NPs were uniformly dispersed on the CNTs and the small average diameter of the particle size was 2.6 ± 0.6 nm, which is consistent with the previously reported results [39,40,45,46].

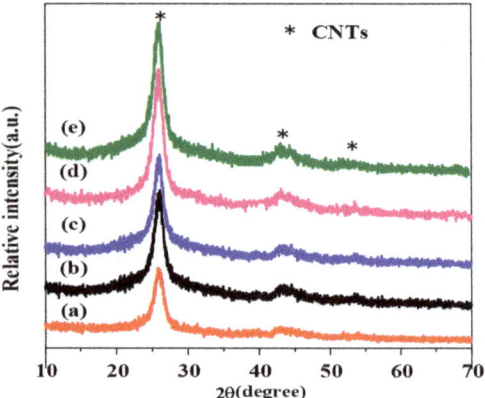

Figure 1. XRD patterns of (**a**) CNTs; Pd/CNTs with different Pd composition (**b**) 0.9 wt%, (**c**) 2.1 wt%, (**d**) 3.0 wt% and (**e**) 4.1 wt%.

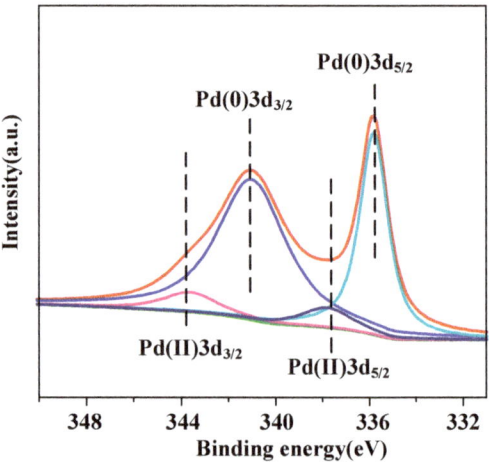

Figure 2. The high-resolution Pd3d peaks in the XPS spectra of 3.0 wt% Pd/CNTs.

Figure 4 shows the hydrogen release of dodecahydro-N-ethylcarbazole over Pd/CNTs with different Pd loadings in the range of 0 wt%–4.1 wt% at 513 K. The hydrogen generation rate significantly relied on the loading of Pd. As shown in Figure 4, 5.6 wt% hydrogen evolved at 90.4, 33.6, and 89.5 min in the presence of the Pd/CNTs with a Pd loading of 2.1 wt%, 3.0 wt% and 4.1 wt%, respectively. Hydrogen evolution catalyzed by 0.9 wt% Pd/CNTs only yielded 4.6 wt% hydrogen even at 97 min. However, no gas was detected for the CNT support, implying that CNTs are inactive for hydrogen production of dodecahydro-N-ethylcarbazole. Obviously, Pd/CNTs with a Pd loading of 3.0 wt% exhibited excellent catalytic activity with a conversion of 96.4%, producing 5.6 wt% H$_2$.

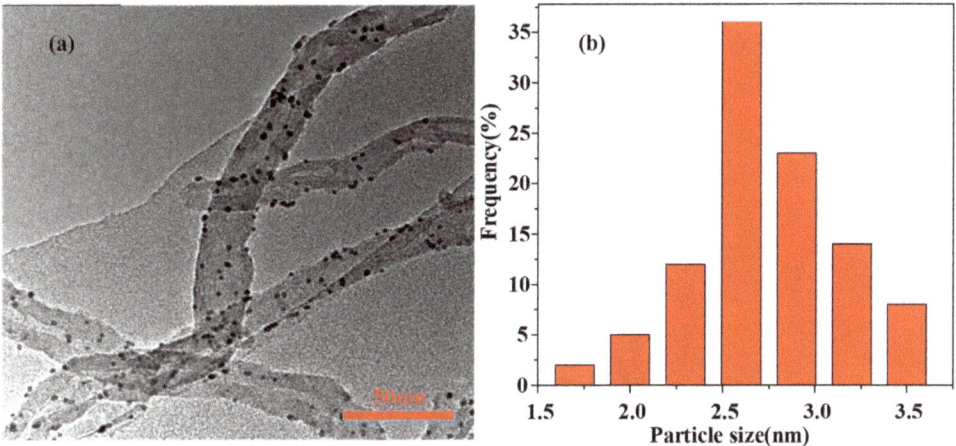

Figure 3. TEM images of (**a**) 3.0 wt% Pd Pd/CNTs and (**b**) particle distribution.

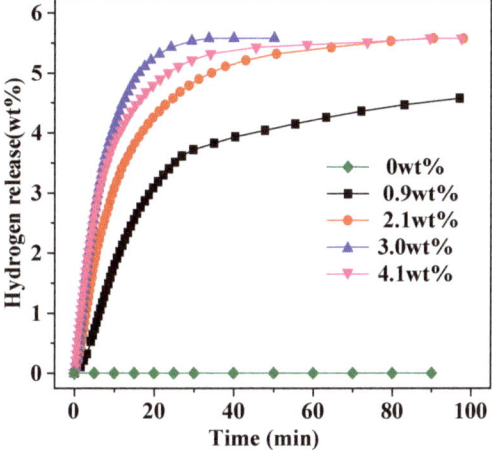

Figure 4. Hydrogen release from dodecahydro-N-ethylcarbazole catalyzed by Pd/CNTs with different Pd loadings at 513 K.

In order to explore the kinetics of the dehydrogenation of dodecahydro-N-ethylcarbazole catalyzed by 3.0 wt% Pd/CNTs, a series of experiments were carried out under varying temperatures. As displayed in Figure 5, when the temperature increased from 453 K to 533 K, hydrogen release increase from 2.7 wt% to 5.8 wt%. It is generally accepted that producing hydrogen from dodecahydro-N-ethylcarbazole is an endothermic reaction, a higher reaction temperature may be favorable for hydrogen generation from dodecahydro-N-ethylcarbazole. It can be seen in Figure 5 that the initial dehydrogenation rate and the amount of hydrogen recovery both increased with an increasing reaction temperature; the higher the reaction temperature, the higher the rate of dehydrogenation. First-order kinetics were established with the concentration of the reactant, dodecahydro-N-ethylcarbazole, measured as a function of time using 3.0 wt% Pd/CNTs catalyst [35,36,38,48,49]. The reaction rate was expressed as:

$$r = dC/dt = kC \tag{1}$$

$$ln(C/C_0) = -kt \tag{2}$$

where C represents the concentration of dodecahydro-N-ethylcarbazole, C_0 denotes the initial concentration of dodecahydro-N-ethylcarbazole.

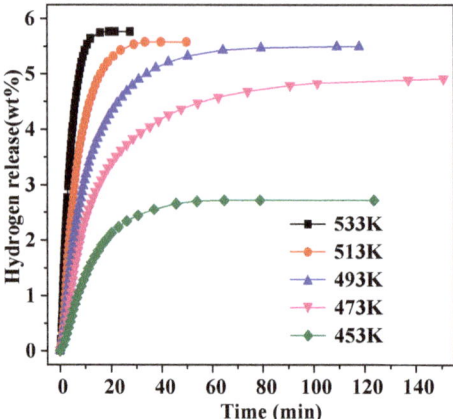

Figure 5. Hydrogen release of dodecahydro-N-ethylcarbazole over 3.0 wt% Pd/CNTs versus time at 453, 473, 493, 513 and 533 K.

On the basis of Figure 5 and the above formula, a linear relation of $ln(C/C_0)$ vs. time is observed in Figure 6a. The values of k under the different temperatures could be acquired, a smooth straight line could be observed by lnk versus $1/T$ plot, as demonstrated in Figure 6b, and its linear correlation coefficient was 99.6%. It is indicated that k and $T(K)$ follow the Arrhenius equation:

$$lnk = -E_a/(RT) + lnk_0. \qquad (3)$$

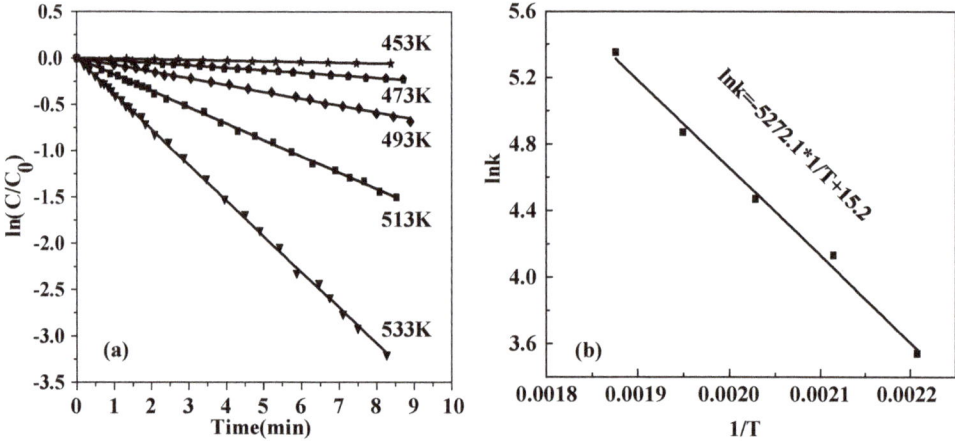

Figure 6. (a) $ln(C/C_0)$ versus time for 3.0 wt% Pd/CNTs at 453, 473, 493, 513 and 533 K; (b) lnk versus $1/T$ for 3.0 wt% Pd/CNTs at 453, 473, 493, 513 and 533 K.

Considering the slope of the straight line, the apparent activation energy of hydrogen production of dodecahydro-N-ethylcarbazole over 3.0 wt% Pd/CNTs was calculated to be 43.8 ± 0.2 kJ/mol.

The durability of the catalyst was of significance for its practical application. Therefore, the reusability of 3.0 wt% Pd/CNTs was investigated at 533 K. As revealed in Figure 7, the catalytic activity of 3.0 wt% Pd/CNTs shows no obvious decrease after five runs for hydrogen generation from dodecahydro-N-ethylcarbazole. The reusability tests revealed that 3.0 wt% Pd/CNTs exhibits activity in consecutive runs in the hydrogen release from dodecahydro-N-ethylcarbazole, demonstrating 96.4% conversion and 5.6 wt% H_2 at the fifth run.

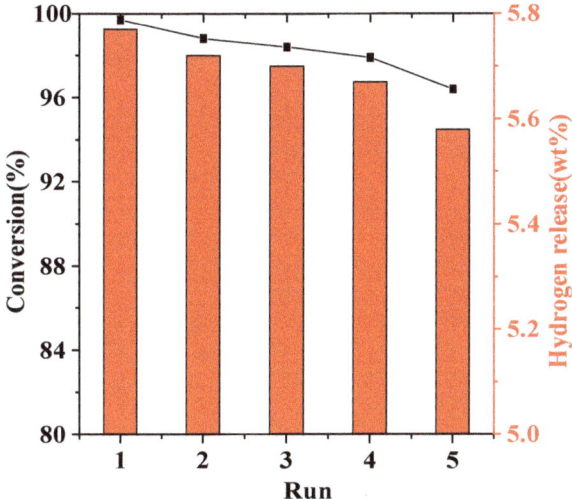

Figure 7. Conversion and hydrogen release of the 3.0 wt% Pd/CNTs in successive runs for the dehydrogenation of dodecahydro-N-ethylcarbazole at 533 K.

3. Materials and Methods

3.1. Materials

All of the chemicals, such as N-ethylcarbazole (purity ≥ 99.5%, Shanghai Infine Chemicals Co., Ltd., Shanghai, China), ultra-high purity hydrogen (99.99999%, Minxing gas company), 5 wt.% Ru/Al$_2$O$_3$ (reduced, Alfa aersa), C$_2$H$_5$OH (AR, Sinopharm Chemical Reagent Co., Ltd., Shanghai, China), Poly (N-vinyl-2-pyrrolidone) (PVP, Sinopharm Chemical Reagent Co., Ltd.), Palladium (II) chloride (AR, Nanjing Chemical Reagent Co., Ltd., Nanjing, China) and carbon nanotubes (CNTs, Φ 20–40 nm, Purity > 97%, Shenzhen Nanotech Port Co., Shenzhen, China), were utilized as purchased without further purification.

3.2. Catalyst Preparation

CNTs were pretreated in a mixture of H$_2$SO$_4$ (90 mL) and HNO$_3$ (60 mL) at 120 °C for 12 h. The treated CNTs were obtained by filtration, washing several times, and vacuum drying at 140 °C for 8 h.

A series of Pd-based catalysts were synthesized through immobilizing the ethanol reduction Pd nanoparticles (Pd NPs) onto the CNTs. In a typical synthesis, PdCl$_2$ was dissolved in an HCl aqueous solution to form an H$_2$PdCl$_4$ aqueous solution. Then, PVP (0.4 g), H$_2$O (40 mL), H$_2$PdCl$_4$ aqueous solution (45 mL) and ethanol (60 mL) were refluxed at 90 °C for 3 h. The foregoing solution was treated through vacuum rotary evaporation and re-dispersed in ethanol to obtain the poly(N-vinyl-2-pyrrolidone)-stabilized Pd nanoparticles (PVP-Pd NPs) solution. Next, an appropriate amount of treated CNTs was added into the above-mentioned PVP-Pd solution under stirring. After another 3 h, the products were stirred to remove the excess solvent in the water bath. The obtained catalyst was dried at 70 °C for 4 h in a vacuum oven and then the X wt% Pd/CNTs catalysts were acquired after calcining at 550 °C for 3 h under a nitrogen atmosphere (X was the nominal Pd loading).

3.3. Hydrogen Generation from Dodecahydro-N-ethylcarbazole

Dodecahydro-N-ethylcarbazole was synthesized via the hydrogenation process using 5 wt% as catalysts for the dehydrogenation of N-ethylcarbazole. The detailed synthesis process has been previously reported [50].

The dehydrogenation of dodecahydro-N-ethylcarbazole was performed in a 25 mL round-bottomed flask in the presence of Pd/CNTs at a temperature ranging from 453 to 533 K. Specifically, 25 mg Pd/CNTs was placed in the round-bottomed flask, which was heated the desired temperature. Then, 5 mL of dodecahydro-N-ethylcarbazole was injected into the reactor under stirring. The evolved gas was measured by recording the displacement of water.

Durability for the catalysts. For testing the recyclability of the Pd/CNTs, after completing the dehydrogenation reaction, Pd/CNTs were separated from the reaction solution through centrifugation and washed with ethanol and water several times. The recovered Pd/CNTs were dried for the next experiment. The dehydrogenation reaction was repeated five times at the designed temperature.

3.4. Characterization

Powder X-ray diffraction (PXRD) patterns were obtained on a Bruker D8-Advance X-ray diffractometer using a Cu Kα radiation source. X-ray photoelectron spectroscopy (XPS) was carried out using an Escalab 250Xi spectrometer with an Al Kα source. BET surface areas were collected from N_2 adsorption/desorption isotherms at 77 K using automatic volumetric adsorption equipment (Micromeritics ASAP2020) after pretreatment under vacuum at 200 °C for 5 h. Transmission electron microscope (TEM) images were recorded on an FEI Tecnai F20 transmission electron microscope with an operating voltage of 200 kV. The metal content of the materials was collected on an inductively coupled plasma-atomic emission spectrometer (ICP-AES, Thermo iCAP6300). The exit gas composition was monitored using a Hiden QIC-20 quadruple mass spectrometer. Liquid samples were analyzed using a Shimadzu QP-2010S GC/MS with a Restek RTX5 30 m 0.25 mm capillary column according to the temperature program (100 °C isotherm for 2 min, then heated to 260 °C with a ramping rate of 10 °C/min).

4. Conclusions

In summary, we have developed an alcohol reduction method for the fabrication of Pd/CNTs with different palladium loadings. The as-synthesized 3.0 wt% Pd/CNTs exhibited outstanding catalytic performance for hydrogen release of dodecahydro-N-ethylcarbazole at 533 K with a conversion rate of 99.7% and 5.8 wt% H_2. Furthermore, the activation energy for producing hydrogen from dodecahydro-N-ethylcarbazole catalyzed by 3.0 wt% Pd/CNTs was found to be 43.8 ± 0.2 kJ mol^{-1}. More importantly, the as-synthesized 3.0 wt% Pd/CNTs possess excellent cycle stability for the dehydrogenation of dodecahydro-N-ethylcarbazole. The reusability tests revealed that 3.0 wt% Pd/CNTs exhibited superior activity even five runs into the hydrogen evolution of dodecahydro-N-ethylcarbazole providing 96.4% conversion and 5.6 wt% H_2 at the fifth run. In addition, this simple synthesis means may provide a new avenue for the noble-metal and CNTs in dehydrogenation reaction.

Supplementary Materials: The following are available online at http://www.mdpi.com/2073-4344/8/12/638/s1, Figure S1: XRD patterns for the synthesized Pd/CNTs with 20 wt% Pd loading, Figure S2: Nitrogen adsorption-desorption isotherms for CNTs and 3.0 wt% Pd/CNTs, Figure S3: Pore size distribution for CNTs and 3.0 wt% Pd/CNTs, Table S1. The content of Pd in Pd/CNTs with different loading based on ICP-AES analysis.

Author Contributions: C.W., Y.A. and L.X. conceived and designed the experiments; M.Z. and M.Y. conducted the experiments, analyzed the data and wrote the manuscript; L.D. analyzed the physicochemical data.

Funding: This research was funded by the National Natural Science Foundation of China.

Acknowledgments: The authors acknowledge Anhui Provincial Natural Science Foundation (1608085QF156), Open Fund of Shaanxi Key Laboratory of Energy Chemical Process Intensification (SXECPI201601), Research Fund for Young Teachers of Anhui University of Technology (QZ201610), the National Natural Science Foundation of China (21376005), and the Scientific Research Foundation of Graduate School of Anhui University of Technology (2016012, 2016017, 2017082, 2017083).

Conflicts of Interest: The authors declare no conflicts of interest.

References

1. Turner, J.A. Sustainable Hydrogen Production. *Science* **2004**, *305*, 972–974. [CrossRef] [PubMed]
2. Li, G.; Kanezashi, M.; Tsuru, T. Catalytic Ammonia Decomposition over High-Performance Ru/Graphene Nanocomposites for Efficient COx-Free Hydrogen Production. *Catalysts* **2017**, *7*, 23. [CrossRef]
3. Li, J.; Li, B.; Shao, H.; Li, W.; Lin, H. Catalysis and Downsizing in Mg-Based Hydrogen Storage Materials. *Catalysts* **2018**, *8*, 89. [CrossRef]
4. Patil, S.P.; Pande, J.V.; Biniwale, R.B. Non-noble Ni–Cu/ACC bimetallic catalyst for dehydrogenation of liquid organic hydrides for hydrogen storage. *Int. J. Hydrogen Energy* **2013**, *38*, 15233–15241. [CrossRef]
5. Xu, L.X.; Liu, N.; Hong, B.; Cui, P.; Cheng, D.G.; Chen, F.Q.; An, Y.; Wan, C. Nickel–platinum nanoparticles immobilized on graphitic carbon nitride as highly efficient catalyst for hydrogen release from hydrous hydrazine. *RSC Adv.* **2016**, *6*, 31687–31691. [CrossRef]
6. Schlapbach, L.; Züttel, A. Hydrogen-storage materials for mobile applications. *Nature* **2001**, *414*, 353–358. [CrossRef] [PubMed]
7. Niaz, S.; Manzoor, T.; Pandith, A.H. Hydrogen storage: Materials, methods and perspectives. *Renew. Sustain. Energy Rev.* **2015**, *50*, 457–469. [CrossRef]
8. Zhu, Q.L.; Xu, Q. Liquid organic and inorganic chemical hydrides for high-capacity hydrogen storage. *Energy Environ. Sci.* **2015**, *8*, 478–512. [CrossRef]
9. Yao, F.; Li, X.; Wan, C.; Xu, L.X.; An, Y.; Ye, M.F.; Lei, Z. Highly efficient hydrogen release from formic acid using a graphitic carbon nitride-supported AgPd nanoparticle catalyst. *Appl. Surf. Sci.* **2017**, *426*, 605–611. [CrossRef]
10. Xia, Z.J.; Liu, H.Y.; Lu, H.F.; Zhang, Z.K.; Chen, Y.F. Study on catalytic properties and carbon deposition of Ni-Cu/SBA-15 for cyclohexane dehydrogenation. *Appl. Surf. Sci.* **2017**, *422*, 905–912. [CrossRef]
11. Bandaru, S.; English, N.J.; Phillips, A.D.; MacElroy, J.M.D. Exploring Promising Catalysts for Chemical Hydrogen Storage in Ammonia Borane: A Density Functional Theory Study. *Catalysts* **2017**, *7*, 140. [CrossRef]
12. Teichmann, D.; Arlt, W.; Wasserscheid, P.; Freymann, R. A future energy supply based on Liquid Organic Hydrogen Carriers (LOHC). *Energy Environ. Sci.* **2011**, *4*, 2767–2773. [CrossRef]
13. Jiang, Z.; Pan, Q.; Xu, J.; Fang, T. Current situation and prospect of hydrogen storage technology with new organic liquid. *Int. J. Hydrogen Energy* **2014**, *39*, 17442–17451. [CrossRef]
14. Markiewicz, M.; Zhang, Y.Q.; Bösmann, A.; Brückner, N.; Thöming, J.; Wasserscheid, P.; Stolte, S. Environmental and health impact assessment of Liquid Organic Hydrogen Carrier (LOHC) systems—challenges and preliminary results. *Energy Environ. Sci.* **2015**, *8*, 1035–1045. [CrossRef]
15. Patil, S.P.; Bindwal, A.B.; Pakade, Y.B.; Biniwale, R.B. On H_2 supply through liquid organic hydrides-Effect of functional groups. *Int. J. Hydrogen Energy* **2017**, *42*, 16214–16224. [CrossRef]
16. Teichmann, D.; Stark, K.; Müller, K.; Zöttl, G.; Wasserscheid, P.; Arlt, W. Energy storage in residential and commercial buildings via Liquid Organic Hydrogen Carriers (LOHC). *Energy Environ. Sci.* **2012**, *5*, 9044–9054. [CrossRef]
17. Wan, C.; An, Y.; Chen, F.Q.; Cheng, D.G.; Wu, F.Y.; Xu, G.H. Kinetics of N-ethylcarbazole hydrogenation over a supported Ru catalyst for hydrogen storage. *Int. J. Hydrogen Energy* **2013**, *38*, 7065–7069. [CrossRef]
18. Eblagon, K.M.; Tam, K.; Yu, K.M.K.; Zhao, S.L.; Gong, X.Q.; He, H.Y.; Ye, L.; Wang, L.C.; Ramirez-Cuesta, A.J.; Tsang, S.C. Study of Catalytic Sites on Ruthenium for Hydrogenation of N-ethylcarbazole: Implications of Hydrogen Storage via Reversible Catalytic Hydrogenation. *J. Phys. Chem. C* **2010**, *114*, 9720–9730. [CrossRef]
19. Preuster, P.; Papp, C.; Wasserscheid, P. Liquid Organic Hydrogen Carriers (LOHCs): Toward a Hydrogen-free Hydrogen Economy. *Acc. Chem. Res.* **2017**, *50*, 74–85. [CrossRef]
20. Sotoodeh, F.; Huber, B.J.M.; Smith, K.J. The effect of the N atom on the dehydrogenation of heterocycles used for hydrogen storage. *Appl. Catal. A* **2012**, *419–420*, 67–72. [CrossRef]

21. Crabtree, R.H. Nitrogen-Containing Liquid Organic Hydrogen Carriers: Progress and Prospects. *ACS Sustain. Chem. Eng.* **2017**, *5*, 4491–4498. [CrossRef]
22. Stark, K.; Emel'yanenko, V.N.; Zhabina, A.A.; Varfolomeev, M.A.; Verevkin, S.P.; Müller, K.; Arlt, W. Liquid Organic Hydrogen Carriers: Thermophysical and Thermochemical Studies of Carbazole Partly and Fully Hydrogenated Derivatives. *Ind. Eng. Chem. Res.* **2015**, *54*, 7953–7966. [CrossRef]
23. Papp, C.; Wasserscheid, P.; Libuda, J.; Steinrück, H.P. Liquid Organic Hydrogen Carriers: Surface Science Studies of Carbazole Derivatives. *Chem. Rec.* **2014**, *14*, 879–896. [CrossRef] [PubMed]
24. Arnende, M.; Gleichweit, C.; Werner, K.; Schernich, S.; Zhao, W.; Lorenz, M.P.A.; Hofert, O.; Papp, C.; Koch, M.; Wasserscheid, P. Model Catalytic Studies of Liquid Organic Hydrogen Carriers: Dehydrogenation and Decomposition Mechanisms of Dodecahydro-N-ethylcarbazole on Pt(111). *ACS Catal.* **2014**, *4*, 657–665.
25. Amende, M.; Schernich, S.; Sobota, M.; Nikiforidis, I.; Hieringer, W.; Assenbaum, D.; Gleichweit, C.; Drescher, H.J.; Papp, C.; Steinruck, H.P. Dehydrogenation Mechanism of Liquid Organic Hydrogen Carriers: Dodecahydro-N-ethylcarbazole on Pd(111). *Chem. Eur. J.* **2013**, *19*, 10854–10865. [CrossRef] [PubMed]
26. Gleichweit, C.; Amende, M.; Schernich, S.; Zhao, W.; Lorenz, M.P.A.; Hofert, O.; Bruckner, N.; Wasserscheid, P.; Libuda, J.; Steinruck, H.P. Dehydrogenation of Dodecahydro-N-ethylcarbazole on Pt(111). *ChemSusChem* **2013**, *6*, 974–977. [CrossRef] [PubMed]
27. Amende, M.; Gleichweit, C.; Schernich, S.; Höfert, O.; Lorenz, M.P.A.; Zhao, W.; Koch, M.; Obesser, K.; Papp, C.; Wasserscheid, P.; et al. Size and Structure Effects Controlling the Stability of the Liquid Organic Hydrogen Carrier Dodecahydro-N-ethylcarbazole during Dehydrogenation over Pt Model Catalysts. *J. Phys. Chem. Lett.* **2014**, *5*, 1498–1504. [CrossRef]
28. Sobota, M.; Nikiforidis, I.; Amende, M.; Zanon, B.S.; Staudt, T.; Hofert, O.; Lykhach, Y.; Papp, C.; Hieringer, W.; Laurin, M. Dehydrogenation of Dodecahydro-N-ethylcarbazole on Pd/Al$_2$O$_3$ Model Catalysts. *Chem. Eur. J.* **2011**, *17*, 11542–11552. [CrossRef]
29. Peters, W.; Eypasch, M.; Frank, T.; Schwerdtfeger, J.; Korner, C.; Bosmann, A.; Wasserscheid, P. Efficient hydrogen release from perhydro-N-ethylcarbazole using catalyst-coated metallic structures produced by selective electron beam melting. *Energy Environ. Sci.* **2015**, *8*, 641–649. [CrossRef]
30. Peters, W.; Seidel, A.; Herzog, S.; Bosmann, A.; Schwieger, W.; Wasserscheid, P. Macrokinetic effects in perhydro-N-ethylcarbazole dehydrogenation and H$_2$ productivity optimization by using egg-shell catalysts. *Energy Environ. Sci.* **2015**, *8*, 3013–3021. [CrossRef]
31. Tarasov, A.L.; Tkachenko, O.P.; Kustov, L.M. Mono and Bimetallic Pt–(M)/Al$_2$O$_3$ Catalysts for Dehydrogenation of Perhydro-N-ethylcarbazole as the Second Stage of Hydrogen Storage. *Catal. Lett.* **2018**, *148*, 1472–1477. [CrossRef]
32. Fei, S.X.; Han, B.; Li, L.L.; Mei, P.; Zhu, T.; Yang, M.; Cheng, H.S. A study on the catalytic hydrogenation of N-ethylcarbazole on the mesoporous Pd/MoO$_3$ catalyst. *Int. J. Hydrogen Energy* **2017**, *42*, 25942–25950. [CrossRef]
33. Sotoodeh, F.; Smith, K.J. Structure sensitivity of dodecahydro-N-ethylcarbazole dehydrogenation over Pd catalysts. *J. Catal.* **2011**, *279*, 36–47. [CrossRef]
34. Wang, Z.H.; Tonks, I.; Belli, J.; Jensen, C.M. Dehydrogenation of N-ethyl perhydrocarbazole catalyzed by PCP pincer iridium complexes: Evaluation of a homogenous hydrogen storage system. *J. Organomet. Chem.* **2009**, *694*, 2854–2857. [CrossRef]
35. Dong, Y.; Yang, M.; Mei, P.; Li, C.G.; Li, L.L. Dehydrogenation kinetics study of perhydro-N-ethylcarbazole over a supported Pd catalyst for hydrogen storage application. *Int. J. Hydrogen Energy* **2016**, *41*, 8498–8505. [CrossRef]
36. Yang, M.; Dong, Y.; Fei, S.X.; Ke, H.Z.; Cheng, H.S. A comparative study of catalytic dehydrogenation of perhydro-N-ethylcarbazole over noble metal catalysts. *Int. J. Hydrogen Energy* **2014**, *39*, 18976–18983. [CrossRef]
37. Kustov, L.M.; Tarasov, A.L.; Kirichenko, O.A. Microwave-activated dehydrogenation of perhydro-N-ethylcarbazol over bimetallic Pd-M/TiO$_2$ catalysts as the second stage of hydrogen storage in liquid substrates. *Int. J. Hydrogen Energy* **2017**, *42*, 26723–26729. [CrossRef]
38. Wang, B.; Chang, T.Y.; Jiang, Z.; Wei, J.J.; Zhang, Y.H.; Yang, S.; Fang, T. Catalytic dehydrogenation study of dodecahydro-N-ethylcarbazole by noble metal supported on reduced graphene oxide. *Int. J. Hydrogen Energy* **2018**, *43*, 7317–7325. [CrossRef]

39. Yazdan-Abad, M.Z.; Noroozifar, M.; Alfi, N.; Modarresi-Alam, A.R.; Saravani, H. A simple and fast method for the preparation of super active Pd/CNTs catalyst toward ethanol electrooxidation. *Int. J. Hydrogen Energy* **2018**, *43*, 12103–12109. [CrossRef]
40. Zhang, J.; Lu, S.F.; Xiang, Y.; Shen, P.K.; Liu, J.; Jiang, S.P. Carbon-Nanotubes-Supported Pd Nanoparticles for Alcohol Oxidations in Fuel Cells: Effect of Number of Nanotube Walls on Activity. *ChemSusChem* **2015**, *8*, 2956–2966. [CrossRef]
41. Xia, Y.; Ye, J.R.; Cheng, D.G.; Chen, F.Q.; Zhan, X.L. Identification of a flattened Pd–Ce oxide cluster as a highly efficient catalyst for low-temperature CO oxidation. *Catal. Sci. Technol.* **2018**, *8*, 5137–5147. [CrossRef]
42. Wang, B.; Yan, T.; Chang, T.Y.; Wei, J.J.; Zhou, Q.; Yang, S.; Fang, T. Palladium supported on reduced graphene oxide as a high-performance catalyst for the dehydrogenation of dodecahydro-N-ethylcarbazole. *Carbon* **2017**, *122*, 9–18. [CrossRef]
43. Wang, A.Q.; Li, J.; Zhang, T. Heterogeneous single-atom catalysis. *Nat. Rev. Chem.* **2018**, *2*, 65–81. [CrossRef]
44. Passaponti, M.; Savastano, M.; Clares, M.P.; Inclán, M.; Lavacchi, A.; Bianchi, A.; García-España, E.; Innocenti, M. MWCNTs-Supported Pd(II) Complexes with High Catalytic Efficiency in Oxygen Reduction Reaction in Alkaline Media. *Inorg. Chem.* **2018**, *57*, 14484–14488. [CrossRef] [PubMed]
45. Teranishi, T.; Miyake, M. Size Control of Palladium Nanoparticles and Their Crystal Structures. *Chem. Mater.* **1998**, *10*, 594–600. [CrossRef]
46. Wu, J.M.; Zeng, L.; Cheng, D.G.; Chen, F.Q.; Zhan, X.L.; Gong, J.L. Synthesis of Pd nanoparticles supported on CeO_2 nanotubes for CO oxidation at low temperatures. *Chin. J. Catal.* **2016**, *37*, 83–90. [CrossRef]
47. Koh, K.; Seo, J.-E.; Lee, J.H.; Goswami, A.; Yoon de, C.W.; Asefa, T. Ultrasmall palladium nanoparticles supported on amine-functionalized SBA-15 efficiently catalyze hydrogen evolution from formic acid. *J. Mater. Chem. A* **2014**, *2*, 20444–20449. [CrossRef]
48. Sotoodeh, F.; Huber, B.J.M.; Smith, K.J. Dehydrogenation kinetics and catalysis of organic heteroaromatics for hydrogen storage. *Int. J. Hydrogen Energy* **2012**, *37*, 2715–2722. [CrossRef]
49. Sotoodeh, F.; Zhao, L.; Smith, K.J. Kinetics of H_2 recovery from dodecahydro-N-ethylcarbazole over a supported Pd catalyst. *Appl. Catal. A Gen.* **2009**, *362*, 155–162. [CrossRef]
50. Wan, C.; An, Y.; Xu, G.; Kong, W. Study of catalytic hydrogenation of N-ethylcarbazole over ruthenium catalyst. *Int. J. Hydrogen Energy* **2012**, *37*, 13092–13096. [CrossRef]

© 2018 by the authors. Licensee MDPI, Basel, Switzerland. This article is an open access article distributed under the terms and conditions of the Creative Commons Attribution (CC BY) license (http://creativecommons.org/licenses/by/4.0/).

Article

Toward the Sustainable Synthesis of Propanols from Renewable Glycerol over MoO_3-Al_2O_3 Supported Palladium Catalysts

Shanthi Priya Samudrala * and Sankar Bhattacharya

Department of Chemical Engineering, Faculty of Engineering, Monash University, Melbourne 3800, Australia; sankar.bhattacharya@monash.edu
* Correspondence: priya.shanthipriya@monash.edu; Tel.: +61-3-9905-8162

Received: 17 August 2018; Accepted: 6 September 2018; Published: 9 September 2018

Abstract: The catalytic conversion of glycerol to value-added propanols is a promising synthetic route that holds the potential to overcome the glycerol oversupply from the biodiesel industry. In this study, selective hydrogenolysis of 10 wt% aqueous bio-glycerol to 1-propanol and 2-propanol was performed in the vapor phase, fixed-bed reactor by using environmentally friendly bifunctional Pd/MoO_3-Al_2O_3 catalysts prepared by wetness impregnation method. The physicochemical properties of these catalysts were derived from various techniques such as X-ray diffraction, NH_3-temperature programmed desorption, scanning electron microscopy, ^{27}Al NMR spectroscopy, surface area analysis, and thermogravimetric analysis. The catalytic activity results depicted that a high catalytic activity (>80%) with very high selectivity (>90%) to 1-propanol and 2-propanol was obtained over all the catalysts evaluated in a continuously fed, fixed-bed reactor. However, among all others, 2 wt% Pd/MoO_3-Al_2O_3 catalyst was the most active and selective to propanols. The synergic interaction between the palladium and MoO_3 on Al_2O_3 support and high strength weak to moderate acid sites of the catalyst were solely responsible for the high catalytic activity. The maximum glycerol conversion of 88.4% with 91.3% selectivity to propanols was achieved at an optimum reaction condition of 210 °C and 1 bar pressure after 3 h of glycerol hydrogenolysis reaction.

Keywords: hydrogenolysis; glycerol; 1-propanol; 2-propanol; palladium catalyst

---

1. Introduction

Alternate sustainable energy resources are vital because of dwindling petroleum reserves and mounting environmental alarms that are allied with fossil fuel exploitation. As a result, alternative bio-based fuels have emerged as the long-standing solution as they are renewable and carbon dioxide neutral [1]. Biodiesel is one such alternative fuel which has received much attention with both demand and production tremendously increased over the last few years. However, along with biodiesel, also known as alkyl esters of long chain fatty acids (C14–C24), a great deal of glycerol as a by-product is also generated, typically equaling 10% of the whole production volume. The crude glycerol, if not handled properly, ends up as a waste product that has low value and is costly to purify in addition to jeopardizing the environmentally friendly nature of the whole biodiesel production process [2]. Valorization of glycerol is, therefore, necessary to enhance the sustainability of the biodiesel industry.

Glycerol, the simplest tri-hydroxy alcohol, is a highly versatile product, with many potential applications. The biodegradability and multi-functional nature of glycerol makes it a promising precursor for the production of high-value bio-renewable fuel/chemical products through various processes involving heterogeneous catalysis, e.g., acetalization [3,4], esterification [5], etherification [6], oxidation [7], dehydration [8], hydrogenolysis [9], and catalytic reforming [10]. The glycerol-derived fuel and chemical products include liquid/gaseous fuels, fuel additives, and chemicals such as solketal, glycerol mono-esters, glyceric acid, 1,3-dihydroxyacetone, epichlorohydrin, glycidol, tartronic acid, lactic acid, acrylonitrile, 1,2-propanediol, and 1,3-propanediol, etc. Catalytic hydrogenolysis of glycerol is a promising approach, resulting in the formation of commodity chemicals such as 1,2-propanediol, 1,3-propanediol, ethylene glycol, propanols (1-propanol and 2-propanol), lower alcohols, and hydrocarbons [11]. Significant efforts have been made to convert glycerol into propanediols, but direct production of propanols via glycerol hydrogenolysis received limited attention.

1-propanol (1-PrOH) and 2-propanol (2-PrOH) are valuable commodity chemicals conventionally produced from fossil-based feedstocks ethylene and propylene by hydroformylation-hydrogenation and hydration reaction processes, respectively [12]. 1-PrOH has potential applications, primarily as a solvent in the pharmaceutical, paint, cosmetics, and cellulose ester industries, organic intermediate for the synthesis of important chemical commodities, and considered as the next-generation gasoline to petroleum substitute. 2-PrOH is extensively used as an industrial solvent and disinfectant, with major applications in the pharmaceutical industry and auto industry [13]. Due to the growing demand of these commodities, the production of 1-PrOH and 2-PrOH based on bio-based glycerol (Scheme 1) appears to be an attractive approach regarding sustainability and energy efficiency compared to the process based on petroleum-derived feedstocks.

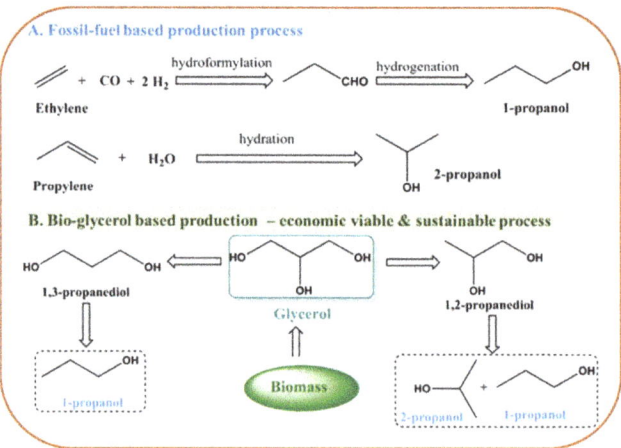

Scheme 1. Synthetic routes to 1-propanol and 2-propanol.

Hydrogenolysis of glycerol generally takes place via the dehydration-hydrogenation pathway over metal-acid bifunctional catalysts [8,14–16]. Noble metal-based catalysts have proven to be highly effective for glycerol hydrogenolysis [17]. Very few catalytic systems based on Ni [18], Ir [19], Rh [20,21], Pt [22,23], and Ru [24,25] based catalysts have so far been reported for the direct synthesis of biopropanols from glycerol, both in batch and fixed bed reactors. Previous research by Zhu et al. [26] on liquid phase glycerol hydrogenolysis to 1-propanol and 2-propanol over Pt-$H_4SiW_{12}O_{40}$/ZrO_2 bifunctional catalysts at 200 °C and 5 MPa H_2 pressure revealed that appropriate metal-acid balance is important to promote the double dehydration-hydrogenation ability of glycerol. In our earlier study [27], we have explored the effect of different heteropolyacids on the glycerol conversion and selectivity to propanols by using a Pt-HPA/ZrO_2 catalytic system and a continuous flow, fixed-bed reactor set-up under ambient pressure. It was found that the metal dispersion and acidity of the catalyst attributed to the high activity. Lin et al. [28] reported the combined use of zeolite and Ni-based catalysts as a sequential two-layer catalyst system and achieved good selectivity of 1-propanol in a fixed-bed reactor.

Despite significant research efforts, most of the reported processes were energy intensive, requiring high hydrogen pressures and the use of organic solvents for the glycerol hydrogenolysis reaction, thus making the process unsustainable. In addition, the conversion of glycerol and selectivities to propanols are still not satisfied with these catalytic systems. Besides, the exact relation between the catalyst acidity and selectivity to propanols in glycerol hydrogenolysis needs to be properly elucidated. Hence, there is a large scope to improve the sustainability of the reaction process by making it energy efficient and increasing the process profitability. It is also most important to find a highly stable, active, and selective catalytic system that favors the production of propanols by one step hydrogenolysis of glycerol under ambient reaction conditions.

In the present investigation, we report a viable catalytic strategy for the direct hydrogenolysis of low-cost glycerol to valued bio propanols over bi-functional Pd/MoO$_3$-Al$_2$O$_3$ catalysts in a fixed-bed reactor without using an organic solvent. A series of 1–4 wt% Pd/10%MoO$_3$-Al$_2$O$_3$ catalysts were prepared, carefully characterized by different characterization techniques, and tested in vapor phase glycerol hydrogenolysis at moderate temperature and atmospheric pressure. The effect of reaction parameters on the catalytic activity was investigated to determine the optimized reaction conditions. The stability of the catalyst has been analyzed to understand the changes in the catalytic activity. The research work reported herein contributes to the development of sustainable biodiesel industry, with the successful first use of alumina-supported palladium-molybdenum catalysts for the glycerol hydrogenolysis to propanols.

2. Results and Discussion

2.1. Characterization Techniques

2.1.1. Structural Characterizations of the Catalysts

The X-ray Diffraction (XRD) patterns of Mo-Al and Pd/Mo-Al catalysts with various palladium contents (1–4 wt%) are shown in Figure 1A, and Figure 1B shows the enlarged view XRD pattern of Mo-Al catalyst. For all catalysts, the diffraction peaks of γ-Al$_2$O$_3$ were identified at 2θ = 45.8° and 67.1° [29] while the peaks at 2θ = 23.4° and 25.7° were assigned to the crystalline MoO$_3$ phase [30]. In addition, other peaks at 2θ = 20.7°, 22.0°, 23.4°, and 25.7° were identified, which are attributed to the Al$_2$(MoO$_4$)$_3$ phase. After the palladium addition, the characteristic diffraction peaks of MoO$_3$ and Al$_2$(MoO$_4$)$_3$ were found to be diminished. This indicates the better dispersion of MoO$_3$ on γ-Al$_2$O$_3$ in amorphous form and saturated monolayer coverage on the catalyst surface. Also, a peak at 2θ = 33° and 59° [31] corresponding to PdO was detected, which was found to be significant at higher Pd loadings and indicates fine dispersion at lower loadings.

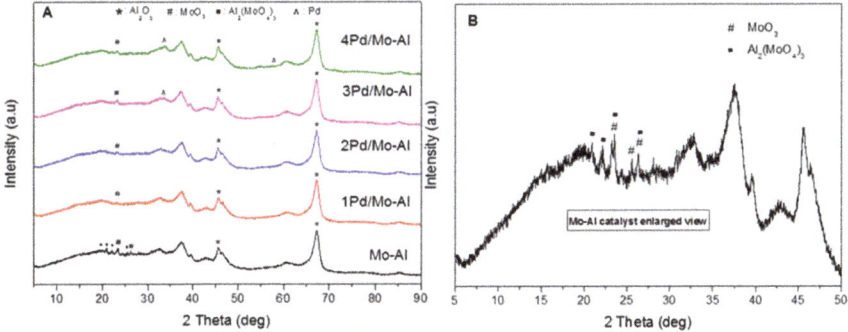

Figure 1. (A) X-ray Diffraction (XRD) patterns of Mo-Al and various Pd/Mo-Al catalysts (B) XRD patterns of Mo-Al catalyst in the enlarged view.

High-temperature X-ray diffraction (HTXRD) of MoO_3/Al_2O_3 support and dried $2Pd/MoO_3$-Al_2O_3 catalysts (the best catalytic system for this process) has been performed at various temperatures up to 550 °C to understand the nature of Pd interaction with the support during calcination, and its role in controlling the catalyst activity. As shown in Figure 2, There were not many significant changes in the XRD patterns of MoO_3/Al_2O_3 support at different temperatures, which clearly indicate the stability of the support at calcination temperature. The peaks appeared at $2\theta = 40°$, $46°$, and $67°$ were assigned to the (hkl) crystalline phases of orthorhombic MoO_3/Al_2O_3. However, the HT-XRD patterns of the dried $2Pd/MoO_3$-Al_2O_3 at different temperatures were found to be slightly different as compared to the HT-XRD profiles of the support. The absence of any diffraction patterns corresponding to metallic Pd indicated that there was no formation of metallic Pd on the support. Relative intensities of the peak, as well as the peak width of these three peaks, corresponding to the support, showed slight changes, which usually happens when a metal ion like Pd is part of the MoO_3-Al_2O_3 crystalline framework. The XRD analysis performed at various temperatures, therefore, enabled us to test the stability of the catalysts and to determine the structure of the catalysts at different temperatures varying from room temperature to 550 °C.

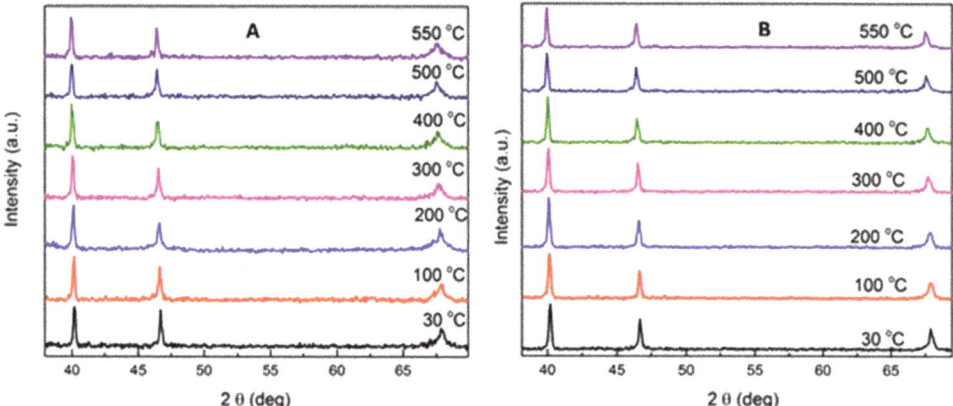

Figure 2. High-temperature X-ray diffraction (HTXRD) patterns of (**A**) uncalcined Mo-Al and (**B**) 2Pd/Mo-Al catalysts.

Scanning electron micrographs (SEM) were acquired to study the topographical morphology of Mo-Al and 2 wt% Pd/Mo-Al catalysts. The semi-quantitative elemental analysis of the catalysts was performed by Scanning Electron Microscopy (SEM) coupled with Energy Dispersive X-ray spectroscopy (EDX). The SEM images and EDX profiles of Mo-Al and 2Pd-Mo-Al catalysts are displayed in Figure 3. As can be seen from Figure 3, the micrographs reveal crystalline aggregates of MoO_3 on γ-Al_2O_3 with a monolayer coverage. In the case of 2Pd/Mo-Al catalyst, both dense and less dense regions of bigger crystallites, along with highly dispersed palladium indicating a good coverage on the support [32], was observed.

The elemental analysis data for Mo-Al and various loadings of Pd/Mo-Al catalysts are presented in Table 1. Al_2O_3 was found to be the major component, with nearly similar atomic percentage of molybdenum in all the catalysts. All Pd/Mo-Al catalysts with varying Pd content (1–4 wt%) showed the presence of palladium, indicating the impregnation of palladium onto MoO_3/Al_2O_3 catalysts. The atomic percentages of Mo and Pd from SEM-EDX results were found to be consistent with the actual loadings used in the catalyst preparation. For a comparison, the Pd contents in a series of 1–4 wt% Pd/Mo-Al catalysts was measured by Inductive Coupled Plasma-Atomic Emission Spectrometer (ICP-AES), and the results are presented in Table 1.

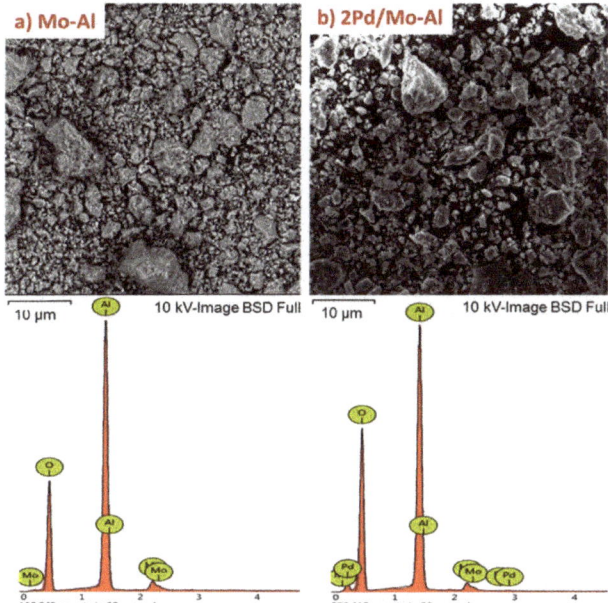

Figure 3. Scanning electron micrograph (SEM) images of Mo-Al and 2 wt% Pd/Mo-Al catalysts.

Table 1. Energy Dispersive X-ray spectroscopy (EDX) and Inductive Coupled Plasma-Atomic Emission Spectrometer (ICP-AES) analysis data of Mo-Al, and various Pd/Mo-Al catalysts.

Catalyst	Weight Percentage (%) [a]					ICP-AES [b] (wt%)
	Mo	Al	O	Pd	N	
Mo-Al	7.83	48.17	44.00	–	–	–
1Pd/Mo-Al	6.99	43.02	48.82	0.67	0.50	0.59
2Pd/Mo-Al	7.12	40.86	50.22	1.22	0.58	1.18
3Pd/Mo-Al	7.01	40.33	49.70	2.41	0.55	2.20
4Pd/Mo-Al	7.24	40.99	48.21	3.07	0.49	2.97

[a] Determined from SEM-EDX analysis; [b] Metal contents determined from ICP-AES analysis.

2.1.2. Physicochemical Properties of Catalysts

The N_2 physisorption results of Mo-Al and various Pd/Mo-Al catalysts are listed in Table 2. The pure Mo-Al catalyst exhibited a total surface area of 189 m^2/g, with a pore diameter of 5.9 nm. After incorporation of palladium into catalysts, the Brunauer–Emmett–Teller (BET) surface area, the average pore volume, and the pore size of all catalysts were found to be smaller than the original Mo-Al catalyst. This change is obvious because of the impregnation of Pd on the support that would block the micropores of the catalysts [30].

Table 2. Physicochemical properties of Mo-Al and various Pd/Mo-Al catalysts.

Catalyst	BET Surface Area (m^2/g)	Pore Size (nm)	Pore Volume (cm^3/g)
Mo-Al	189	5.9	0.29
1Pd/Mo-Al	181	5.8	0.28
2Pd/Mo-Al	170	5.6	0.28
3Pd/Mo-Al	162	5.3	0.26
4Pd/Mo-Al	158	5.3	0.23

2.1.3. Acidity Measurement by NH_3–TPD Analysis

Temperature programmed desorption (TPD) experiments of γ-Al_2O_3, Mo-Al and 1–4 wt% Pd/Mo-Al catalysts were performed to determine the amount and strength of acid sites. The number of acid sites calculated from the desorption TPD peak area is presented in Table 3. Agreeing with the literature [33], three ranges of NH_3 desorption temperature should be taken into account: 50–200, 200–350, and 350–550 °C, which correspond to weak, moderate, and strong acid sites, respectively. The pure γ-Al_2O_3 catalyst exhibited weak to moderate acidity in the temperature regions of 150–350 °C. The total acidity decreased with the addition of Mo, with a significant decrease in moderate acid sites. This could be due to a decrease in surface area and blockage of acidic sites of γ-Al_2O_3. After incorporation of Pd in Mo-Al catalyst, weaker to moderate acid sites remained in all catalysts, with no generation of new acid sites. However, it is to be noted that there was an increase in the moderate acid site in all Pd loaded catalysts compared to Mo-Al catalyst. This indicates that weak to moderate acid centers of catalyst had a crucial role in facilitating the glycerol hydrogenolysis reaction.

Table 3. Acidities of γ-Al_2O_3, Mo-Al, and 1–4 wt% Pd/Mo-Al catalysts from Temperature programmed desorption of ammonia (NH_3-TPD) analysis.

Catalyst	NH_3 Uptake (µmol/g)			Total NH_3 Uptake (µmol/g)
	Weak	Moderate	Strong	
γ-Al_2O_3	298.7	426.2	–	724.9
Mo-Al	323.5	350.6	–	674.1
1Pd/Mo-Al	297.3	361.2	–	658.5
2Pd/Mo-Al	273.3	372.2	–	645.5
3Pd/Mo-Al	194.2	369.0	–	563.2
4Pd/Mo-Al	195.0	364.9	–	559.9

2.1.4. ^{27}Al NMR Spectroscopy

Solid state ^{27}Al nuclear magnetic resonance (NMR) spectroscopy is a non-invasive and non-destructive technique vital to studying the structural modifications and coordination of aluminium nuclei within the catalyst subjected to impregnation and calcination. The chemical analysis and comparison of ^{27}Al NMR spectra of Mo-Al and 1–4 wt% Pd/Mo-Al catalyst samples is presented in Figure 4. The spectrum of Mo-Al consists of a single peak at a chemical shift of 54.1 ppm from octahedral aluminium, characteristic of as-prepared catalysts. Similar results were obtained for all the Pd/Mo-Al catalysts, indicating that palladium impregnation did not alter the basic aluminium framework. However, upon impregnation of palladium onto Mo-Al catalyst, it was observed that there was a slight change in the chemical shift of octahedral Al peak. At higher loadings, the peak was markedly broadened, and the intensity of the peak was decreased. This is probably due to the palladium interaction with the support material that causes the expulsion or distortion of aluminium atoms from the framework sites [34].

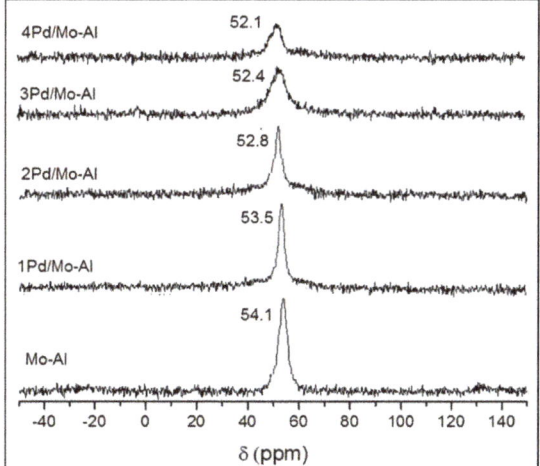

Figure 4. ^{27}Al nuclear magnetic resonance (^{27}Al NMR) spectra of Mo-Al and 1–4 wt% Pd/Mo-Al catalysts.

2.2. Catalytic Studies

Glycerol hydrogenolysis is a complex reaction (Scheme 2) and is generally believed to proceed via dehydration-hydrogenation route in two distinct pathways, which involve the formation of intermediates (acetol and 3-hydroxypropanaldehyde (3-HPA)) by acid-catalyzed dehydration and subsequent formation of propanediols (1,2-PDO and 1,3-PDO) over metal sites [35–39]. Thereafter, C-O hydrogenolysis of 1,2-PDO results in the formation of 1-propanol (1-PrOH), while C-O hydrogenolysis of 1,3-PDO gives rise to 2-propanol (2-PrOH). Another possible pathway is the formation of glyceraldehyde and 2-hydroxy acrolein by the dehydrogenation-dehydration-hydrogenation mechanism, in which acetol can be indirectly formed by this pathway. Direct hydrogenolysis of glycerol to propanols in one step using a single catalyst, though highly challenging, would be consistent and more preferential over a two-step process in terms of energy efficiency.

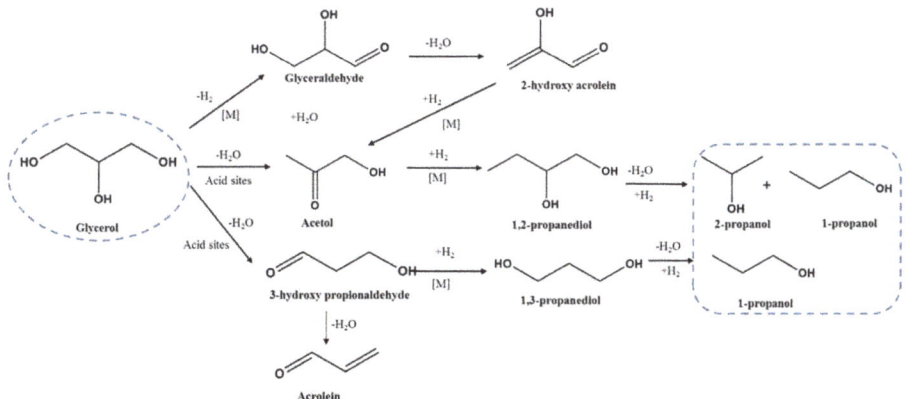

Scheme 2. Series and parallel network of reactions in hydrogenolysis of glycerol.

The present work focused on the investigation of hydrogenolysis of glycerol over highly efficient Pd/MoO$_3$-Al$_2$O$_3$ catalysts with various Pd loadings (1–4 wt%) in a continuous flow, fixed-bed reactor at 210 °C under atmospheric pressure. The reaction successfully progressed to give propanols (1-PrOH

+ 2-PrOH) as major products. Interestingly, the presence of weak to moderates acid sites in the Pd/Mo-Al catalyst promoted the double dehydration and hydrogenation of glycerol to 1-propanol and 2-propanol, respectively. Further analysis and optimization of the reaction process helped to achieve the best glycerol conversion and selectivity to propanols. Optimisation of the reaction process included adjusting various reaction parameters such as the effect of Pd loading, reaction temperature, and hydrogen flow rate. All Pd/Mo-Al catalysts showed good activity in terms of glycerol conversion and propanol selectivity in glycerol hydrogenolysis. However, the activity differed upon altering the reaction parameters, which facilitated determining the best catalyst that resulted in the highest activity.

2.2.1. Effect of Pd Loading

Defining the ideal metal loading of a catalyst for glycerol hydrogenolysis is important, as metal sites play a crucial role in the reaction mechanism. Therefore, the glycerol hydrogenolysis was performed using Pd/Mo-Al catalysts with varying Pd loading (1–4 wt%), and the results obtained are presented in Figure 5. In the absence of palladium over 10 wt% MoO_3-Al_2O_3, only about 52% glycerol conversion was observed, with 76% selectivity to total propanol. The incorporation of Pd into the catalyst resulted in significant increase in the glycerol conversion and product selectivity. As can be seen from Figure 5, a maximum of 88.4% glycerol conversion with a 91.3% selectivity to total propanol was attained over 2 wt% Pd catalyst. Further, increase in Pd loading to 3 and 4 wt% showed a drop in activity, which concludes that 2 wt% Pd/Mo-Al catalyst with an appropriate number of Pd sites influenced the direct hydrogenolysis of glycerol to propanols. Thus, 2 wt% Pd catalyst has been chosen to be the optimal catalyst for further investigations.

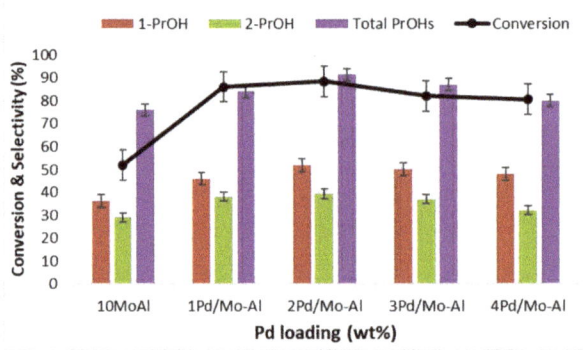

Figure 5. Effect of Pd loading on glycerol hydrogenolysis to propanols. Reaction conditions: Reaction temperature: 210 °C, 0.1 MPa H_2; H_2 flow rate: 100 mL/min; 10 wt% aqueous glycerol; 0.5 g catalyst; Reaction time: 6 h; 1-PrOH: 1-propanol; 2-PrOH: 2-Propanol; Total PrOHs: Total propanol.

2.2.2. Effect of Reaction Temperature

To assess the influence of reaction temperature on glycerol conversion and propanol selectivity, the glycerol hydrogenolysis reaction was performed over 2 wt% Pd/Mo-Al catalyst at different reaction temperatures going from 170 to 250 °C, and the results are shown in Figure 6. It is obvious that the reaction temperature had a positive influence on the glycerol conversion, as suggested by previous studies [40,41]. With the increase in the temperature from 170 °C to 230 °C, the glycerol conversion increased steadily from 70.6% up to 92.6%. The highest total propanol selectivity of 91.3% was obtained at a reaction temperature of 210 °C. Further increase of temperature to 250 °C caused a decrease in glycerol conversion and 1-PrOH selectivity. However, it is noteworthy that the selectivity to 2-PrOH elevated at a higher temperature. Therefore, the results suggest that a reaction temperature of 210 °C promoted excessive C-O hydrogenolysis of glycerol to produce the highest amounts of 1-propanol and

2-propanol, in addition to confirming that reaction temperature had a significant effect on the product distribution of 1-PrOH and 2-PrOH.

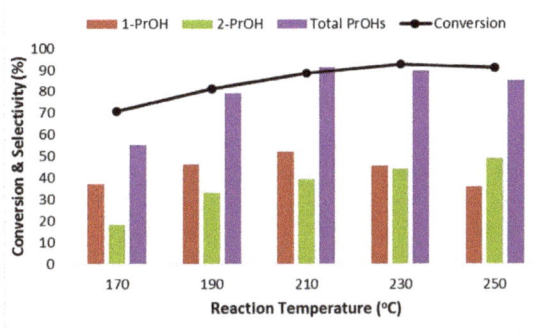

Figure 6. Effect of Reaction temperature on glycerol hydrogenolysis to propanols. Reaction conditions: Reaction temperature = 170–250 °C, Reduction temperature = 350 °C; 0.1 MPa H_2; H_2 flow rate = 100 mL/min; 10 wt% aqueous glycerol; 0.5 g catalyst; Reaction time: 6 h; 1-PrOH: 1-propanol; 2-PrOH: 2-Propanol; Total PrOHs: Total propanols.

2.2.3. Effect of Hydrogen Flow Rate

Hydrogen is a key reactant in hydrogenolysis reaction, and it is important to understand the influence of the hydrogen flow rate on glycerol hydrogenolysis. Optimising the hydrogen flow rate will add value to the process and will balance the higher costs of high-pressure apparatus. Hence the reaction was investigated by varying the hydrogen pressures at 60, 80, 100, and 120 mL/min over 2 wt% Pd/Mo-Al catalyst at 210 °C under atmospheric pressure. As can be observed from Figure 7, the glycerol conversion and selectivity of 1-PrOH increased with an increase in hydrogen flow rate from 60 to 120 mL/min. However, the total PrOH selectivity was found to be the highest at a hydrogen flow rate of 100 mL/min. While when the hydrogen flow rate was 120 mL/min, total PrOH selectivity slightly dropped, in spite of distinct increment in glycerol conversion, to 94%. This is probably because at higher hydrogen flow rates excessive hydrogenolysis occurs, leading to the formation of degradative products. A similar observation was made in previous studies [42]. Thus, 100 mL/min hydrogen flow rate was considered to be optimum for this reaction, resulting in a maximum total PrOHs selectivity of 91.3% at 88.4% glycerol conversion.

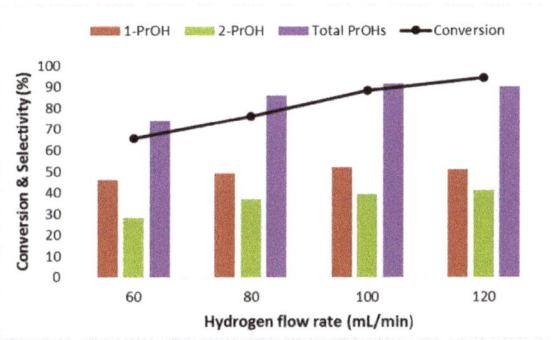

Figure 7. Effect of Hydrogen flow rate on glycerol hydrogenolysis to propanols. Reaction conditions: Reaction temperature = 210 °C; H_2 flow rate = 60, 80, 100 & 120 mL/min, 0.1 MPa H_2; 10 wt% aqueous glycerol; 0.5 g catalyst; Reaction time: 6 h; 1-PrOH: 1-propanol; 2-PrOH: 2-Propanol; Total PrOHs: Total propanols.

2.2.4. Effect of Glycerol Concentration

The glycerol concentration serves as one of the most important parameters which has a significant effect on the activity and selectivity of glycerol hydrogenolysis reaction. It is known that a low concentration of glycerol is favorable to increase the glycerol conversion [17,43]. Figure 8 shows the activity results at different glycerol concentrations (5–20 wt%), which was performed over 2 wt% Pd/Mo-Al catalyst to identify the best activity. From the results, it was observed that by increasing the glycerol concentration from 5 wt% to 20 wt%, both the glycerol conversion and selectivity to total propanols declined. As the glycerol content increased, the glycerol conversion and the selectivity towards propanols decreased considerably. With 5 wt% glycerol feed, 88% of glycerol converted, with about 80% selectivity to total propanols. At 10 wt% glycerol feed, the selectivity to total propanols was improved to 91.3% at nearly the same glycerol conversion, indicating the excessive hydrogenolysis of glycerol to propanols. At higher glycerol concentration (15 and 20 wt%), it was observed that both conversion and selectivity dropped considerably, which is ascribed to the fact that the reaction rate is lowered due to high viscosity of the glycerol feed and the imbalance between the catalytically active sites and the excess glycerol available to react at higher concentrations. Consequently, 10 wt% glycerol feed was considered the optimal concentration to attain the highest glycerol conversion of glycerol and selectivity to total propanols.

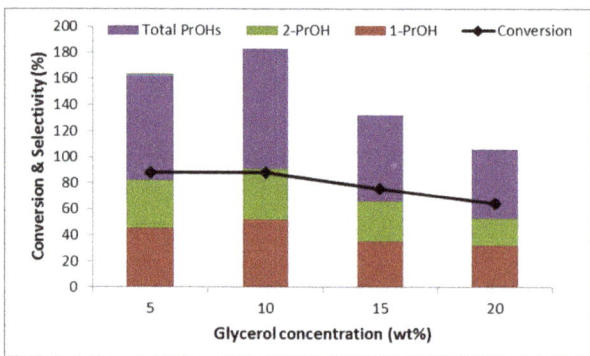

Figure 8. Effect of glycerol concentration on glycerol hydrogenolysis to propanols. Reaction conditions: Reaction temperature = 210 °C; H_2 flow rate = 100 mL/min, 0.1 MPa H_2; 5–10 wt% glycerol feed; 0.5 g catalyst; Reaction time: 6 h; 1-PrOH: 1-propanol; 2-PrOH: 2-Propanol; Total PrOHs: Total propanols.

2.2.5. Effect of the Partial Pressure of Glycerol

Figure 9 shows the effect of partial pressure of glycerol on glycerol conversion, and total propanols selectivity for the 2 Pt/Mo-Al catalyst at 210 °C, 100 mL/min H_2 flow rate, 10 wt% glycerol solution, and at different glycerol feed flow rates of 0.5, 1.0, 1.5, and 2.0 mL/h. With the increase in partial pressure of glycerol from 8.3–13.0 mm Hg, the conversion of glycerol decreased from 88.4% to 71%, and the selectivity to total propanols decreased almost two-fold. This decrease in conversion and selectivity can be attributed to a decrease in the number of active sites due to the increase in the flow of glycerol feed [44].

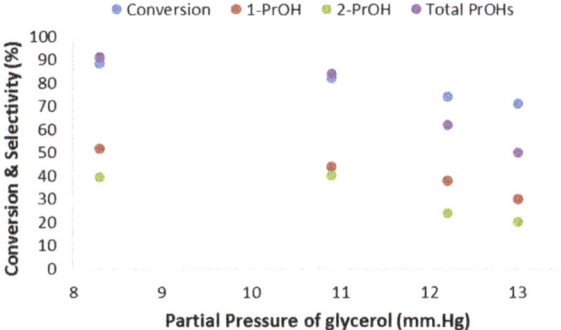

Figure 9. Effect of the partial pressure of glycerol on glycerol hydrogenolysis to propanols. Reaction conditions: Reaction temperature = 210 °C; H$_2$ flow rate = 100 mL/min, 0.1 MPa H$_2$; 10 wt% aqueous glycerol; 0.5, 1.0, 1.5 and 2.0 mL/h glycerol feed; 0.5 g catalyst; Reaction time: 6 h; 1-PrOH: 1-propanol; 2-PrOH: 2-Propanol; Total PrOHs: Total propanols.

2.2.6. Effect of Contact Time (W/F)

The dependence of the rate of glycerol hydrogenolysis on contact time (W/F) in the range of 0.4 to 2.0 g mL^{-1} h at 210 °C, 0.5 mL/h 10 wt% glycerol feed was studied over 2Pd/Mo-Al catalyst by varying the weight of the catalyst. As can be noted from Figure 10, the glycerol conversion increases steadily from 60% to 88.4% with an increase in the contact time. Similarly, the selectivity to total propanols was also found to increase substantially from 68% to 91.3% upon increasing the contact time. This observation suggests that greater contact time enables the strong adsorption of glycerol on to the catalytically active sites and facilitates the excessive hydrogenolysis reaction of glycerol to produce double dehydration-dehydrogenation products 1-propanol and 2-propanol, which is in good agreement with the previous studies [27].

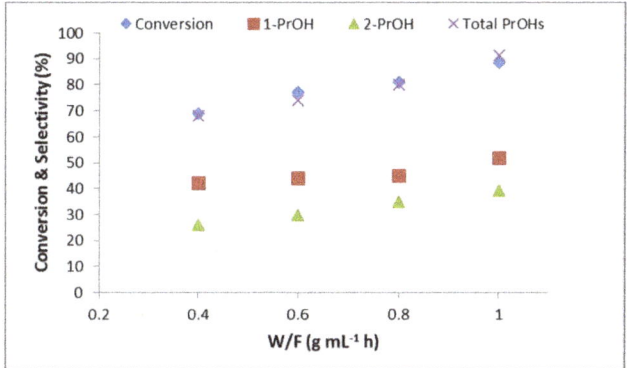

Figure 10. Effect of contact time on glycerol hydrogenolysis to propanols. Reaction conditions: Reaction temperature = 210 °C; H$_2$ flow rate = 100 mL/min, 0.1 MPa H$_2$; 10 wt% aqueous glycerol; 0.5 mL/h glycerol feed; 0.2–0.5 mg catalyst; Reaction time: 6 h; 1-PrOH: 1-propanol; 2-PrOH: 2-Propanol; Total PrOHs: Total propanols.

From all the above sequence of experiments, the optimal set of reaction conditions to accomplish the best possible glycerol conversion and total propanols selectivity over 2Pd/Mo-Al catalyst were identified as 210 °C, 100 mL/min H$_2$ flow rate, 10 wt% glycerol concentration, 8.3 mm Hg partial pressure of glycerol, and at 1.0 g mL^{-1} h contact time. Under these best set of conditions, the most active 2Pd/Mo-Al catalysts was tested for stability and reusability, and the results are analyzed.

2.2.7. Time on Stream Experiment

Time on stream experiment for 8 h was conducted over the best catalyst 2 wt% Pd/Mo-Al under the similar reaction conditions to investigate the stability and activity of catalyst (Figure 11). The conversion of glycerol progressively increased from 82% and reached a maximum of 88.4% at 3 h and remained constant up until 6 h. Also, the selectivity to total PrOH also maximized at 3 h, remained the same until 5 h, and presented a slight decline during 6–8 h. This decline in activity might be probably due to the fact that the catalyst suffers slow deactivation during the reaction due to carbon deposition. A similar finding was reported in our previous studies [8,35].

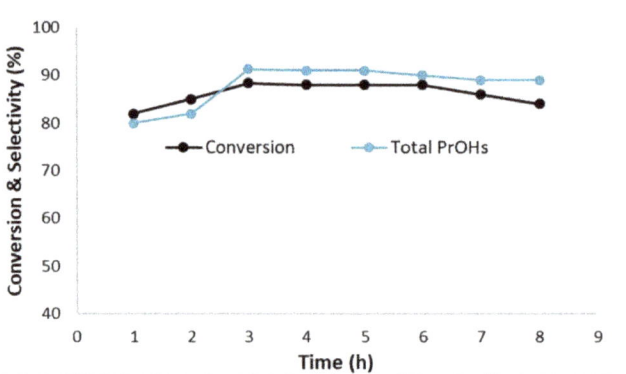

Figure 11. Time-on-stream experiment over fresh 2 wt% Pd/Mo-Al$_2$O$_3$ catalyst. Reaction conditions: Reaction temperature = 210 °C; 0.1 MPa H$_2$; H$_2$ Flow rate = 100 mL/min, 10 wt% aqueous glycerol; 0.5 g catalyst; Total PrOHs: Total propanols.

2.3. Reusability of the Catalyst

To further understand the stability and reusability of catalyst, the spent 2Pd/Mo-Al catalyst was regenerated by activating the catalyst in air flow (100 mL/min) at 500 °C for 2 h and then tested in glycerol hydrogenolysis time-on-stream experiment under the similar reaction conditions used for the fresh catalyst. The performance of 2Pd/Mo-Al catalyst upon re-use is illustrated in Figure 12A. The results were reproduced over the used catalyst and demonstrated a similar pattern to that of the fresh catalyst during 1–6 h. However, there is a slight decline in glycerol conversion and a significant drop in the selectivity to total propanols at later hours. The cause for the decrease in activity might be due to the catalyst deactivation over time due to agglomeration or carbon deposition. The used and reactivated catalysts were further analyzed by various characterization techniques, and the results are presented in Table 4 & Figure 12.

Table 4. Studies on the used 2Pd/Mo-Al catalyst.

Catalyst	Conversion (%)	Selectivity of Total PrOHs (%)	BET Surface Area (m^2/g)	Total Acidity (NH$_3$ µmol/g)
Fresh	88.4	91.3	170	645.5
Used	87.0	90.0	162	594.0
Reactivated	88.0	90.9	168	617.2

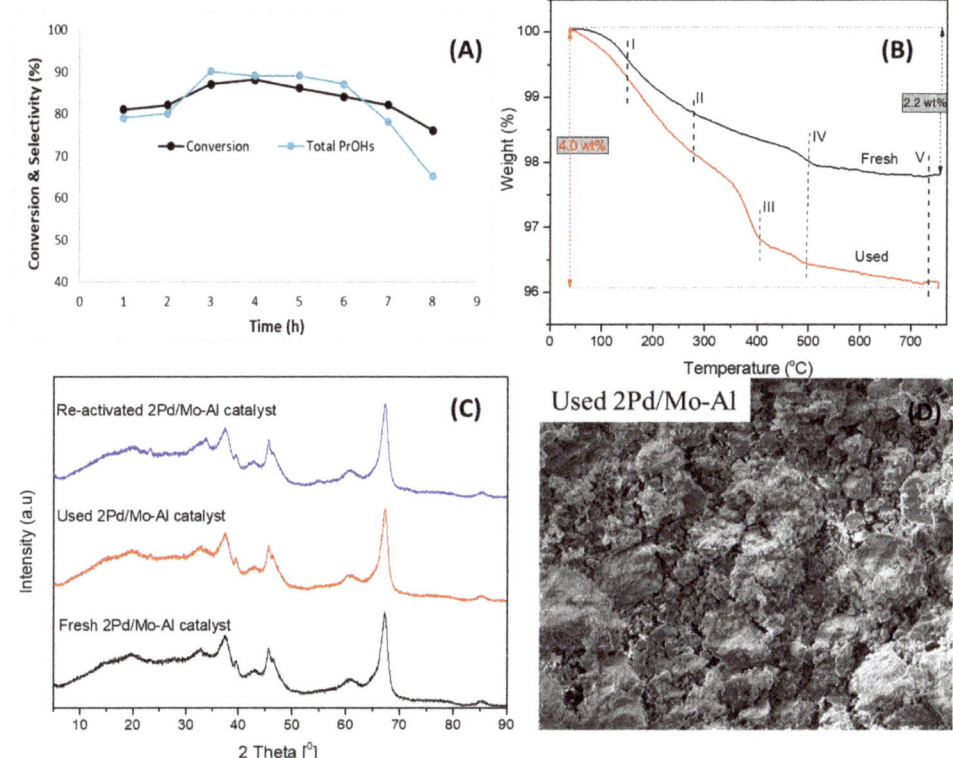

Figure 12. (**A**) Time-on-stream experiment over used 2 wt% Pd/Mo-Al catalyst; (**B**) thermogravimetric analysis (TGA) profile of fresh and used catalyst; (**C**) XRD pattern of fresh, used and reactivated catalyst; (**D**) SEM image of used 2Pd/Mo-Al catalyst.

The XRD pattern, SEM image, BET SA, and total acidity from NH_3-TPD analysis of the used 2Pd/Mo-Al catalyst showed no significant changes, indicating that the catalytic structure remained intact during the course of the reaction. Thermogravimetric analysis (TGA) was used to characterize the used catalyst to understand the mode of deactivation. The TGA measurements were performed over a temperature range of 50 °C to 750 °C at a heating rate of 10 °C/min under a constant flow of nitrogen (20 mL/min). The recorded TG curves of fresh and used catalysts presented as Figure 12B consist of four mass loss processes common in both fresh and used catalysts and an additional mass loss process observed in used catalyst. The first step (I) between 100 and 130 °C corresponds to removal of physisorbed water; the second mass loss region (II) around 250 °C indicates the loss of coordinated water molecules; the fourth weight loss step (IV) between 450–500 °C is attributed to the removal of strongly bound hydroxyl groups from alumina matrix; and a small weight loss (V) above 700 °C is due to the sublimation of molybdena [45]. In addition, a large weight loss region (III) extended above 380–500 °C in the used catalyst corresponds to the loss of carbonaceous materials [46], indicating the deactivation of the catalyst by carbon deposition. The total weight loss due to heating at the temperature range 50–750 °C for the used catalyst was found to be 4.0 wt% while it was just 2.2 wt% for the fresh catalyst, which clearly indicates that the excess weight loss is due to the deposition of carbon species and undesired materials on the catalyst surface during the reaction.

3. Materials and Methods

3.1. Catalyst Preparation

A series of palladium catalysts on 10 wt% MoO_3/Al_2O_3 with varying palladium loading from 1.0–4.0 wt% was prepared by wet impregnation method. Tetraammine palladium (II) nitrate solution (10 wt% in H_2O, produced by Sigma Aldrich Co., Ltd., St. Louis, MO, USA) was used as a precursor on the support. 10 wt% MoO_3/Al_2O_3 (2–5 mm pellets) was obtained from Riogen. The prepared catalysts were dried overnight at 100 °C and subsequently calcined at 500 °C for 2 h in air. The prepared catalysts were labelled as xPd/Mo-Al, where x refers to Pd loading, Mo refers to MoO_3, and Al refers to Al_2O_3.

3.2. Catalyst Characterization

The X-ray Diffraction (XRD) of catalysts was performed on a Miniflex X-ray diffractometer (Rigaku, Neu-Isenburg, Germany) with Ni-filtered Cu-Kα radiation (λ = 1.392 Å). The angles of scanning were from 2° to 90° with a rate of 2°/min, with the beam voltage and a beam current of 30 kV and 15 mA, respectively. High Temperature X-ray diffraction (HTXRD) profiles of the catalysts as a function of temperature were carried out using Bruker D8 Advance X-ray diffractometer equipped with the Anton-Parr heating accessory.

Nitrogen physisorption analysis of the catalysts was carried at −196 °C under liquid N_2 with a Quantachrome Autosorb 1 instrument. As a pretreatment of the N_2 physisorption analysis, each catalyst sample is degassed under vacuum for 6 h at 250 °C. The multi-point Brunauer–Emmet–Teller (BET) method was used to calculate the specific surface areas of each catalyst and the Barrett–Joyner–Halenda model (BJH) method was used to measure the average pore diameter and pore volumes.

Scanning electron microscopy (SEM) was used to study the morphology of the catalyst samples, and the elemental identification of the catalysts was performed by Energy Dispersive X-Ray Analyzer coupled to Phenom XL scanning electron microscopy (SEM-EDX). Before analysis by SEM, each sample was mounted on an aluminum support using double adhesive carbon tape. At 10 kV beam voltage and 5000× magnification, the micrographs of the catalysts were captured using a backscatter electron detector (BSD). The elemental analysis was performed at high resolution (15 kV of beam voltage) and high vacuum pressure (1 Pa) with a secondary electron detector (SED), where the point and mapping analysis for element identification was performed using a Phenom Pro Suite software.

The amount of Pd in all Pd/Mo-Al catalysts was quantitatively analyzed by Inductively coupled plasma atomic emission spectrometry (Agilent Technologies-4200MP-AES, Santa Clara, CA, USA). The samples were prepared by acid digestion of catalyst (~10 mg in 2 mL aquaregia) at 60 °C followed by dilution to desired concentration.

Temperature programmed desorption of ammonia (NH_3-TPD) experiments were conducted on AutoChem 2910 (Micromeritics, Norcross, GA, USA) instrument. In a typical experiment, 100 mg of oven-dried sample was pretreated by passage of high purity (99.995%) helium (50 mL min^{-1}) at 200 °C for one hour. After pretreatment, the sample was saturated with highly pure anhydrous ammonia (50 mL min^{-1}) with a mixture of 10% NH_3–He at 80 °C for 1 h and subsequently flushed with He flow (50 mL min^{-1}) at 80 °C for 30 min to remove physisorbed ammonia. TPD analysis was carried out from ambient temperature to 700 °C at a heating rate of 10 °C min^{-1}. The amount of NH_3 desorbed was calculated using the GRAMS/32 software.

Solid state ^{27}Al Nuclear Magnetic Resonance Spectroscopy (^{27}Al NMR) technique was used to identify the basic aluminium framework structure of catalysts on DD2 Oxford Magnet AS-500MHz spectrometer (Agilent Technologies) using aluminium oxide (Aldrich) as a probe. The chemical shifts (δ) are revealed in ppm.

Thermo gravimetric analysis (TGA) was performed using a Perkin Elmer TGA-7 from 35 °C to 700 °C with a heating rate of 10 °C per minute under the flow of nitrogen.

3.3. Catalyst Testing

The glycerol hydrogenolysis experiments were conducted in the vapour phase under atmospheric pressure in a vertical fixed bed quartz reactor (40 cm length, 9 mm i.d.) using 0.5 g of catalyst. Before the reaction, the catalysts were pretreated at 350 °C for 2 h in flowing H_2 (60 mL min^{-1}). After cooling down to the reaction temperature (210 °C), an aqueous solution of 10 wt% glycerol was introduced into the reactor using a feed pump along with the flow of hydrogen (100 mL/min). The reaction products were condensed in an ice–water trap and collected hourly for analysis on a gas chromatograph GC-2014 (Shimadzu, Kyoto, Japan) equipped with a DB-wax 123-7033 (Agilent) capillary column (0.32 mm i.d., 30 m long) and flame ionization detector. The conversion of glycerol (1) and selectivity (2) of products were calculated as follows:

$$Conversion\ (\%) = \frac{moles\ of\ glycerol\ (in) - moles\ of\ glycerol\ (out)}{moles\ of\ glycerol\ (in)} \times 100$$

$$Selectivity\ (\%) = \frac{moles\ of\ one\ product}{moles\ of\ all\ products} \times 100$$

4. Conclusions

A simple and highly efficient protocol of direct hydrogenolysis of glycerol to 1-propanol and 2-propanol over a series of 1–4 wt% Pd/MoO$_3$-Al$_2$O$_3$ catalysts has been demonstrated. Pd/Mo-Al catalysts were prepared by wetness impregnation method and characterized by XRD, SEM, NH$_3$-TPD, ^{27}Al NMR, and BET to evaluate the physical and chemical properties of catalysts. The hydrogenolysis of glycerol was performed in vapour phase under atmospheric pressure. Among the catalysts tested, 2 wt% Pd/Mo-Al catalyst established high activity and selectivity in the glycerol hydrogenolysis, with 91.3% total propanol selectivity at 88.4% glycerol conversion at best reaction conditions of 210 °C, 1 bar pressure, 100 mL/min H_2 flow rate, 10 wt% aq. Glycerol, and 8.3 mm Hg partial pressure of glycerol. The weak to moderate acidity of the catalyst was found to affect the degree of glycerol dehydroxylation, and promoted double dehydration of glycerol to form 1-propanol and 2-propanol. A comprehensive study on reaction parameters revealed the optimized reaction conditions and it was found that the reaction temperature, partial pressure of glycerol, and contact time had a substantial influence on the glycerol conversion and propanol selectivity. Further, the time on stream studies demonstrated that the catalyst remained stable for a longer period without great loss in the conversion and selectivity to products. The activity results were well correlated with the results obtained from catalyst characterization. The study signifies a sustainable technology that can not only develop the opportunity of biodiesel industry, but also deliver a green chemical production route from biomass-derived glycerol.

Author Contributions: Conceptualization, Methodology, Formal Analysis, Investigation, Writing-Original Draft Preparation, S.S.P.; Supervision, and Draft Review, S.B.

Funding: This research received no external funding.

Acknowledgments: S.S.P. would like to acknowledge Industry Connect Seed Fund support from the Monash Energy Materials and Systems Institute and Discovery Seed Fund Scheme from Monash Engineering.

Conflicts of Interest: The authors declare no conflict of interest.

References

1. Werpy, T.; Petersen, G.; Aden, A.; Bozell, J.; Holladay, J.; White, J.; Manheim, A.; Eliot, D.; Lasure, L.; Jones, S. *Top Value Added Chemicals from Biomass. Results of Screening for Potential Candidates from Sugars and Synthesis Gas*; US Department of Energy: Southwest Washington, DC, USA, 2004; Volume 1.
2. Zhou, C.H.C.; Beltramini, J.N.; Fan, Y.X.; Lu, G.Q.M. Chemoselective catalytic conversion of glycerol as a biorenewable source to valuable commodity chemicals. *Chem. Soc. Rev.* **2008**, *37*, 527–549. [CrossRef] [PubMed]

3. Priya, S.S.; Selvakannan, P.R.; Chary, K.V.R.; Kantam, M.L.; Bhargava, S.K. Solvent-Free Microwave-Assisted Synthesis of Solketal from Glycerol Using Transition Metal Ions Promoted Mordenite Solid Acid Catalysts. *J. Mol. Catal. A Chem.* **2017**, *434*, 184–193. [CrossRef]
4. Li, R.; Song, H.; Chen, J. Propylsulfonic Acid Functionalized SBA-15 Mesoporous Silica as Efficient Catalysts for the Acetalization of Glycerol. *Catalysts* **2018**, *8*, 297. [CrossRef]
5. Bossaert, W.D.; De Vos, D.E.; Rhijn, W.M.V.; Bullen, J.; Grobet, P.J.; Jacobs, P.A. Mesoporous Sulfonic Acids as Selective Heterogeneous Catalysts for the Synthesis of Monoglycerides. *J. Catal.* **1999**, *182*, 156–164. [CrossRef]
6. Karinen, R.S.; Krause, A.O.I. New biocomponents from glycerol. *Appl. Catal. A Gen.* **2006**, *306*, 128–133. [CrossRef]
7. Demirel, S.; Lucas, M.; Warna, J.; Salmi, T.; Murzin, D.; Claus, P. Reaction kinetics and modeling of the gold catalysed glycerol oxidation. *Top. Catal.* **2005**, *44*, 299–305. [CrossRef]
8. Kim, Y.T.; Jung, K.D.; Park, E.D. Gas-phase dehydration of glycerol over ZSM-5 catalysts. *Microporous Mesoporous Mater.* **2010**, *131*, 28–36. [CrossRef]
9. Priya, S.S.; Kandasamy, S.; Bhattacharya, S. Turning Biodiesel Waste Glycerol into 1,3-Propanediol: Catalytic Performance of Sulphuric acid-Activated Montmorillonite Supported Platinum Catalysts in Glycerol Hydrogenolysis. *Nat. Sci. Rep.* **2018**, *8*, 7484.
10. King, D.L.; Zhang, L.; Xia, G.; Karim, A.M.; Heldebrant, D.J.; Wang, X.; Peterson, T.; Wang, Y. Aqueous phase reforming of glycerol for hydrogen production over Pt–Re supported on carbon. *Appl. Catal. B Environ.* **2010**, *99*, 206–213. [CrossRef]
11. Sun, D.; Yamada, Y.; Sato, S.; Ueda, W. Glycerol hydrogenolysis into useful C3 chemicals. *Appl. Catal. B Environ.* **2010**, *193*, 75–92. [CrossRef]
12. Unruh, J.D.; Pearson, D. *Kirk-Othmer Encyclopedia of Chemical Technology*; John Wiley & Sons: New York, NY, USA, 2000.
13. Logsdon, J.E.; Loke, R.A. *Kirk-Othmer Encyclopedia of Chemical Technology*; John Wiley & Sons: New York, NY, USA, 2000.
14. Dam, J.T.; Djanashvili, K.; Kapteijn, F.; Hanefeld, U. Pt/Al_2O_3 Catalyzed 1,3-Propanediol Formation from Glycerol using Tungsten Additives. *ChemCatChem* **2013**, *5*, 497–505.
15. Arundhathi, R.; Mizugaki, T.; Mitsudome, T.; Jitsukawa, K.; Kaneda, K. Highly Selective Hydrogenolysis of Glycerol to 1,3-Propanediol over a Boehmite-Supported Platinum/Tungsten Catalyst. *ChemSusChem* **2013**, *6*, 1345–1347. [CrossRef] [PubMed]
16. Zhu, S.; Qiu, Y.; Zhu, Y.; Hao, S.; Zheng, H.; Li, Y. Hydrogenolysis of glycerol to 1,3-propanediol over bifunctional catalysts containing Pt and heteropolyacids. *Catal. Today* **2013**, *212*, 120–126. [CrossRef]
17. Priya, S.S.; Kumar, V.P.; Kantam, M.L.; Bhargava, S.K.; Chary, K.V.R. Catalytic performance of $Pt/AlPO_4$ catalysts for selective hydrogenolysis of glycerol to 1,3- propanediol in the vapour phase. *RSC Adv.* **2014**, *4*, 51893–51903. [CrossRef]
18. Ryneveld, E.V.; Mahomed, A.S.; Heerden, P.S.V.; Green, M.J.; Friedrich, H.B. A catalytic route to lower alcohols from glycerol using Ni-supported catalysts. *Green Chem.* **2011**, *13*, 1819–1827. [CrossRef]
19. Tamura, M.; Amada, Y.; Liu, S.; Yuan, Z.; Nakagawa, Y.; Tomishige, K. Promoting effect of Ru on Ir-ReOx/SiO_2 catalyst in hydrogenolysisof glycerol. *J. Mol. Catal. A Chem.* **2014**, *388–399*, 177–187. [CrossRef]
20. Furikado, I.; Miyazawa, T.; Koso, S.; Shimao, A.; Kunimori, K.; Tomishige, K. Catalytic performance of Rh/SiO_2 in glycerol reaction under hydrogen. *Green Chem.* **2007**, *9*, 582–588. [CrossRef]
21. Amada, Y.; Koso, S.; Nakagawa, Y.; Tomishige, K. Hydrogenolysis of 1,2-Propanediol for the Production of Biopropanols from Glycerol. *ChemSusChem* **2010**, *3*, 728–736. [CrossRef] [PubMed]
22. Ryneveld, E.V.; Mahomed, A.S.; Heerden, P.S.V.; Friedrich, H.B. Direct Hydrogenolysis of Highly Concentrated Glycerol Solutions over Supported Ru, Pd and Pt Catalyst Systems. *Catal. Lett.* **2011**, *141*, 958–967. [CrossRef]
23. Li, C.; He, B.; Ling, Y.; Tsang, C.W.; Lian, C. Glycerol hydrogenolysis to n-propanol over Zr-Al composite oxide-supported Pt catalysts. *Chin. J. Catal.* **2011**, *39*, 1121–1128. [CrossRef]
24. Wang, M.; Yang, H.; Xie, Y.; Wu, X.; Chen, C.; Ma, W.; Dong, Q.; Hou, Z. Catalytic transformation of glycerol to 1-propanol by combining zirconium phosphate and supported Ru catalysts. *RSC Adv.* **2016**, *6*, 29769. [CrossRef]

25. Schlaf, M.; Ghosh, P.; Fagan, P.J.; Hauptman, E.; Bullock, R.M. Catalytic Deoxygenation of 1,2-Propanediol to Give n-Propanol. *Adv. Synth. Catal.* **2009**, *351*, 789–800. [CrossRef]
26. Zhu, S.; Zhu, Y.; Hao, S.; Zheng, H.; Mo, T.; Li, Y. One-step hydrogenolysis of glycerol to biopropanols over Pt–$H_4SiW_{12}O_{40}$/ZrO_2 catalysts. *Green Chem.* **2012**, *14*, 2607–2616. [CrossRef]
27. Priya, S.S.; Kumar, V.P.; Kantam, M.L.; Bhargava, S.K.; Periasamy, S.; Chary, K.V.R. Metal–acid bifunctional catalysts for selective hydrogenolysis of glycerol under atmospheric pressure: A highly selective route toproduce propanols. *Appl. Catal. A Gen.* **2015**, *498*, 88–98. [CrossRef]
28. Lin, X.; Lv, Y.; Xi, Y.; Qu, Y.; Phillips, D.L.; Liu, C. Hydrogenolysis of Glycerol by the Combined Use of Zeolite and Ni/Al_2O_3 as Catalysts: A Route for Achieving High Selectivity to 1-Propanol. *Energy Fuels* **2014**, *28*, 3345–3351. [CrossRef]
29. Zhang, Y.; Han, W.; Long, X.; Nie, H. Redispersion effects of citric acid on CoMo/-Al_2O_3 hydrodesulfurization catalysts. *Catal. Commun.* **2016**, *82*, 20–23. [CrossRef]
30. Meng, D.; Wang, B.; Yu, W.; Wang, W.; Li, Z.; Ma, X. Effect of Citric Acid on MoO_3/Al_2O_3 Catalysts for Sulfur-Resistant Methanation. *Catalysts* **2017**, *7*, 151. [CrossRef]
31. Halasz, I.; Brenner, A. Preparation and characterization of PdO-MoO & Al_2O_3 catalysts. *Appl. Catal. A Gen.* **1992**, *82*, 51–63.
32. Del Arco, M.; Carrazan, S.R.G.; Martin, C.; Martin, I.; Rives, V.; Maletb, P. Surface dispersion of molybdena supported on silica, alumina and titania. *J. Mater. Chem.* **1993**, *3*, 1313–1318. [CrossRef]
33. Yang, X.L.; Dai, W.L.; Gao, R.; Fan, K. Characterization and catalytic behavior of highly active tungsten-doped SBA-15 catalyst in the synthesis of glutaraldehyde using an anhydrous approach. *J. Catal.* **2007**, *249*, 278–288. [CrossRef]
34. Barras, J.; Klinowski, J. 27Al and 29Si Solid-state NMR Studies of Dealuminated Mordenite. *J. Chem. Soc. Faraday Trans.* **1994**, *90*, 3719–3723. [CrossRef]
35. Priya, S.S.; Bhanuchander, P.; Kumar, V.P.; Deepa, D.; Selvakannan, P.R.; Kantam, M.L.; Bhargava, S.K.; Chary, K.V.R. Platinum supported on H-Mordenite: A highly efficient catalyst for selective hydrogenolysis of glycerol to 1,3-PDO. *ACS Sustain. Chem. Eng.* **2016**, *4*, 1212–1222. [CrossRef]
36. Oh, J.; Dash, S.; Lee, H. Selective conversion of glycerol to 1,3-propanediol using Pt-sulfated Zirconia. *Green Chem.* **2011**, *13*, 2004–2007. [CrossRef]
37. Priya, S.S.; Kumar, V.P.; Kantam, M.L.; Bhargava, S.K.; Srikanth, A.; Chary, K.V.R. High Efficiency Conversion of Glycerol to 1,3-Propanediol Using a Novel Platinum−Tungsten Catalyst Supported on SBA-15. *Ind. Eng. Chem. Res.* **2015**, *54*, 9104–9115. [CrossRef]
38. Priya, S.S.; Bhanuchander, P.; Kumar, V.P.; Bhargava, S.K.; Chary, K.V.R. Activity & Selectivity of Platinum-Copper bimetallic catalysts supported on mordenite for glycerol hydrogenolysis to 1,3-propanediol. *Ind. Eng. Chem. Res.* **2016**, *55*, 4461–4472.
39. Priya, S.S.; Kumar, V.P.; Kantam, M.L.; Bhargava, S.K.; Chary, K.V.R. Vapour-Phase Hydrogenolysis of Glycerol to 1,3-Propanediol Over Supported Pt Catalysts: The Effect of Supports on the Catalytic Functionalities. *Catal. Lett.* **2014**, *144*, 2129–2143. [CrossRef]
40. Amada, Y.; Shinmi, Y.; Koso, S.; Kubota, T.; Nakagawa, Y.; Tomishige, K. Reaction mechanism of the glycerol hydrogenolysis to 1,3-propanediol over Ir-ReO_x/SiO_2 catalyst. *Appl. Catal. B Environ.* **2011**, *105*, 117–127. [CrossRef]
41. Huang, L.; Zhu, Y.; Zheng, H.; Ding, G.; Li, Y. Direct conversion of glycerol into 1,3-propanediol over Cu-$H_4SiW_{12}O_{40}$/SiO_2 in vapor phase. *Catal. Lett.* **2009**, *131*, 312–320. [CrossRef]
42. Cavani, F.; Guidetti, S.; Trevisanut, C.; Ghedini, E.; Signoretto, M. Unexpected events in sulfated zirconia catalyst during glycerol-to-acrolein conversion. *Appl. Catal. A Gen.* **2011**, *409–410*, 267–278. [CrossRef]
43. Miyazawa, T.; Kusunoki, Y.; Kunimori, K.; Tomishige, K. Glycerol conversion in the aqueous solution under hydrogen over Ru/C + an ion-exchange resin and its reaction mechanism. *J. Catal.* **2006**, *240*, 213–221. [CrossRef]
44. Kumar, V.P.; Priya, S.S.; Harikrishna, Y.; Kumar, A.; Chary, K.V.R. Catalytic Functionalities of Nano Ruthenium/γ-Al_2O_3 Catalysts for the Vapour Phase Hydrogenolysis of Glycerol. *J. Nanosci. Nanotechnol.* **2016**, *16*, 1952–1960. [CrossRef]

45. El-Shobakya, H.G.; Mokhtarb, M.; Ahmeda, A.S. Effect of MgO-doping on solid-solid interactions in MoO_3/Al_2O_3 system. *Thermochim. Acta* **1999**, *327*, 39–46. [CrossRef]
46. Toniolo, F.S.; Barbosa-Coutinho, E.; Schwaab, M.; Leocadio, I.C.L.; Aderne, R.S.; Schmal, M.; Pinto, J.C. Kinetics of the catalytic combustion of diesel soot with MoO_3/Al_2O_3 catalyst from thermogravimetric analyses. *Appl. Catal. A Gen.* **2008**, *342*, 87–92.

© 2018 by the authors. Licensee MDPI, Basel, Switzerland. This article is an open access article distributed under the terms and conditions of the Creative Commons Attribution (CC BY) license (http://creativecommons.org/licenses/by/4.0/).

Article

Effect of Microwave Drying, Calcination and Aging of Pt/Al$_2$O$_3$ on Platinum Dispersion

Xavier Auvray [1,2,*] and Anthony Thuault [3]

1. Department of Chemical Engineering, NTNU—Norwegian University of Science and Technology, NO-7491 Trondheim, Norway
2. Competence Center for Catalysis, Chemical Engineering, Chalmers University of Technology, SE-412 96 Göteborg, Sweden
3. LMCPA—Laboratoire des Matériaux Céramiques et Procédés Associés, Université de Valenciennes et du Hainaut Cambrésis, Boulevard Charles de Gaulle, 59600 Maubeuge, France; anthony.thuault@univ-valenciennes.fr
* Correspondence: auvray@chalmers.se; Tel.: +46-031-772-3039

Received: 11 August 2018; Accepted: 24 August 2018; Published: 26 August 2018

Abstract: The effect of heating method employed for drying and calcination during the synthesis of 1 wt% Pt/Al$_2$O$_3$ catalyst was investigated. Conventional heating (CH) in resistive oven and microwave heating (MW) in single mode were applied, and the Pt dispersion and Brunauer-Emmett-Teller (BET) surface area were measured to characterize the samples. It was evidenced that the fast and homogeneous heating offered by the microwave heating led to higher Pt dispersion. However, this benefit was only achieved when the subsequent calcination was performed in a conventional oven. The aging in microwave oven of conventionally prepared—as well as MW-prepared—catalysts demonstrated the great ability of microwave irradiation to accelerate platinum sintering. After 1 h at 800 °C under microwave, catalysts showed a dispersion of 5%. Therefore, microwave treatment should be considered for accelerated catalyst aging but should be avoided as a calcination technique for the synthesis of highly dispersed Pt/Al$_2$O$_3$.

Keywords: microwave; catalyst synthesis; Pt/Al$_2$O$_3$; aging; platinum dispersion; drying; DOC

---

1. Introduction

A common method to prepare supported noble metal catalyst is wet impregnation. It consists in impregnating a porous metal oxide, with a high surface area, with a solution of noble metal precursor. The solvent is then evaporated by a slow drying step while the remaining compounds from the metal complex are decomposed by high temperature calcination. These two steps are crucial for the resulting metal distribution and metal particle size. Drying is a complex process in which the solvent is transferred from the pores to the liquid-gas interface, where it evaporates. The diffusion of solvent from the inside to the outside of the support particle induces the transport of the dissolved metal towards the exterior of the support particles, where it becomes more concentrated, especially when the metal adsorption on the support is weak [1,2]. If the metal cannot diffuse back inward in the last stage of the drying process, the formation of larger particles located within the outer layer of the catalyst particle in a "egg-white" or "egg-shell" configuration occurs. Exposure to high temperature during calcination brings about metallic particles mobility and coalescence, which also results in the formation of large particles. To control the metal deposition uniformity and the metal particle size it is necessary to limit the transport phenomena observed during drying in a conventional furnace based on resistive heating, also called Joule heating. With this aim, other ways of drying impregnated catalysts have been used such as spray drying [3] and freeze-drying [4,5], yielding high initial dispersion. The specificity of microwave heating is its homogeneity. It has been shown that microwave drying leads to a more uniform distribution of metal particles than regular oven drying in the case of Ni/Al$_2$O$_3$ catalyst [2,6,7]. These results were explained by the unique volumetric heating provided by microwaves. Microwave heating is also appreciated in zeolite synthesis [8–11] due to shorter time required, energy saving and narrower particle size distribution. In ceramics science, microwave (MW) sintering can lead to higher material density at lower temperature or after a shorter time than conventional oven sintering [12,13].

Although the impact of microwave treatment during catalyst preparation has been addressed, little is known on the effect of microwave-assisted drying and calcination on the resulting particle size. Moreover, the studies reporting on the effect of microwaves on metal distribution employed household multimode pulsed microwave ovens which limit the electromagnetic field intensity and does not allow temperature measurement. In the present study, microwave-assisted synthesis of 1 wt% Pt/Al$_2$O$_3$, widely used in automotive catalysis to carry out oxidation reactions, was performed. The impact of microwave drying and calcination on the resulting platinum dispersion was compared with conventional drying and calcination.

2. Materials and Methods

2.1. Catalyst Preparation

Platinum was deposited by incipient wetness method onto a pre-calcined γ-Al$_2$O$_3$ support to obtain a Pt content of 1 wt%. Pt(NO$_3$)$_2$ (Sigma-Aldrich, St. Louis, MO, USA) precursor was used and diluted in deionized water to obtain an aqueous solution, the volume of which corresponded to the pore volume of the support. The samples were then dried either in a conventional oven at 110 °C overnight or in a microwave oven at 210 °C for 15 min (see the description of the set-up below). The MW drying temperature was chosen for practical reasons as it was the temperature reached by bare alumina irradiated with the minimum controlled power allowed by the generator (60 W). The drying time was determined by MW drying of wetted alumina. 15 min appeared sufficient to evaporate all water from the alumina pores, which was controlled by weighing the alumina after drying. A short drying time was sought to study solely the drying stage and limit the catalyst exposure to MW. Calcination of dried samples was carried out at 600 °C for 2 h and 30 min in conventional and MW oven, respectively. Similarly, the MW calcination duration was determined during a calcination test and kept as short as possible to limit unnecessary aging. The color change of the catalyst indicated the decomposition of nitrate and the oxidation of platinum. The usual calcination was conducted under air flow in a vertical tubular quartz reactor surrounded by a furnace. Aging was performed in the same MW oven at 700 and 800 °C. The catalysts were maintained at these respective temperatures for 1 h.

2.2. Microwave Set-Up

The apparatus used was a Sairem microwave oven (Neyron, France), equipped with a 2.45 GHz cavity magnetron, a 2-kW generator and a single-mode microwave applicator (Figure S1). In this study, the transverse electric mode (TE105) was used in order to heat the sample by direct interaction between the matter and the electric field. The microwaves produced were in resonance within a standard WR340 rectangular waveguide. The generated power spanned from 60 to 2000 W. The catalyst was contained in a low-absorption alumina crucible (height = 14 mm, Ø = 25 mm) surrounded by an in-house made SiC ring acting as susceptor and a sample holder and thermal insulation made of fibrous alumina as described in previous studies [14]. The sample holder was placed in a chamber where the intensity of the electromagnetic field could be fine-tuned. This system allows to reach high temperature rapidly and with low input power due to the resonance phenomenon. The temperature was monitored by a pyrometer placed above the sample and pointing down to the sample. Two pyrometers with various temperature range were used: 60–400 °C and 350–2000 °C. The pyrometers were calibrated with conventional furnace and the emissivity used was 0.8, corresponding to alumina. Preliminary tests were performed on alumina or wet alumina to determine the power settings required to heat the sample at a steady desired temperature.

2.3. Pt Dispersion Measurement

The platinum dispersion was measured by chemisorption of probe molecules on an ASAP2020 (Micromeritics, Norcross, GA, USA). Ca. 400 mg of catalyst was used to get sufficient available metal to perform reliable measurement. After in situ oxidation (30 min at 350 °C in air) and subsequent reduction (2 h at 450 °C in H_2), the sample was evacuated, and stepwise CO adsorption was performed at 35 °C to obtain an isotherm of the adsorbed quantity from 1 to 400 mmHg. The adsorption was repeated after evacuation to quantify the amount of physisorbed CO. The difference of both isotherms is a constant line, the value of which corresponds to the quantity of chemisorbed CO. An example of CO adsorption isotherm is provided in Figure S2. The stoichiometry of CO adsorption on Pt was set to 1 although the presence of bridged CO adsorbates would decrease the CO:Pt ratio and lead to underestimation of the dispersion. H_2–O_2 titration was also employed to validate the CO chemisorption results. The same instrument was used, and the pretreatment protocol was identical as for CO chemisorption. The amount of H_2 consumed by both reaction with oxygen adsorbed on platinum and dissociative adsorption of H_2 on platinum was obtained by extrapolation of the isotherm at $P = 0$.

2.4. Surface Area Measurements

The surface area of the materials was measured by physisorption of N_2 at 77 K with a Micromeritics Tristar apparatus (Norcross, GA, USA). The catalysts were previously degassed for 4 h at 250 °C. The surface area was calculated according to the BET method.

3. Results

To study the impact of microwave thermal treatment on Pt dispersion, a set of 1 wt% Pt/Al_2O_3 samples was prepared with different combinations of drying and calcination methods. Table 1 summarizes the synthesis conditions of each sample, as well as their aging treatment.

Table 1. Summary of 1 wt% Pt/Al_2O_3 samples synthesized.

	1	2	3	4	5	6	7
Drying	MW	MW	MW	MW	CH	CH	CH
Calcination	CH	MW	MW	MW	CH	CH	CH
Aging (MW)	–	–	700 °C	800 °C	* 600 °C	700 °C	800 °C

MW: Microwave Heating. CH: Conventional Heating. Samples 5, 6 and 7 come from the same batch, therefore having the same dispersion after calcination. * MW treatment at 600 °C for 30 min, like MW calcination.

Some samples were entirely prepared with the help of a conventional oven, operating under gas flow, where the heat was transferred to the sample via the surrounding atmosphere. Those samples were then aged under microwave radiation at various temperatures in order to study the effect of microwaves on already prepared catalysts. It should be noted that the normal aging time of 1 h, used at 700 and 800 °C, was reduced to 30 min in the case of the aging at 600 °C to reproduce the same treatment as the microwave-assisted calcination. One singular sample was dried under microwave and calcined in conventional oven to investigate the effect of MW drying. Additional catalysts were both dried, calcined and, when applicable, aged in MW oven. This addresses the effect of MW calcination and allows to compare the impact of MW aging on MW-prepared and conventionally prepared 1 wt% Pt/Al_2O_3 catalysts.

3.1. Effect of Microwave Drying

The dispersion presented by the two fresh catalysts calcined in usual oven and dried in a different way is shown in Figure 1. It can be seen that both CO chemisorption and H_2–O_2 titration reveal higher Pt dispersion for the MW-dried sample. H_2–O_2 titration gave consistently higher dispersion values than CO chemisorption which can be attributed to the underestimation of the dispersion measured by CO chemisorption due to the chosen CO:Pt adsorption stoichiometric ratio. In addition, both samples had similar BET surface area (147 m^2/g), as reported in Table 2. In light of this result, one can conclude that MW drying yields higher platinum dispersion than conventional resistive drying.

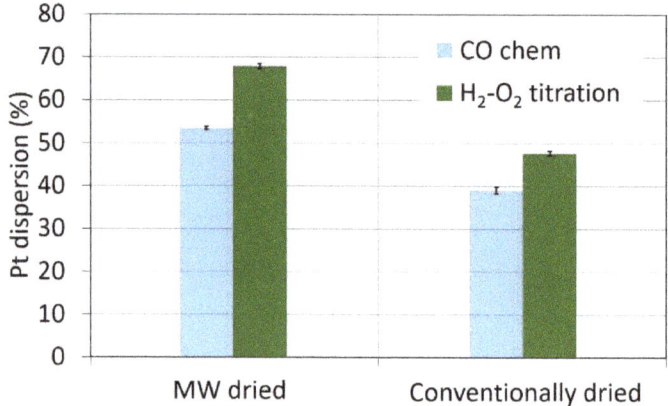

Figure 1. Comparison of Pt dispersion of samples conventionally dried (110 °C overnight) and dried by microwave irradiation (210 °C, 15 min). The dispersion was measured by CO chemisorption and H_2–O_2 titration. Both samples were calcined in conventional furnace.

Table 2. BET surface area of all prepared samples.

Catalyst	Drying	CH	MW	MW	CH	CH	CH	MW
	Calcination	CH	CH	MW	CH	CH	CH	MW
	Aging (°C)	–	–	–	MW600	MW700	MW800	MW800
BET Surface Area (m^2/g)		147	147	151	144	144	138	137

The higher dispersion measured on MW-dried catalyst can be attributed to the higher drying temperature (210 °C vs. 110 °C) or to a specific effect of microwave irradiation and uniform heating. Vallee and Conner investigated the sorption and sorption selectivity of various compounds on an oxide surface under MW irradiation [15,16]. They found that adsorption of compounds depends on the properties of the adsorbates, the adsorbents and the surface of adsorption. Desorption of compounds with high permittivity, which reflects the ability to absorb microwave energy, was much greater under MW heating than in the conventional oven. In the case of a mixture, this property leads to the possibility of selective heating and selective adsorption/desorption depending on the compounds' permittivity [16,17]. Since water has a significantly higher permittivity than Pt, water desorption is enhanced in favor of Pt adsorption under MW irradiation. In addition, the fast and intense evaporation of the solvent due to MW heating provokes deposition of the solute in situ [18] leading to the high platinum dispersion measured. The higher dispersion measured after a fast drying indicates either that the diffusion of platinum is extremely slow [19] or its interaction with the surface is strong [18]. As reported by Bond et al. [20], microwave drying of a Ni/Al$_2$O$_3$ catalyst after wet impregnation leads

3.2. Effect of Microwave Calcination

After impregnation and drying, a calcination step is necessary to decompose the counterions or ligands of the metal precursor. It consists in a thermal treatment beyond the decomposition temperature of the precursor for a sufficient time. However, care must be taken of metal migration and aggregation due to high temperature. The use of microwave to perform this step was compared to the resistive heating. In Figure 2, the effect of the calcination method can be compared on samples dried with MW. It is clear that MW calcination is detrimental for platinum dispersion despite the shorter time of exposure to 600 °C. During MW calcination, the catalyst lost near 30% of available platinum sites. This result is found for both dispersion measurement techniques. As the H_2 titration stoichiometry is known ($H/Pt_{surface} = 1$), the amount of Pt surface atoms can be inferred for each catalyst and subsequently the CO chemisorption stoichiometry can be assessed. Our results show consistently a CO/Pt ratio in the range 0.71–0.79 for all samples, regardless of the dispersion or preparation method. This stoichiometry corresponds to a ratio of linearly bonded CO and bridged CO of around 1. The BET surface area was not decreased by MW aging (151 m^2/g for the sample both dried and calcined by MW in Table 2). MW heating presents therefore the ability to rapidly sinter Pt at relatively low temperature. This implies great mobility of Pt nanoparticles on alumina surface during MW irradiation. It could be due to a specific effect of the electromagnetic field on metallic particles and on their interaction with the support [21], which enhances their mobility. Local hot spots in the catalyst bed are difficult to detect but are likely to occur [22,23] and would cause platinum sintering. The elucidation of the enhanced sintering mechanism will require further study.

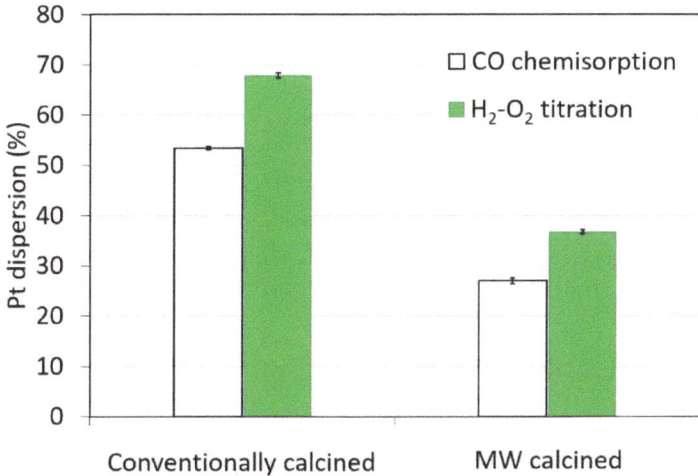

Figure 2. Comparison of Pt dispersion of samples conventionally calcined (600 °C, 2 h) and calcined by microwave irradiation (600 °C, 30 min). The dispersion was measured by CO chemisorption and H_2–O_2 titration. Both samples were dried by microwave irradiation.

3.3. Effect of Microwave Aging

In practical applications, catalysts are often subjected to high temperatures or thermal treatment that leads to morphological changes which are often studied in the laboratory. However, this leads to noble metal sintering and/or loss of surface area due to support phase changes. Thermal and chemical aging is often simulated in the laboratory to elucidate these mechanisms and evaluate the impact of aging conditions on catalytic activity [4,24,25]. Here, we attempted to use MW irradiation to provoke accelerated aging of Pt/Al_2O_3. Alumina has very low dielectric constant (9) and dielectric loss (0.0063) at 25 °C [26] which makes this material transparent to MW. However, these electrical parameters increase with temperature and impurities in the material [27]. As a metal, Pt reflects MW and can induce arcing. However, the dielectric parameters are temperature dependent and a SiC susceptor was used in our experiment to ignite the temperature raise. Unlike for drying, where water filled the alumina pores, and for calcination, where nitrate species could absorb and convert MW into heat, Pt/Al_2O_3 was the sole medium to interact with the MW during aging. Ca. 1 g of sample was used for aging and the temperature was reached, in all cases, in less than 20 min. Firstly, at the temperatures tested, it was noted that alumina easily generated heat from MW absorption and consequently little extra input power was needed to increase the temperature from 700 to 800 °C. Secondly, arcing was not observed during MW aging due to the susceptor presence. High temperature MW treatment of Pt/Al_2O_3 was therefore successfully performed. The dispersion measured after MW aging of MW and CH prepared catalysts is depicted in Figure 3. It should be noted that the CH-prepared sample, MW-aged 30 min at 600 °C, was previously CH-calcined for 2 h at 600 °C, while the MW-prepared sample was only MW-calcined for 30 min at 600 °C. The CH sample has therefore been exposed to 600 °C for a longer time (2.5 h) than the MW sample (0.5 h) and yet presents a slightly greater dispersion. This highlights the strong effect of MW calcination at 600 °C on Pt dispersion.

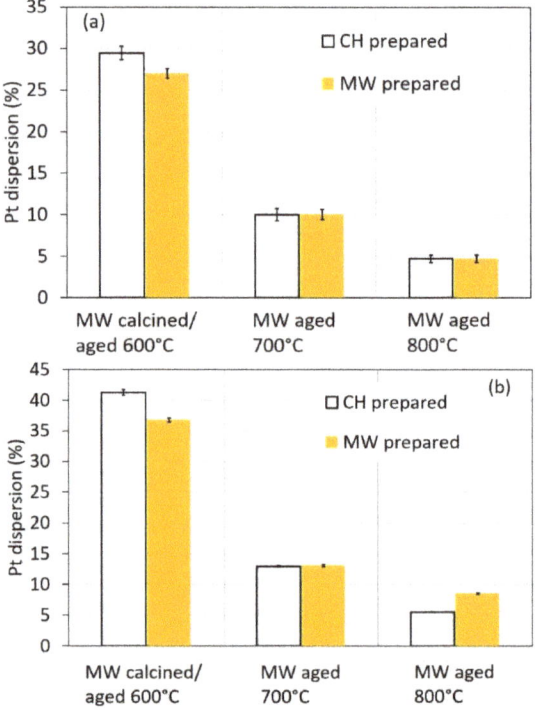

Figure 3. Platinum dispersion of various MW-aged Pt/Al_2O_3 catalysts (30 min at 600 °C and 1 h at 700 and 800 °C). (**a**) CO chemisorption and (**b**) H_2–O_2 titration results.

The BET surface area of Pt/Al_2O_3 decreased moderately due to microwave aging. It dropped from 147 to 138 m^2/g and from 151 to 137 m^2/g after 1 h at 800 °C (Table 2) for the conventionally prepared catalyst and the microwave prepared catalyst, respectively. This indicates that γ-alumina

4. Conclusions

The utilization of microwave heating during the synthesis of Pt/Al$_2$O$_3$ catalyst was investigated and compared to conventional heating (CH), with focus on resulting platinum dispersion. Higher dispersion was obtained when drying of the impregnated alumina was performed under microwave irradiation than under conventional heating. However, this result was only obtained when subsequent calcination was done in a conventional oven. Indeed, MW calcination led to lower dispersion after both MW and CH drying. Aging at 700 and 800 °C under microwave irradiation was also studied and the results revealed platinum sintering on both MW- and CH-prepared catalysts. A 1 h exposure to 700 °C yielded a dispersion of ca. 10% while at 800 °C, the dispersion dropped to ca. 5%, which is a very low value based on the initial dispersion and the rather short aging time. Microwaves are capable of accelerating Pt sintering. The BET surface area, however, was moderately affected by microwaves.

Supplementary Materials: The following are available online at http://www.mdpi.com/2073-4344/8/9/348/s1, Figure S1: Description of the single-mode microwave apparatus (Sairem). Figure S2: Example of a typical CO adsorption isotherm: 1st isotherm corresponds to both physisorbed and chemisorbed CO, 2nd isotherm, measured after evacuation, corresponds to physisorbed CO and the difference corresponds, therefore, to the amount of chemisorbed CO.

Author Contributions: X.A. conceived and performed the experiments, acquired funding, analyzed the data and wrote the original draft. A.T. provided the microwave equipment and expertise in microwave irradiation, performed the microwave experiments and reviewed and edited the draft.

Funding: This research received no external funding.

Acknowledgments: The Norges teknisk-naturvitenskapelige universitet (NTNU) is gratefully acknowledged for financial support.

Conflicts of Interest: The authors declare no conflicts of interest.

References

1. Van Den Berg, G.H.; Rijnten, H.T. The Impregnation and Drying Step in Catalyst Manufacturing. *Stud. Surf. Sci. Catal.* **1979**, *3*, 265–277.
2. Lekhal, A.; Glasser, B.J.; Khinast, J.G. Impact of drying on the catalyst profile in supported impregnation catalysts. *Chem. Eng. Sci.* **2001**, *56*, 4473–4487. [CrossRef]
3. Santiago, M.; Restuccia, A.; Gramm, F.; Pérez-Ramírez, J. Spray deposition method for the synthesis of supported catalysts with superior metal dispersion. *Microporous Mesoporous Mater.* **2011**, *146*, 76–81. [CrossRef]
4. Auvray, X.; Pingel, T.; Olsson, E.; Olsson, L. The effect gas composition during thermal aging on the dispersion and NO oxidation activity over Pt/Al$_2$O$_3$ catalysts. *Appl. Catal. B* **2013**, *129*, 517–527. [CrossRef]
5. Eggenhuisen, T.M.; Munnik, P.; Talsma, H.; de Jongh, P.E.; de Jong, K.P. Freeze-drying for controlled nanoparticle distribution in Co/SiO$_2$ Fischer–Tropsch catalysts. *J. Catal.* **2013**, *297*, 306–313. [CrossRef]
6. Liu, X.; Khinast, J.G.; Glasser, B.J. Drying of supported catalysts for low melting point precursors: Impact of metal loading and drying methods on the metal distribution. *Chem. Eng. Sci.* **2012**, *79*, 187–199. [CrossRef]
7. Vergunst, T.; Kapteijn, F.; Moulijn, J.A. Monolithic catalysts—Non-uniform active phase distribution by impregnation. *Appl. Catal. A* **2001**, *213*, 179–187. [CrossRef]
8. Choi, K.Y.; Tompsett, G.; Conner, W.C. Microwave assisted synthesis of silicalite–Power delivery and energy consumption. *Green Chem.* **2008**, *10*, 1313–1317. [CrossRef]
9. Tompsett, G.A.; Conner, W.C.; Yngvesson, K.S. Microwave synthesis of nanoporous materials. *Chem. Phys. Chem.* **2006**, *7*, 296–319. [CrossRef] [PubMed]
10. Brar, T.; France, P.; Smirniotis, P.G. Control of Crystal Size and Distribution of Zeolite A. *Ind. Eng. Chem. Res.* **2001**, *40*, 1133–1139. [CrossRef]

11. Gharibeh, M.; Tompsett, G.A.; Conner, W.C. Microwave reaction enhancement: The rapid synthesis of SAPO-11 molecular sieves. *Top. Catal.* **2008**, *49*, 157–166. [CrossRef]
12. Xie, Z.; Yang, J.; Huang, Y. Densification and grain growth of alumina by microwave processing. *Mater. Lett.* **1998**, *37*, 215–220. [CrossRef]
13. Presenda, Á.; Salvador, M.D.; Penaranda-Foix, F.L.; Catalá-Civera, J.M.; Pallone, E.; Ferreira, J.; Borrell, A. Effects of microwave sintering in aging resistance of zirconia-based ceramics. *Chem. Eng. Process.* **2017**, *122*, 404–412. [CrossRef]
14. Heuguet, R.; Marinel, S.; Thuault, A.; Badev, A. Effects of the Susceptor Dielectric Properties on the Microwave Sintering of Alumina. *J. Am. Ceram. Soc.* **2013**, *96*, 3728–3736. [CrossRef]
15. Vallee, S.J.; Conner, W.C. Microwaves and Sorption on Oxides: A Surface Temperature Investigation. *J. Phys. Chem. B* **2006**, *110*, 15459–15470. [CrossRef] [PubMed]
16. Vallee, S.J.; Conner, W.C. Effects of microwaves and microwave frequency on the selectivity of sorption for binary mixtures on oxides. *J. Phys. Chem. C* **2008**, *112*, 15483–15489. [CrossRef]
17. Jobic, H.; Santander, J.E.; Conner, W.C.; Wittaker, G.; Giriat, G.; Harrison, A.; Ollivier, J.; Auerbach, S.M. Experimental evidence of selective heating of molecules adsorbed in nanopores under microwave radiation. *Phys. Rev. Lett.* **2011**, *106*, 157401. [CrossRef] [PubMed]
18. Lee, S.-Y.; Aris, R. The Distribution of Active ingredients in Supported Catalysts Prepared by Impregnation. *Catal. Rev. Sci. Eng.* **1985**, *27*, 207–340. [CrossRef]
19. Nelmark, A.V.; Kheifez, L.I.; Fenelonov, V.B. Theory of Preparation of Supported Catalysts. *Ind. Eng. Chem. Prod. Res. Dev.* **1981**, *20*, 439–450. [CrossRef]
20. Bond, G.; Moyes, R.B.; Pollington, S.D.; Whan, D.A. The Advantageous use of Microwave Radiation in the Preparation of Supported Nickel Catalysts. *Stud. Surf. Sci. Catal.* **1993**, *75*, 1805–1808.
21. Olevsky, E.A.; Maximenko, A.L.; Grigoryev, E.G. Ponderomotive effects during contact formation in microwave sintering. *Modell. Simul. Mater. Sci. Eng.* **2013**, *21*, 055022. [CrossRef]
22. Stuerga, D.; Gaillard, P. Microwave heating as a new way to induce localized enhancements of reaction rate. Non-isothermal and heterogeneous kinetics. *Tetrahedron* **1996**, *52*, 5505–5510. [CrossRef]
23. Zhang, X.; Hayward, D.O.; Mingos, D.M.P. Microwave dielectric heating behavior of supported MoS$_2$ and Pt catalysts. *Ind. Eng. Chem. Res.* **2001**, *40*, 2810–2817. [CrossRef]
24. Auvray, X.P.; Olsson, L. Sulfur dioxide exposure: A way to improve the oxidation catalyst performance. *Ind. Eng. Chem. Res.* **2013**, *52*, 14556–14566. [CrossRef]
25. Mihai, O.; Fathali, A.; Auvray, X.; Olsson, L. DME, propane and CO: The oxidation, steam reforming and WGS over Pt/Al$_2$O$_3$. The effect of aging and presence of water. *Appl. Catal. B* **2014**, *160–161*, 480–491. [CrossRef]
26. Durka, T.; van Gerven, T.; Stankiewicz, A. Microwaves in heterogeneous gas-phase catalysis: Experimental and numerical approaches. *Chem. Eng. Technol.* **2009**, *32*, 1301–1312. [CrossRef]
27. Atlas, L.M.; Nagao, H.; Nakamura, H.H. Control of Dielectric Constant and Loss in Alumina Ceramics. *J. Am. Ceram. Soc.* **1962**, *45*, 464–471. [CrossRef]

© 2018 by the authors. Licensee MDPI, Basel, Switzerland. This article is an open access article distributed under the terms and conditions of the Creative Commons Attribution (CC BY) license (http://creativecommons.org/licenses/by/4.0/).

Review

Perovskite Structure Associated with Precious Metals: Influence on Heterogenous Catalytic Process

Guilhermina Ferreira Teixeira [1], Euripedes Silva Junior [2], Ramon Vilela [1], Maria Aparecida Zaghete [2] and Flávio Colmati [1,*]

[1] Laboratório de Bio-eletrocatálise e Células Combustíveis (LABEL-FC)-Instituto de Química, Universidade Federal de Goiás (UFG), 74690-900 Goiânia-Goiás, Brazil
[2] Laboratório Interdisciplinar de Eletroquímica e Cerâmica (LIEC)-Centro de Desenvolvimento de Materiais Funcionais (CDMF), Instituto de Química-UNESP, 14800-060 Araraquara-SP, Brazil
* Correspondence: colmati@ufg.br; Tel.: +55-623-521-1097

Received: 29 July 2019; Accepted: 23 August 2019; Published: 27 August 2019

Abstract: The use of perovskite-based materials and their derivatives can have an important role in the heterogeneous catalytic field based on photochemical processes. Photochemical reactions have a great potential to solve environmental damage issues. The presence of precious metals in the perovskite structure (i.e., Ag, Au, or Pt) may improve its efficiency significantly. The precious metal may comprise the perovskite lattice as well as form a heterostructure with it. The efficiency of catalytic materials is directly related to processing conditions. Based on this, this review will address the use of perovskite materials combined with precious metal as well as their processing methods for the use in catalyzed reactions.

Keywords: perovskite; photochemical; photodegradation; precious metals

1. Introduction

Perovskite-based materials have been extensively used in photocatalytic systems. The incorporation of a precious metal on perovskite lattice or the use of these elements to modify the surface of perovskite particles result in better light absorptions, which promotes the photochemical behavior required for the catalytic process, such as photocatalysis and electrocatalysis; silver (Ag) and gold (Au) are the most used precious metals for this purpose. In order to understand the advantages of using perovskite structures associated with precious metals as photocatalysis in the catalytic process, this review addresses considerations on photocatalysis principles, perovskite structure, and synthesis methods to obtain a catalyst.

2. General Considerations on Photocatalysis

According to the IUPAC Gold Book [1], photocatalysis is defined as a change in the rate of a chemical reaction or its initiation under the action of ultraviolet, visible or infrared radiation in the presence of a substance—the photocatalyst—that absorbs light and is involved in the chemical transformation of the reaction partners. Thereby, the photo term comprises the light (UV-vis, IR, etc.), while the catalysis ones represent the process through which a catalyst compound or a substance changes the reaction rate of chemical transformation by increasing the reaction kinetics without modifying the overall standard Gibbs energy (ΔG^0) alteration in the reaction [1]. Considering the catalyst aspect, a photocatalyst is defined according to the IUPAC as a compound or a substance able to produce, upon absorption of light, chemical transformations of the reaction partners. The excited state of the photocatalyst repeatedly interacts with the reaction partners forming reaction intermediates and regenerates itself after each cycle of such interactions [1]. Therefore, a photocatalytic process can be understood as a synergic process in which light radiation and the catalyst act jointly to support and speed up a chemical reaction.

The photocatalytic processes are included in a broader and general category of chemical and photochemical reactions known as advanced oxidation processes (AOPs), which apply powerful oxidizing agents in order to degrade persistent organic pollutants as well as remove certain inorganic pollutants. Through such process, the pollutants are converted partially or totally into simpler and less toxic species resulting in easily degraded substances through the application of common technologies [2–5]. The development of the AOPs started in the 1980s from new approaches for potable water treatment that dispose of hydroxyl radicals ($^\bullet$OH) as major oxidizing agent [3–5]

Because of their non-selective nature, hydroxyl radicals may easily lead to the mineralization of a wide range of potentially toxic organic species and depending on the operating conditions (temperature, pH, pollutant concentration, etc.) transform them into less complex and harmful intermediate products [5–7]. Although molecular ozone is a selective chemical oxidant and has a good oxidizing capability, hydroxyl radicals can reach reaction rates of 1 million to 1 billion times faster than those found in chemical oxidants and consequently display rate constants ranging from 10^6 to 10^{10} (M^{-1} s^{-1}) [2,4,7]. Even though fluorine is a strong oxidizing agent, its use in drink water treatment or to degrade organic pollutants is constantly intensively discussed due to the risk that fluoride derivates may represent to the human health and biological systems [8,9], in addition to being an expensive treatment technology on a large scale [10,11].

Most industrial waste and wastewater treatment processes employ classical treatment methods from different physical, chemical, and biological processes, which only carry over organic compounds from water to another phase generating hazardous sludge that results in secondary pollution [12–16]. Despite usually being high-cost technologies, AOPs emerge as an interesting solution to wastewater treatment for presenting attractive characteristics such as versatile treatment —enabled before or after conventional treatment or even during the main stage; mineralization of several chemical pollutants, as well as intermediates; no generation of sludge hazardous (eco-friendliness technologies) as it occurs in the physical, chemical, and/or biological processes; high reaction rate, and satisfactory cost-effectiveness [13,14,17,18].

From the point of view of a chemical reaction, applying AOPs in wastewater treatments comprises basically the following steps: (i) firstly, the generation of strong oxidizing agents ($^\bullet_0$OH, $HO_2^\bullet$, $O_2^{\bullet-}$, etc.) from the species into medium reactional (ions, molecules, catalyst, etc.); (ii) followed by the oxidizing species reacting with organic contaminants molecules in solution, generally converting them into easily degradable intermediary compounds; (iii) finally, the oxidation of these intermediates leading to complete mineralization in water, carbon dioxide, and inorganic salts [3,4,18–20]. In contrast, AOPs exhibit complex oxidation mechanisms due to the very large number of reactions that may occur; therefore, it is laborious to predict all reaction pathways of products formed [5,18–20]. In this regard, although AOPs display fast oxidation rates for many persistent organic pollutants, controlling the formation of reaction products becomes impossible. In particular cases, depending on medium conditions of the supports reaction, these products can be more toxic than parent molecules due to a fast attack to the water contaminant molecules by some free radicals, just as other intermediate species available in solution, such as short-lived species like hydroxyl radicals [5,19,20].

Although AOPs show many advantages in relation to other conventional oxidation processes, an important consequence of most AOPs—regarded as a remarkable disadvantage to any oxidation process based on hydroxyl radical attack—consists of the scavenging of $^\bullet_0$OH radicals by scavenger agents presents in the solution (HCO_3^-, CO_3^{2-}, excess H_2O_2 and pH conditions) [2,5,18,21–23]. In this regard, the scavengers trap the $^\bullet_0$OH radicals decreasing their availability in the reactive medium and consequently compromising the degradation efficiency, which lead others species to emerge with much lower oxidizing power than the hydroxyl radical. It is worth highlighting that scavenger species (ions, radicals, charge carriers, etc.) can act either in a favorable or harmful manner during the degradation process.

AOPs are classified as homogeneous and heterogeneous systems; in both cases, the hydroxyl radicals are generated with or without the presence of a stimulus from an external source (UV, ultrasound, microwave, etc.) [17,19,24]. In the homogeneous systems, a single phase is formed between the catalyst and the substrate, whereas in the heterogeneous systems the substrate and the catalyst present more than one phase, in which the catalyst can be found as a solid phase [19].

2.1. Heterogeneous Photocatalysis: TiO$_2$ vs. Perovskite-based Materials

In the context of the AOPs systems, heterogeneous photocatalysis based on a semiconductor photocatalyst is among the most explored procedures in the area of water and wastewater treatment. The first experiments regarding AOPs were performed in the 1930s and used TiO$_2$ in aqueous suspension under UV irradiation in order to degrade dyes [25–27]. Nonetheless, the applications of semiconductor photocatalysts are not limited to degradation of dyes and other potential pollutants that are harmful to human health and the environment. Its application also involves the production of hydrogen, purification of air, and antibacterial activity [12,26–30]. The evolution of photocatalysis progressed in parallel to advances in researches on the investigation of TiO$_2$ properties applied to photoelectrochemical reactions. The advantages of using TiO$_2$ include higher photoreactivity and physical-chemical stability as well as lower cost and toxicity in relation to other photocatalyst candidates [25,26,29,30]. Researchers investigating the use of TiO$_2$ in photocatalytic processes were boosted in the year of 1972 when Honda and Fujishima pioneered and reported the decomposition of water into H$_2$ and O$_2$ through means of a photochemical experiment under sunlight irradiation [26,30].

In the photoactivation of the semiconductor, electrons located in the valence band maximum (VBM) are photoexcited under $hv \geq E_g$ toward conduction band minimum (CBM) leaving positive holes in the valence band. Subsequently, these charge carriers may have different destinations in the photocatalytic process [23,25,31–35]. A photocatalyst will oxidize a species upon a less positive oxidation potential (E_{ox}) than the potential edge conduction band (CBM) and analogously reduces it upon a more positive reduction potential (E_{red}) than the potential edge valence band (VBM) [36]. Thus, the positions of the electronic band-edge structures of the semiconductor can easily promote both oxidation and reduction of the species presents in solution since the valence/conduction band gap (E_g) is sufficiently large for encompassing the redox potentials of some species that are essential to the photocatalytic process [25,33–37].

TiO$_2$-based photocatalysts are undeniable the most studied material in the context of photocatalytic applications. Nevertheless, its use as pristine TiO$_2$ bulk is restricted to the UV region since the TiO$_2$ anatase electronic structure absorbs at ~3.2 eV and [25,26,38]. Complex oxides systems—such as perovskite-based photocatalysts—emerge as an interesting alternative to photocatalytic processes [39–41]. Among the different advantages of the photophysical properties of perovskite, we can highlight: (i) their structures can be formed from a wide variety of elements, although their basic structures are similar; (ii) their valency, stoichiometry, and vacancy can be easily changed from the simple and non-laborious proceedings; (iii) possibility of a satisfactory prediction of their surface properties since their bulk structures are very well characterized; (iv) their crystal structures generally provide an appropriate electronic structure that shifts the band gap energy to visible-light absorption; (v) their crystal structural arrangements allow lattice distortions, strongly affecting the separation of photogenerated charge carriers and avoiding the recombination processes [41,42]. Therefore, the possibility of controlling the physical-chemical properties of perovskite structures allows unraveling the relationship between structural properties and photocatalytic activity, which makes this material a good photocatalyst alternative to TiO$_2$ [42].

2.2. Perovskite Structure

The perovskite-type structure term comes from mineral calcium titanate (CaTiO$_3$), also called perovskite. Materials under perovskite-type structure have a general chemical formula of type ABX$_3$, in which atoms A and B are metal cations and X atoms are non-metallic anions, so that upon the occupation of the site X with oxygen, this structure is called perovskite oxides (ABO$_3$), as illustrated in Figure 1. In addition, the A atom can be monovalent, divalent or trivalent, while the B atom can be trivalent, tetravalent or pentavalent [43,44]. In a typical perovskite oxide structure, the lattice is formed by small B cations that are six-fold coordinated to oxygen (BO$_6$), while larger A cations are twelve-fold coordinated by oxygen (AO$_{12}$), so that BO$_6$ octahedra are corner connected and AO$_{12}$ clusters are arranged between the eight BO$_6$ octahedra, resulting in a crystal structure that extends in three dimensions and has excellent flexibility depending on the valence and ionic radii of the A and B cations [43–45].

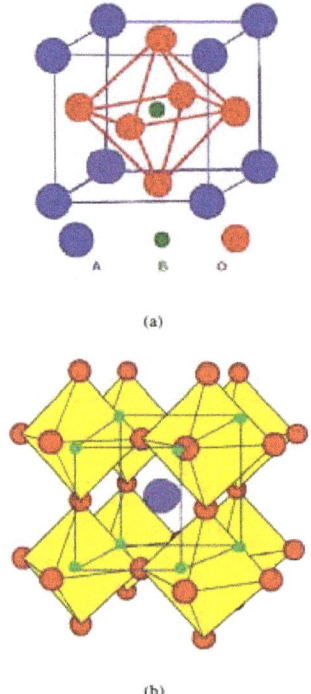

Figure 1. (a) ABO$_3$ ideal cubic perovskite structure and (b) the perovskite framework Mishra, A.; Prasad, R. Preparation and application of perovskite catalyst for diesel soot emission control: an overview. *Catal Rev.* 2014, *56*, 57–81. Copyright 2014 Taylor & Francis [46].

An ideal, fully ordered perovskite oxide structure shows a cubic crystal structure with space group $Pm\overline{3}m$; however, there are several cases in which the ideal structure is distorted. The distortions observed in ABO$_3$ structures can derive from (i) distortions in the BO$_6$ octahedra, (ii) displacement of the B-site cation within the octahedron, (iii) displacement of the A-site cation within the lattice, and (iv) tilting of the corner-sharing octahedrons [47,48]. Therefore, the A and B cations can promote strain, stress, and/or distortions under the perovskite structure according to the distortion degree, also known as tolerance factor (t), depending on the ionic radii of the A and B cations [49,50].

Nevertheless, in general terms, the tolerance factor is not solely the unique variable able to determine the formability and stability of a general perovskite structure since other non-geometric factors, such as bond valence and chemical stability, should also be considered [51,52]. Thereby, the size of the atoms constituting the structure must be adequate to make the perovskite structure geometrically robust as well as less reactive under ambient conditions, for example, in the presence of moisture and oxygen [51].

Figure 2 shows two correlated metals with similar perovskite structures, whereby the difference between the ionic radii of cations Sr^{2+} (118 pm) and Ca^{2+} (100 pm) promoted distinct degrees of BO_6 octahedral distortions, resulting in a more "distorted" crystal structure of $CaVO_3$ in relation to the $SrVO_3$. Despite being similar perovskite structures, because of the variation in cation radii, the crystal structure of $SrVO_3$ is cubic (c = 3.842 Å), whereas $CaVO_3$ is orthorhombically distorted (c = 3.770 Å) [53].

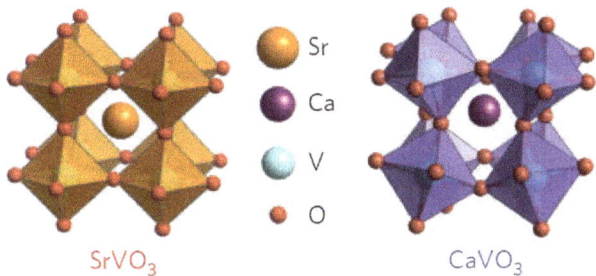

Figure 2. Crystal structure of $SrVO_3$ and $CaVO_3$. Reprinted (adapted) with permission from Zhang, L.; Zhou, Y.; Guo, L.; Zhao, W.W.; Barnes, A.; Zhang, H.-T.; Eatom, C.; Zheng, Y.; Brahlek, M.; Haneef, H.F.; Podroza, N.J.; Chan, M.H.W.; Gopalan, V.; Rabe, K.M.; Engel-Herbert, R. Correlated metals as transparent conductor. *Nat. Mater.* 2016, *15*, 204–210. Copyright 2015 Spring Nature [53].

In terms of photocatalytic process, there are several factors assigned to structural properties that affect the photocatalytic performance of ABO_3-based photocatalysts. The main factors are related to electron-hole separation as well as the carrier transport mechanisms within the crystal lattice, such as exciton lifetime and diffusion length, exciton binding energy, and electron/hole effective mass [39,54]. Thereby, the lattice distortion and structural defects may change the electronic structure of perovskite oxide materials remarkably, consequently affecting the carrier transport mechanisms [39,44,47,48].

The electronic structure in perovskite oxides is defined by a lattice of BO_6 octahedra clusters, whereby their B-O-B bonds build up an electrical conduce way along the crystal structure through which the electrons and/or holes "flow" [43,45,55]. The lattice distortion degree of the pristine ABO_3 induced by ion doping may induce structural defects on the crystal structure generating and/or rearranging energy state levels that change the electronic structure strongly. These lattice distortions affect the separation of charge carriers avoiding the recombination processes and/or shifting the band gap energy to enable visible-light absorption and tuning the band edge potentials to meet the requirement of specific photocatalytic reactions [39,56,57].

2.3. Perovskite-based Photocatalysts Supported by Precious Metals: A Synergistic Effect on Photocatalytic Performance

Aiming at achieving the best of the photocatalytic process, innumerous strategies can be applied. Although other challenges are yet to be overcome in the photocatalytic universe, hampering the electron-hole pair recombination is an essential strategy to ensure optimum photocatalytic efficiency. In this regard, our review addresses one of the most interesting strategies to improve the photocatalytic activity of perovskite-based semiconductor. This strategy is based on the modification of perovskite structure by inserting a precious metal into its crystalline lattice through doping process or obtaining precious metal/perovskite composite through modification of perovskite particles surface with a precious metal in order to decrease and/or retard the recombination process of charge carriers, apart from shifting the band gap values to visible-light absorption region.

Precious metals integrate a class of dopants widely used in the development of new photocatalysts since their optical properties may allow the activity in visible light as well as decrease the electron-hole pair recombination promoting a synergistic effect on the photocatalytic activity [58,59]. On the whole, individual precious metals nanoparticles (Ag, Au, Pt, Rh, Ir, etc.) or precious metals supported on the photocatalyst surface exhibit a specific behavior when interacting with light at plasma frequency, which enables the visible light absorption by the surface plasmon resonance (SPR) effect [60,61]. The plasmon term can be understood as a wave coming from the collective oscillation of electrons on the surface of metals. Basically, the electric field associated with light exerts a force on the outermost electrons of the conduction band of metal redistributing the spatial electron density and generating intense electric fields on the surface metal. In a given oscillation pattern at plasma frequency, when the depth of light penetration is approximately equal to the size of the metallic nanoparticles, such oscillation resonates with incident light resulting in a strong oscillation of the surface electrons that change the absorption profile of metal [61,62].

The key point of the SPR is the frequency of incident radiation, which must be resonant with the frequency of oscillation of the electrons on the metal surface. However, if the incident radiation is lower than plasma frequency, it will be reflected because the electrons of metal filter the electric field incident. In contrast, if the incident radiation is higher than the plasma frequency, the light will be transmitted because the electrons are incapable of responding quickly to the electric field oscillation to filter it [61].

The oscillations of electrons and electromagnetic fields under resonance frequency are defined as located at the surface plasmonic resonance (LSPR), whereby electrons would oscillate with the maximum amplitude. Nevertheless, for a same noble metal, the resonance frequency can be attuned by means of change in size, shape, and nature of the surrounding medium and type of matrix (photocatalyst) when the precious metal is a dopant on the photocatalyst surface [58,59,63]. In this respect, silver and gold nanoparticles are good examples of metals that present an SPR effect due to decreased particles size. A smaller particle size allows silver to absorb light at a visible frequency and gold to absorb light at IR frequency [61].

Thus, the great interest in precious metals nanostructured for photocatalytic applications arise from the localized surface plasmon resonance mechanisms since it is well-known that such plasmonic effect can increase the cross-section absorption in photocatalyst through the process of strong field enhancement, consequently shifting the light absorption to larger wavelengths, in addition to enhancing the separation of charge carriers photogenerated in photocatalyst [58,60].

In this regard, for metallic nanoparticles supported on the semiconductor surface, the energy generates by the electron oscillation via LSPR effect can be transferred to the semiconductor as electromagnetic energy or as excited electrons. In the first case, the energy absorbed in the excitation of the plastron band is quickly converted to heating, and then this energy is transferred to the semiconductor via photothermal effect [58–60,62]. The second case involves a complex charge of transfer mechanisms in which the surface plasmon resonance acts by amplifying the light absorption of the metallic nanoparticles directly injecting photogenerated electrons from the metallic nanoparticles to the semiconductor conduction band via direct transfer mechanism and/or through the resonance electron transfer mechanism. The local electromagnetic field generated by LSPR, in turn, facilitates the generation of electron-hole pairs close to semiconductor surface, in addition to creating a Schottky barrier suppressing the recombination processes since the photogenerated electrons are quickly transferred to the semiconductor conduction band [58,59,62,63]. In the case of our work, the semiconductors are the perovskite compounds used as the catalyst.

Figure 3 illustrates a scheme presenting both direct electron transfer and resonance electron transfer which have been proposed to explain the visible light activity of metallic nanoparticles on the photocatalyst. In the scheme, Au is the metallic nanoparticles and TiO_2 is the photocatalyst semiconductor. However, this scheme can be applied to illustrate these electronic transfer effects whenever a perovskite structure is used as a semiconductor. Details on the species formed during the photocatalytic process can be found further in the reference section [63].

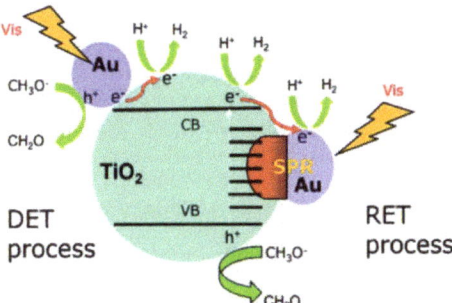

Figure 3. The visible light activity of Au–TiO$_2$ explained by two electron transfer mechanisms: resonance energy transfer process (right) and direct electron transfer process (left). Reprinted with permission from Nie, J.; Scheneider, J.; Sieland, F.; Zhou, L.; Xia, S.; Bahnemann. New insights into the surface plasmon resonance (SPR) driven photocatalytic H$_2$ production of Au-TiO$_2$. *RSC Adv.* 2018, *8*, 25881–25887. Copyright 2018 The Royal Society of Chemistry [63].

On the plasmonic point of view, the decision on the noble metals to be used should consider those able to support a strong surface plasmon resonance effect at the desired resonance wavelength. The ability of a metal to support an SPR depends on their dielectric properties, i.e., dielectric function (ε), including its real (ε_r) and an imaginary (ε_i) part, as both of which vary according to the excitation frequency [60]. Therefore, a resonance in the absorption occurs at the plasmon frequency when the real part of the dielectric function reaches zero [61]. Thereby, among the noble metals, silver is undoubtedly the most employed one in plasmonic applications for its unique electrical and optical properties, in addition to low cost in relation to Au and Pt [60], for example.

In a hybrid structure applied in plasmonic photocatalysis, the charge-transfer mechanism between metallic nanoparticles and semiconductors displays a fundamental role for noble metals to present an intense SPR effect in the range of energy level states of the semiconductor conduction band in order to enable the electron transfer between metal and semiconductor [64–66]. Generally, interband transitions between Ag/semiconductor hybrid structures occur at high frequencies, therefore, silver nanoparticles present an intense SPR effect in the range of 400–450 nm [65], whereas, for Au and Cu, these transitions limit their SPR excitation to wavelengths above 500 and 600 nm, respectively [67]. It is worth mentioning that these interband transitions from the SPR effect in metal/semiconductor hybrid structures strongly depend on the electronic band structure and mainly on the minimum conduction band.

Therefore, plasmonic photocatalysis applications emerge as new and upgraded approaches on the photocatalytic universe and combined with the structure-property relationship of perovskite materials generate new structures of photocatalysts by blending the unique properties of perovskite compounds and the optical and electrical properties of precious metals nanostructured highlighted by the SPR effect.

The plasmonic effect caused by precious metals on the perovskite structure not only improves the photocatalytic behavior of the material, the incorporation of metals such as Pt, Pd, and Rh into a perovskite structure, also stabilizes the metal against sintering, metal-support interaction, and volatilization. The catalytic activity and selectivity of perovskite in the exhaust cleaning process are enhanced even if small amounts of precious metals are combined with the perovskite structure [68]. As aforementioned and illustrated in Figure 4, the distortion of BO$_6$ octahedral of perovskite is promoted by ionic radii of cations present in the crystalline lattice resulting in a more distorted structure. Thus, the same precious-doped perovskite matrix may suffer different distortion degrees according to the kind of precious metal present in its structure. Thereby, the lattice distortion may change the electronic structure of perovskite oxide materials remarkably and consequently affect the carrier transport mechanisms [39,44,47,48,55].

Based-perovskite plasmonic photocatalytic materials can be processed as powder or film. Our work approaches the chemical synthesis method (bottom-up method) most commonly used to produce perovskite powders. Through this method, it is possible to control the stoichiometry, shape, surface area, and size of the particles, among other relevant features.

3. Chemical Synthesis for Perovskite Photocatalyst Obtaining

3.1. Solid-State Reaction

The solid-state reaction method is the oldest method to produce inorganic powders. It is the simplest and most widely used method in the synthesis of compounds with perovskite structure. This reaction consists of the combined heating of two or more non-volatile solids that react to form the desired product. The reactants in the stoichiometric ratios should be very well mixed at the level of individual particles on a scale around 1 μm or 10^{-3} mm [69]. An efficient mix process ensures the maximum contact among the reactants minimizing the diffusion distance among them. The mix is performed using a mortar or a mill for the precursors to be subsequently heated at high temperature in a furnace for several hours in an alumina crucible [70]. The reaction can be slow and its speed increases upon a higher temperature promoting the faster diffusion rate of the ions that compose the final product. The reaction occurs at the interface of the solids (reactants); after the surface layer reacts, the reaction continues as the reactant diffuse from the bulk to the interface [69,70].

Iwashina and Kudo obtained $SrTiO_3$ powders doped with Rhodium at Ti sites ($SrTiO_3$:Rh). We performed the airborne calcination of precursor mixture according to the following two steps: firstly, at 1173 K for 1 hour and then at 1373 K for 10 hours. X-ray diffraction patterns revealed that the single phase of $SrTiO_3$: Rh is obtained until 4 atom% of dopant. Small amounts of impurity phase are obtained for 7 and 10% of atom dopant. After powder preparation, this one was employed in photoelectrochemical water splitting by covering of an ITO electrode. Undoped $SrTiO_3$ is an n-type semiconductor and the doping process transforms it in a p-type semiconductor. The optimum material was the one that doped with 7 atom% of Rhodium. The dopant contributes to the visible light absorber and a recombination center for photogenerated electrons and holes because these factors are responsible for the device efficiency in the photocatalysis process. Through diffuse reflectance spectra, we found that $SrTiO_3$:Rh (7 atom%) had visible light absorption bands that did not occur for the non-doped material. The Faradaic efficiency for H_2 and O_2 evolution were 100% within experimental error and stable cathodic photocurrent could be observed for a long period of time [71].

The $(Ag_ySr_{1-y})(Ti_{1-y}Nb_y)O_3$ was produced seeking to evaluate its efficiency in the photocatalytic oxidation through the decomposition of gaseous 2-propanol under UV and visible light irradiations. The products present a cubic crystalline structure with homogeneous $SrTiO_3$ crystalline phase at least up to y = 0.1. The band gap energy value (3.18 eV) was similar to $(Ag_ySr_{1-y})(Ti_{1-y}Nb_y)O_3$ and pure $SrTiO_3$. The density of states calculations for $(Ag_{0.25}Sr_{0.75})(Ti_{0.75}Nb_{0.25})O_3$, Nb 4d contributes to the conduction band, but did not contribute to the valence band. In contrast, Ag 4d contributed only to the valence band, meanwhile, we may consider that Ag 4d did not mix with valence band composed of O 2p, but Ag 4d forms an isolated band in the forbidden above the valence band. We evaluated the photocatalytic activity in the decomposition of gaseous 2-propanol by monitoring the acetone and CO_2 concentration on the subjection of the photocatalyst under UV and visible lights. After adsorbed into the semiconductor surface, the 2-propanol concentrations in presence of $(Ag_{0.25}Sr_{0.75})(Ti_{0.75}Nb_{0.25})O_3$ were 18 and 2.5 ppmv, respectively, showing that acetone and CO_2 are generated when absorbing either UV or visible light. In addition, the color of $(Ag_{0.25}Sr_{0.75})(Ti_{0.75}Nb_{0.25})O_3$ did not change at all after photocatalytic test indicating the stability of the compound under light irradiation. The effective work of photocatalyst at decomposing organic compounds occurs if photoexcited electrons are consumed in the oxygen reduction. In the case of $(Ag_{0.25}Sr_{0.75})(Ti_{0.75}Nb_{0.25})O_3$, the Ag^+ was attributed to the electronic structure of matrix valence band and not to that of its conduction band. If the Ag^+ were attributed to the electronic structure of the matrix conduction band, photoexcited electrons could not be consumed in the oxygen reduction. Despite $(Ag_{0.25}Sr_{0.75})(Ti_{0.75}Nb_{0.25})O_3$ presenting photocatalytic activity under both UV and visible light irradiation, the formation of the Ag 4d isolated band above the valence band composed of O 2p is responsible for making the efficiency of the compound under UV light superior to that under visible light [72].

Saadetnejad and Yildirim studied the photocatalytic hydrogen production by water splitting over $Au/Al-SrTiO_3$. The Al-doped $SrTiO_3$ was obtained from solid-state reaction and loaded with Au via the homogeneous deposition-precipitation method. XRD patterns indicated that samples were indeed $SrTiO_3$ with good crystallinity, which contributes to an efficient migration of charge carriers in the water-splitting process. The best performing catalyst was obtained by using a material with composition equal to 1% Al doped, 0.25% Au loaded $SrTiO_3$, in the presence of methanol as a sacrificial agent, while 1% Al doped 0.50% Au was the best for the isopropyl medium. The relative performance of methanol and isopropyl medium seems to depend on the alcohol concentration since the isopropyl performs better at 10% and 30%, while methanol is better at 20%. Furthermore, the distinguished results obtained from differences among concentrations and different alcohol types may be related to the carbon number present in the alcohol molecules, which may result in different byproducts, the redox potential of alcohol, and their OH scavenging behavior [73].

$AgTaO_3/AgBr$ heterojunction presents intense visible light absorption and consequently better photocatalyst perform than individual $AgTaO_3$ and AgBr in the methyl orange degradation. $AgTaO_3$ showed no efficiency at dye degradation under visible light irradiation. After the heterojunction formation, the photocatalytic activity efficiency significantly improves due to the enhanced interfacial charge separation efficiency between $AgTaO_3$ and AgBr. The heterojunction induced an efficient transfer of photogenerated electrons from AgBr conduction band to $AgTaO_3$ conduction band leading to a better charge separation improving photocatalytic behavior. We applied solid-state reaction to obtain Perovskite $AgTaO_3$ and solution method to produce AgBr. Composites were processed by mixing individual $AgTaO_3$ and AgBr suspensions containing poly(ethyleneglycol)-block-poly(ethylene glycol) $NH_3.H_2O$. After a photocatalytic test, an additional weak diffraction peak regarding cubic Ag was observed in $AgTaO_3/AgBr$ heterostructure indicating the formation of the ternary $AgTaO_3/AgBr/Ag$. XPS results indicated that partial Ag^+ ions present in the heterojunction were reduced to metallic Ag nanoparticles to form the ternary $AgTaO_3/AgBr/Ag$ during the photocatalytic experiment. It was evidenced that the photogenerated electrons may migrate from the conduction band of AgBr to the conduction band of $AgTaO_3$ [74].

Konta and co-workers investigated the photocatalytic application of $SrTiO_3$:M(0.5%) (M = Mn, Ru, Rh, Pd, Ir, and Pt ions) powders in H_2 evolution from methanol solution and O_2 evolution from an aqueous solution nitrate solution under visible light irradiation. All material exhibited pure phase of $SrTiO_3$, except for $SrTiO_3$:Pd(0.5%), since its Pd^{2+} ion radius are much larger than Ti^{4+} ion radius. The diffuse reflectance spectra of materials revealed that all samples presented visible light absorption. The spectra of the sample doped with Mn, Ru, Rh, and Ir, we found an absorption band as a shoulder in addition to the absorption band regarding $SrTiO_3$ indicating that the dopants had created discontinuous levels in the forbidden band. The presence of two absorption bands as shoulders in the diffuse reflectance spectra of $SrTiO_3$:Rh powders result from some doping levels formed in the forbidden band due to different oxidation numbers (Rh^{3+} and Rh^{5+}). $SrTiO_3$:Rh showed the best photocatalytic activity in H_2 evolution, while $SrTiO_3$:Ru proved the best for O_2 evolution. In contrast to the higher O_2 evolution for $SrTiO_3$:Mn and $SrTiO_3$:Ru, their low photocatalytic activity for H_2 evolution would occur due to the surface states generated by Mn and Ru dopants below the conduction bands, resulting in difficulty for the H_2 evolution. Because $SrTiO_3$:Rh showed the highest activity for H_2 evolution, further investigations should be carried out to approach this compound. We found the occurrence of at least two different species of doped Rh (Rh^{3+} and probably Rh^{5+}). Rh with higher oxidation number acts as an electron acceptor because they can be easily reduced to Rh^{3+} at an early stage of the photocatalytic reaction; in addition, during the photocatalytic process, the Rh^{3+} acts as an electron donor. The $SrTiO_3$:Rh visible light response resulted from the electronic transition from the electron donor level formed by Rh ions to the Ti $3d$ orbitals present in the conduction band of $SrTiO_3$ matrix. Despite its excellent performance in H_2 evolution, the $SrTiO_3$:Rh presented no activity to O_2 evolution. One of the reasons for no O_2 evolution using $SrTiO_3$:Rh as photocatalyst is the kinetic limitation of the lack of active sites. The photocatalytic reaction may be efficient after reducing the Rh with high oxidation number. However, these species cannot be reduced in an aqueous silver nitrate solution medium because they act as recombination center in the absence of a hole scavenger, which is another reason for the absence of photocatalytic activity in O_2 evolution [75]. Table 1 shows both H_2 and O_2 evolution when using $SrTiO_3$:M(0.5%) under visible light absorption as well as the respective band-gap values of semiconductors.

Table 1. Photocatalytic activities of $SrTiO_3$:M(0.5%) for H_2 and O_2 evolution from the aqueous solution containing sacrificial reagents under visible light irradiation [a]. Reprinted with permission from Konta, R.; Ishii, T.; Kato, H.; Kudo, A. Photocatalytic activities of noble metal ion doped $SrTiO_3$ under visible light irradiation. J. Phys. Chem. B 2004, 108, 8992–8995. Copyright 2004 American Chemical Society [75].

M	E_g/eV	Activity/µmol h^{-1}	
		H_2 [b]	O_2 [c]
Mn	2.7	0.2	2.7
Ru	1.9	1.7	3.9
Rh	1.7	17.2	0
Ir	2.3	8.6	0.4

[a] Catalyst, 0.3 g; reactant solution, 150 mL; light source, 300 W Xe lamp with cutoff filter (λ > 440nm). [b] From 10 vol% aqueous MeOH; cocatalyst, Pt (0.5 wt %). [c] From 0.05 mol.L^{-1} aqueous $AgNO_3$.

3.2. Sol–Gel Method

Sol–gel method also represents a widely used approach to produce perovskite particles. This process consists in the transition of a solution system from a liquid known as "sol", most often a colloid liquid, into a solid phase named "gel". Inorganic metal salts or metalorganic compounds are usually employed as starting materials for the sol–gel method. To form the desired product, it initially occurs the hydrolysis of a precursor metal to produce the metal hydroxide solution. Subsequently, an immediate condensation occurs providing a three-dimensional gel that is dried by removing the solvent. Depending on the way the solvent is removed, the formation of the Xerogel or Aerogel occurs. If the solvent is removed slowly at room temperature, a Xerogel is formed. In contrast, if the drying process is performed under supercritical conditions, the Aerogel is obtained [76–78].

Since the ion–codoped semiconductor has been improving photocatalytic activities under visible-light irradiation, Zhang and co-workers studied the effect of monovalence silver and trivalence lanthanum codoping $CaTiO_3$ in the water-splitting process. The sol–gel method was chosen to provide the perovskite material and the ultrasonic dispersing technique was used to enhance the properties of the materials. XRD patterns showed the pure $CaTiO_3$ orthorhombic phase even for doped samples. Diffuse reflection spectra showed a shift to a longer wavelength when replacing Ca^{2+} by Ag^+ with La^{3+}. Such behavior may result from the electronic transition between the electrons present in the Ag $4d5s$ orbital to the O $2p$ + Ti $3d$ hybrid orbital. The Ag–La codoped $CaTiO_3$ introduces a band in the visible light region due to the transition of Ag $4d5s$ electrons to the conduction band. The photocatalytic efficiency increased until the amount of dopant equal to 3% and is much higher than the value obtained from pure $CaTiO_3$. Upon a doped concentration lower than the optimum value, the photocatalytic activity increased because of the absence of enough capture traps of charge carriers in the photocatalyst particles. In contrast, the recombination rate increases when the doped concentration is higher than the optimal value, which promotes lower photocatalyst activity [79].

ZnTiO$_3$ is UV light active photocatalyst and the formation of a composite between this perovskite and plasmonic metals (e.g., Au) represent satisfactory applications of this perovskite in visible light assisted photocatalytic reactions. Gold is considered an efficient metal for this purpose due to its great properties, such as surface plasmon resonance characteristics, electron storage effect, chemical stability, and high catalytic activity. In this sense, Hemalata Reddy and co-workers obtained ZnTiO$_3$ through sol-gel auto combustion method and subsequently loaded the particles with Au nanoparticles using the precipitation-deposition method forming Au/ZnTiO$_3$ nanocomposites. The material was applied in H$_2$ generation from methanol solution under UV and visible light irradiation. Pure ZnTiO$_3$ has absorption in UV region and the addition of Au on ZnTiO$_3$ surface promotes an additional absorption in the visible region due to the surface plasmon absorption corresponding to the gold nanoparticles. The photocatalytic tests were performed by using Au/ZnTiO$_3$ nanocomposite with Au concentration varying from 0.5 to .5 wt%. We achieved the best results for H$_2$ evolution using the nanocomposites with 1 wt% of Au under visible light irradiation, while pure ZnTiO$_3$ efficiency was practically insignificant. In contrast, through UV as the radiation source, the photocatalytic efficiency increases with a higher Au content up to 1.5% wt due to the separation of charge carriers in the interface between Au nanoparticles and perovskite particles through the Schottky barrier. In addition, the interfacial defect sites in the nanocomposites act as Au-induced charge separation effect that balances the recombination effect of Au/ZnTiO$_3$ in photocatalytic reaction from UV light irradiation [80].

The physicochemical properties and photocatalytic activity for the degradation of methylene blue under simulated solar light were investigated in novel Au-induced nanostructured BiFeO$_3$ homojunctions (Au$_x$-BFO, x = 0, 0. 6, 1.2, 1.8, 2.4 wt%) obtained through in situ synthesis, despite of a simple reduction method using spinning techniques and post-thermal processing. Among the Au$_x$-BFO structures, the Au$_{1.2}$-BFO sample proved the best photocatalytic activity (85.76%) after 3h of irritation, being much higher than pristine BFO samples (49.49%) considering the same experimental conditions. The authors attributed the notorious improvement of photocatalytic activity primarily to the SPR effect of Au NPs in the hierarchical of Au-BFO nanostructured homojunction and secondly to the structural defects (Fe^{2+}/Fe^{3+} pairs and oxygen vacancy) [81].

3.3. Hydrothermal Method

Another widely used synthesis route to grow perovskite structures is the hydrothermal method. This reaction occurs in an aqueous solution into a close system under high pressure and temperature above the boiling point. When the materials synthesis is performed in a nonaqueous solvent at relatively high temperature, the process is known as the solvothermal method. In the hydrothermal/solvothermal method, the temperature is relatively high, but lower than the temperature employed in the methods explained above. In these methods, the crystallization process occurs directly in the solution, which excludes the calcination step and contributes to the growth of particles with very well-controlled morphology [82].

CaTiO$_3$ orthorhombic nanocuboids with a size around 0.3–0.5 µm in width and 0.8–1.1 µm in length were produced using the conventional hydrothermal method. After the CaTiO$_3$ synthesis, the hybridized Au nanoparticles were deposited on perovskite surface obtaining Au@CaTiO$_3$ nanocomposite with an Au mass fraction of 4.3%. XPS spectrum indicated that Au species exist in the metallic state on perovskite surface. We performed photocatalytic tests under simulated-sunlight, ultraviolet, and visible light using Rhodamina B as pollutant model. After 120 minutes of simulated-sunlight irradiation, the degradation of Rhodamine reached 76.4% using pure CaTiO$_3$ and 99.9% when the photocatalyst was the composite. Under UV light irradiation, the efficiency of CaTiO$_3$ reached 95.7%, while the efficiency of Au@CaTiO$_3$ was 99.9%. By irradiating the system with a visible light source, the CaTiO$_3$ efficiency had the worst result (15.1%), whereas the composite presented 46.1% of efficiency. The presence of Au onto CaTiO$_3$ enhanced visible-light absorption; decreased charge-transfer resistance enhanced the photocatalytic performance of the material. CaTiO$_3$ is excited under UV irradiation producing electrons in its conduction band and holes in its valence band. Even though Au nanoparticles could not be excited under UV irradiation, they can act as electron sinks to capture the photogenerated electrons in $CaTiO_3$. The electronic transfer between the conduction band of $CaTiO_3$ to Au nanoparticles promotes the spatial separation of the electron/hole pair in perovskite. For this reason, more holes in the valence band of $CaTiO_3$ are available in the photocatalytic process. $CaTiO_3$ cannot be directly excited to produce electron/hole pairs under visible light irradiation. However, the located surface plasmon resonance (LSPR) of Au nanoparticles is induced by visible-light absorption stimulating the generation of electron/hole pairs in $CaTiO_3$ due to electromagnetic field caused by LSPR effect. This is why Au@$CaTiO_3$ nanocomposites also present visible-light photocatalytic degradation of Rhodamine B. When simulated sunlight is used as irradiation source, Au nanoparticles act as electron sinks and are responsible for the LSPR effect in the Au@$CaTiO_3$ composites, improving the photocatalytic performance of nanocomposite under simulated sunlight irradiation [83].

Malkhasian studied the photocatalytic performance of Pt/$AgVO_3$ nanowires under visible light irradiation in the photooxidation of atrazine. The author used the hydrothermal method to synthesize β-$AgVO_3$ nanowires posteriorly used to obtain the Pt/$AgVO_3$ compound. For this purpose, Pt metal was deposited on β-$AgVO_3$ surface under UV-light irradiation using a photo-assisted deposition (PAD) method. XRD patterns reveal that only the β-$AgVO_3$ peaks are present in the samples indicating that Pt content was well dispersed on the perovskite surface, which was confirmed through TEM analyses. In contrast, higher Pt percentage decreases the crystalline size of β-$AgVO_3$. Higher Pt weight percentage of Pt from 0.2% to 0.6% also leads to improved photocatalytic efficiency. When using 0.6 g/L of the 0.6%wt Pt/β-$AgVO_3$, the oxidation of atrazine was 99% after a 60-minute reaction time. The oxidation time of atrazine decreases from 60 minutes to 40 minutes upon the use of 0.9 g/L of photocatalyst. This behavior results from the higher number of available active sites to oxidize atrazine with higher photocatalyst concentration. In addition, the photocatalyst can be used five times without losing its stability [84].

Me-loaded $NaNbO_3$ catalysts (Me = Fe, Ni, Co, and Ag) were prepared, aiming at producing H_2 through the water-splitting process. The material was obtained by impregnating $NaNbO_3$ (previously synthesized through hydrothermal method) in an aqueous solution of cation nitrates followed calcination. The formation of colored powders indicated the formation of metal oxidized on the surface of perovskite particles. The samples presented similar morphologies composed by irregular structures without significant modification caused by metal loading on $NaNbO_3$. Photocatalyst results showed that different metal oxides loaded with $NaNbO_3$ have a different effect in the photocatalysis for H_2 evolution. Interestingly, only Ag/$NaNbO_3$ proved superior efficiency than pristine $NaNbO_3$ in the performed tests. This behavior was attributed to the fact that the difference between the electronegativity of Ag and Nb is higher than the difference between the electronegativity of Nb and the other metals employed on the surface of $NaNbO_3$. Additionally, the Ag species on the perovskite surface acts as active site for proton reduction, which can explain the highest photocatalytic activity of Ag/$NaNbO_3$ [85]. Ruthenium particles supported on hierarchical nanoflowers assembled by cubic phase $NaNbO_3$ have been applied to degrade Rhodamine B. The surface modification of $NaNbO_3$ with Ruthenium promotes the photosensitized degradation of dye, while Ru-dopping $NaNbO_3$ inhibits the photocatalytic effect. The photocatalyst activity improved with surface decoration due to efficienct electron transfer and charge separation achieved by composite. The best results occurred for $NaNbO_3$ with 0.5% of Ru species and the degradation of Rhodamine B follows pseudo-first order kinetics. In addition, the photocatalyst can be used at least five times successively without losing its photocatalytic efficiency. Interestingly, the degradation time reduces after reusing photocatalyst. After the catalyst process, both crystalline structure and morphology of the perovskite do not change, which may indicate the stability of photocatalyst posteriorly confirmed through the XPS analysis [86].

The presence of Ag nanoparticles onto $KNbO_3$ nanowires drastically improved the photoreactivity of the perovskite and the photodecomposition of Rhodamine B under UV and visible light irradiation. The material begins to lose its activity after three cycles of aqueous organic compound degradation under UV or visible light irradiation. The use of p-benzoquinone, $_t$-butyl alcohol, and ammonium oxalate scavengers revealed that under UV illumination, $\cdot O_2^-$ and $\cdot OH$ are the major active species in the reaction. When visible light is used as an irradiation source, the major active species are $\cdot O_2^-$ and H^+. During the photocatalytic process under UV irradiation, electrons present in the valence band of $KNbO_3$ are excited to the conduction band with hole generation in the valence band. Due to the energy gap of $Ag/KNbO_3$ composites, the photoexcited electrons could transfer from $KNbO_3$ perovskite to Ag nanoparticles, enabling the separation between photogenerated electrons and holes in $KNbO_3$ to promote the photocatalytic reaction. However, the excess of Ag on perovskite surface can act as a recombination center and partly avoid the UV absorption of perovskite decreasing the photocatalytic activity of the material. Visible-light irradiation promotes the photoexcitation of Ag nanoparticles electrons through intraband excitations within the sp band. Furthermore, the photocatalytic process under visible-light is driven by plasmonic sensitization due to an electron electron-transfer process from Ag nanoparticles to $KNbO_3$ via localized surface plasmonic resonance. In addition, researchers revealed that the $Ag/KNbO_3$ composites proved efficient not only for organic dye decomposition, but also for application in other areas, like water splitting [87]. More recently, Xing et al. studied the new application of $Ag/KNbO_3$ nanocomposites in photocatalytic NH_3 synthesis by converting N_2 into NH_3. The presence of Ag on $KNbO_3$ surface significantly enhances the photocatalytic behavior of perovskite. The highest performance for NH_3 generation is achieved when the Ag content in the sample is 0.5%, which reaches about four times more than pure $KNbO_3$. NH_3 generation is related to Ag content since the ammonium production increases with higher Ag and then reduces. Pristine $KNbO_3$ has a poor response on visible light irradiation due to its high band gap value and the formation of composite with Ag nanoparticles significantly improves the performance of the material under visible light exposure due to the photosensitization effect of Ag nanoparticles. In addition, the composites also proved efficient at H_2 generation under simulated sunlight radiation and photocatalytic Rhodamine B degradation [88].

Ag-decorated $ATaO_3$ nanocubes (A = K, Na) were prepared using the hydrothermal method in order to evaluate their photocatalytic water-splitting activity under simulated and pure sunlight. The photocatalytic water-splitting results revealed that both Ag-decorated $KTaO_3$ and $NaTaO_3$ nanocubes exhibited a rate for H_2 evolution from aqueous CH_3OH solutions up to 185.60 and 3.54 µmol/hg under simulated sunlight, receptively, which were more than two times of pristine $NaTaO_3$ and $KTaO_3$ in the same experimental conditions. In contrast, under purely visible light illumination, the photocatalytic performance of both Ag-decorated $ATaO_3$ materials was lower, therefore, the highest H_2 evolution of 25.94 and 0.83 µmol/h g for Ag-decorated $KTaO_3$ and $NaTaO_3$, respectively. The authors conclude that the surface plasmonic effect of Ag nanoparticles was responsible for increasing appreciably the quantum efficiency photocatalytic water-splitting activity of the Ag-decorated K, $NaTaO_3$ materials [89].

As observed, several perovskites structures applied in photocatalysis are obtained from hydrothermal synthesis. One of the most interesting advantages in the use of this method to obtain semiconductor materials is the possibility of producing the same material with different morphologies. Spherical shaped $SrTiO_3$ with sponge-like mesoporous morphology [90], cubic-like $SrTiO_3$ [91,92], $SrTiO_3$ dendritic particles, among other $SrTiO_3$ morphologies, can be obtained from hydrothermal method [93] (Figure 4). In this sense, the syntheses parameters can be controlled in order to produce a material with different morphology. The work conducted by Kalyani et al. is a good example of how synthesis parameters influence the product obtained since the authors developed a comprehensive study on the $SrTiO_3$ growth mechanism as a function of the synthesis conditions [93]. After the synthesis, the surface of the particles can be easily decorated with a precious metal through a chemical solution process; in some cases, the metal deposition process may be photo-assisted [83,84]. Figure 5 from the work published by Kumar et al. exemplifies an easy chemical solution route to deposit precious metal nanoparticles on the perovskite surface [93]. Kumar addressed the fabrication of photoanodes of $NaNbO_3$ nanorods and Ag decorated $NaNbO_3$ nanorods across hydrothermal method and chemical solution method. Due to the visible plasmonic effect, Ag nanoparticles are responsible for reducing the material optical band gap. Additionally, the Ag decorated NaNbO$_3$ nanorods presented the best photocurrent density because the surface plasmon mediated electron transfer from metallic nanoparticles decorated NaNbO$_3$ nanorods [94].

Based on the processing routes discussed here, any pure perovskite structure composed of precious metal or precious metal/perovskite hybrid structure can be obtained aiming at the applicability in a catalyst test.

Figure 4. (**a**) FE-SEM image of spherical shaped SrTiO$_3$ with sponge-like mesoporous morphology. Reprinted (adapted) with permission from Jayabal, P.; Sasirekha, V.; Mayandi, J.; Jeganathan, K.; Ramakrishnan, V. A facile hydrothermal synthesis of SrTiO$_3$ for dye-sensitized solar cell application. *J. Alloys Compd.* 2014, *586*, 456–461. Copyright (2013) Elsevier B.V. [90]; (**b**) SEM image of cubic-shaped SrTiO$_3$. Reprinted (adapted) with permission from Gao, H.; Yang, H.; Wang, S. Hydrothermal synthesis, growth mechanism, optical properties and photocatalytic activity of cubic SrTiO$_3$ particles for the degradation of cationic and anionic dyes. *Optik* 2018, *175*, 237–249. Copyright (2018) Elsevier GmbH [91]; (**c**) TEM image of 0.5%Ag-SrTiO3 nanocomposites (the arrows indicate Ag nanoparticles). Reprinted (adapted) with permission from Zhang, Q.; Huang, Y.; Xu, L.; Cao, J.; Ho, W.; Lee, S.C. Visible-light-active plasmonic Ag-SrTiO$_3$ nanocomposites for the degradation of NO in air with high selectivity. *ACS Appl. Mater. Interfaces* 2016, *8*, 4165–4178. Copyright (2016) American Chemical Society [92]; and (**d**) SEM image of SrTiO3 dendritic particles. Reprinted (adapted) with permission from Kalyani, V.; Vasile, B.S.; Ianculescu, A.; Testino, A.; Carino, A.; Buscaglia, M.T.; Buscaglia, V.; Nanni, P. Hydrothermal synthesis of SrTiO$_3$: role of interfaces. *Cryst Growth Des.* 2015, *15*, 5712–5725. Copyright (2015) American Chemical Society [93].

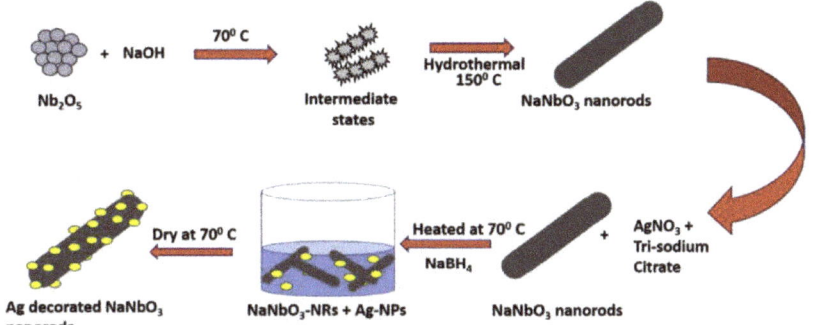

Figure 5. Schematic representation of the different steps involved in the formation of perovskite particles from the hydrothermal method and Ag decoration pervoskite surface by the chemical solution. Reprinted with permission from Kumar, D.; Singh, S.; Khare, N. Plasmonic Ag nanoparticles decorated NaNbO$_3$ nanorods for efficient photoelectrochemical water splitting. *Int. J. Hydrogen Energ.* 2018, 43, 1–8. Copyright (2018) Hydrogen Energy Publications LLC. Published by Elsevier Ltd [94].

These materials can be also processed as film. Jeong and co-workers [91] developed a fully solution-deposited nanocomposite photoanode from the impregnation of Ag NPs in the BiVO$_4$ films prepared through spin coating and post-annealing (500 °C for 2h) to evaluate the photocatalytic water-splitting activity under simulated sunlight of 1 sun (100 mW/cm^2) with Xe lamp. Ag nanoparticles-incorporated BiVO$_4$ nanocomposite, for optimum Ag concentration of 4.0 mM, exhibited an appreciable improvement of photocatalytic performance at low potential (0.4 V), in addition to generating a saturated photocatalytic current density 3.3 times higher than the pristine BiVO$_4$ film. Jeong et al. explain that these results can be associated with the SPR-induced enhancement mechanism, once Ag NPs impregnated in BiVO$_4$ films are responsible for promoting enhanced carrier generation and separation. The performance of the films enhances the current density at low potentials and improves the kinetics of the carrier generation and separation due to the efficient charge carrier generation [95].

Zhang et al. [96] was approached by the role of charge transfer (CT) on the degradation of Rhodamine B under UV (265 nm) and UV–Vis light (Xe lamp 500 W with a 420 nm optical filter) using Ag-nanoparticle-dispersed BaTiO$_3$ (Ag/BTO) composite films prepared through the sol-gel and spin-coating methods. Among the BTO materials obtained, only the monolithic BTO did not exhibit the absorption peak in the visible wavelength region of 450–600 nm, which is the peak associated with the SPR effect of the Ag NPs. The optical absorption spectra revealed that as Ag content increases until 25 mol%, a red-shift was observed, whereas a blue shift occurs with Ag content of 30 mol%. The photocatalytic results showed that Ag$_{25}$/BTO film was the best Ag/BTO composite films; therefore, its photocatalytic efficiency was better under visible (62%) rather than UV–Vis (42%) light irradiation. The authors attributed this improvement in photocatalytic activity of the Ag$_{25}$/BTO film to the SPR effect of the Ag nanoparticles derived from the maximum absorption profile of the band with slightly red-shift trend from 517 to 523 nm. XPS and PL measurements provide great evidence on the synergetic effect between the charge transfer from perovskite to Ag nanoparticles under UV light and charge transfer from Ag nanoparticles to BaTiO$_3$ under visible light irradiation [96].

Until now, we approached the photocatalysis as environmental contamination solution, however, the electrocatalysis also represents an alternative to this purpose. Both processes involve oxidation and reduction mechanisms from photochemical reactions. As explained, in the photocatalysis the redox reaction is promoted by light absorption, while in the case of electrocatalysis, the redox process occurs through the flow of an electric current through an external circuit in the reaction system, as in further explanations below.

4. Application of Perovskites in Electrocatalysis

Electrocatalysis is the study area concerning chemical reactions that occur on electrode surfaces. It is strategical to develop devices and methodologies applied to produce clean energy and benefit a response to the increasing worldwide energetic demand [97]. The most efficient electrocatalysts derive from noble metals, e.g., Pt, Ru, Rh, and Ir based nanomaterials, but the high cost of these materials makes the large-scale production of technologies aiming at sustainable energy production, conversion, and storage a goal difficult to be achieved [98,99]. To overcome such issue, perovskites have been gaining attention as promising materials to act as electrocatalysts for their exceptional thermal stability, ionic conductivity, electron mobility, and redox behavior [100]. The perovskite electrocatalysts can be applied to plenty of electrochemical reactions, such as oxygen reduction, oxygen evaluation, hydrogen evaluation, alcohol oxidation, carbon dioxide reduction, water splitting, among others [101]. In this section, we present some insights on the perovskites electrocatalysts containing precious metals and their applications aiming at the development of energy generation devices.

The oxygen reduction reaction (ORR) and oxygen evaluation reaction (OER) have great importance in the development of electrochemical energy-conversion devices such metal-air batteries, polymer electrolyte membrane fuel cells (PEMFC), and other devices, e.g., in oxygen sensors. The large overpotential associated with these processes, due their slow reactions kinetics, is one of the major challenges to be overcome to develop high-performance catalysts [102]. Taking as an example the PEMFC, which accounts for one of the most representative applications of oxygen electrocatalysis, the oxygen could react through 2-electron or 4-electron pathways, as shown in the chemical equations below [103]:

ORR in alkaline medium	
$O_2 + 2H_2O + 4e^- \rightarrow 4OH^-$ (4-electrons process)	−1
$O_2 + H_2O + 2e^- \rightarrow HO_2^- + OH^-$	(2a)
$H_2O + HO_2^- + 2e^- \rightarrow 3OH^-$ (2-electrons process)	(2b)
ORR in acid medium	
$O_2 + 4H^+ + 4e^- \rightarrow 2H_2O$ (4-electrons process)	−3
$O_2 + 2H^+ + 2e^- \rightarrow H_2O_2$	(4a)
$H_2O_2 + 2H^+ + 2e^- \rightarrow 2H_2O$ (2-electrons process)	(4b)

Currently, the most efficient catalysts for both anodic and cathodic electrochemical reactions in PEMFC are the Pt-based electrocatalysts. However, the high costs involved in large scale production represent an enormous problem. In this scenario, the perovskites structures arise as one alternative to obtain cheaper electrocatalysts that exhibit the same features of the Pt electrocatalysts, like outstanding catalytic and electrical properties and superior resistant characteristics to corrosion [104]. Recently, perovskite-type oxides have gained noticeable popularity as cost-effective bifunctional oxygen catalysts with promising catalytic activity and stability for the ORR and OER in an alkaline medium for their excellent electrical properties and abundant active sites [105].

Zhang et al. [106] provide an example of a precious metal-containing perovskite catalyst applied to ORR by using thermal treatments incorporated Ag-decorating nanoparticles into $PBMO_5$ perovskites nanofibers matrix, followed by a second thermal treatment, which allowed the Ag nanoparticles to be exsolved from the crystal structure, establishing a strong interaction with the perovskite matrix and generating the Ag-$(PrBa)_{0.95}Mn_2O_{5+\delta}$ (Ag-$PBMO_5$) catalyst, as illustrated in Figure 6a. This catalyst has favorable ORR activity enhanced in relation to the $PBMO_5$ and higher durability than the state-of-art Pt/C catalyst in alkaline solution. The ORR electrocatalytic tests were performed using linear sweep voltammetry (LSV) technique and the results revealed that Ag-$PBMO_5$ catalyst demonstrated considerably higher catalytic activity in comparison to the its predecessors with ORR E_{onset} and $E_{1/2}$ of Ag-$PBMO_5$ catalyst of 0.92 V and 0.81 V vs. RHE, and the ORR E_{onset} and $E_{1/2}$ of $PBMO_5$ catalyst of 0.84 V and 0.74 V vs. RHE, respectively. The use of a Koutecky–Levich analysis based on the LSV experiments performed with a rotating electrode revealed that the Ag-PBMO$_5$ electrocatalyst reduces the oxygen completely and directly to OH$^-$ though a 4-electron reaction pathway, achieving the highest energy efficiency in theory. The outstanding catalytic features of Ag-PBMO$_5$ arises from a sum of factors, like crystal structure of reconstructed perovskite with oxygen defects internally ordered, which facilitates the charge transfer during the ORR through the perovskite structure and the resulting Mn^{3+} valence with e^g occupancy, which moderates the intermediate binding strength on Mn sites and contributes to electrocatalytic activity of PBMO$_5$. The contribution of exsolved Ag nanoparticles was investigated using DFT calculations, which revealed a significant ligand effect between the nanoparticles and PBMO$_5$ and an increase in electron occupancy in s, p, and d orbitals on Mn sites, which results in a charge transfer from Ag to Mn sites. In this case, the d-band orbitals of Ag atoms are not completely filled and vacancies are generated, which consequently narrows the d-orbital of Ag and up-shifts the d-band center of Ag atom. The Ag-atom with and up-shifted d-band center is expected to adsorb O$_2$ more strongly and split the O–O bond more efficiently. Therefore, charge transfer plays a crucial role in the enhancement of the ORR activity.

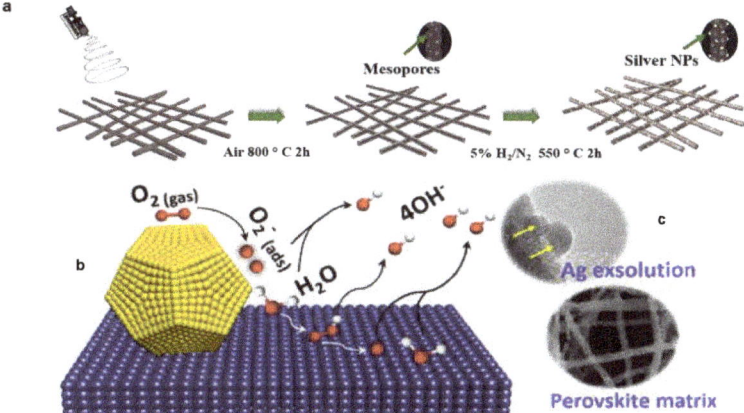

Figure 6. (a) Synthesis steps for the obtention of Ag-PBMO$_5$ electrocatalyst; (b) proposed scheme for ORR at the electrocatalyst surface; c) HRTEM images of exsolved Ag$_2$O nanoparticles bonded to the perovskite surface and FESEM images of PBAMO$_3$ nanofibers. Reprinted with permission from Zhang, Y.Q.; Tao, H.B.; Liu, J.; Sun, Y.F.; Chen, J.; Hua, B.; Thundat, T.; Luo, J.L. A rational design for enhanced oxygen reduction: Strongly coupled silver nanoparticles and engineered perovskite nanofibers. *Nano Energy*, 2017, 38, 392–400. Copyright (2017) Nano Energy Publications LLC. Published by Elsevier Ltd. [106].

Zhu et al. [107] reported another example of perovskite electrocatalyst with exsolved Ag nanoparticles. The heterostructured Ag nanoparticle-decorated perovskite (denoted by e-SANC) was obtained from the precursor Sr$_{0.95}$Ag$_{0.05}$Nb$_{0.1}$Co$_{0.9}$O$_{3-\delta}$ (SANC) through an easy exsolving process. The ORR activity was measured using the electrochemical impedance spectroscopy (EIS) technique and data were calculated from impedance loops the area-specific resistance (ASR). Electrocatalysts with lower ASR values show higher ORR activity. The e-SANC electrocatalyst exhibit very low area-specific resistance (0.214 Ω cm^2 at 500 °C), which is much lower than its precursor (0.363 Ω cm^2 at 500 °C). To illustrate the effect of Ag atoms in the perovskite structure, the results were compared to an electrode prepared without Ag (SNC0.95). In this case, the e-SANC electrocatalyst also showed ASR values at least 50% lower in addition to lower energy activation for ORR for the entire temperature range studied. The analysis of the circuit model used for EIS experiments can offer some insights on the ORR mechanism. The high-frequency resistance (R$_{E1}$) is associated with the charge-transfer process and the results indicate that this feature is not affected in a significant way when comparing the catalysts e-SANC and SNC0.95. The low-frequency resistance (R_{E2}) is related to oxygen surface process, like adsorption, desorption, dissociation at gas–cathode interface, and surface diffusion of intermediate species. The R_{E2} value for e-SANC decreased greatly when compared to SNC0.95, indicating that the Ag nanoparticles increase the velocity of the oxygen surface processes, an effect also found in k_{ex} (surface exchange coefficient) measurements using electrical conductivity relaxation (ECR) technique. The e-SANC electrocatalyst performance was evaluated in a solid oxide fuel cell (SOFC), in which the material was used as a cathode; the system showed the power-density peak of 1116 mW cm^{-2}, having remained stable for 140 h at 500 °C. Furthermore, the cathode tolerance to CO_2 was improved in relation to the precursor. It is possible to find that the exsolving methods have great potential to prepare precious metal nanoparticle-modified perovskites oxides as efficient catalysts towards fuel cells applications [108].

Taking into account oxygen electrocatalysis, there are some cases of bifunctional catalysts, which means that they act in ORR and OER. The bifunctional electrocatalysts are fundamental to the development of regenerative fuel cells, which are devices that produce energy and electrolytically regenerate their reactants by using stacks of electrochemical cells [109]. Retuerto and co-workers [110] prepared the functional perovskite electrocatalyst $La_{1.5}Sr_{0.5}NiMn_{0.5}Ru_{0.5}O_6$ (denoted as LSNMR) from the mixture of La_2O_3, the Ru, Mn and Ni oxides, and $SrCO_3$ at stoichiometric proportions. The efficiency of LSNMR electrocatalyst towards ORR and OER was measured in alkaline medium using the bifunctional index (BI) proposed by Schuhmann et al. [111]. The BI value determined to LSNMR bifunctional electrocatalyst was approximately 0.83 V, while an ideal ORR-OER electrocatalyst must show a BI value close to 0 V and the state-of-art bifunctional electrocatalysts shows values around 0.9 V. To the best of our knowledge, the LSNMR electrocatalyst appears to be the unique example of bifunctional catalyst for ORR and OER containing a precious metal into a perovskite structure. For comparison purposes, we can highlight the BI values of some bifunctional electrocatalysts, such as $La_{0.58}Sr_{0.4}Co_{0.2}Fe_{0.8}O_3/FeN_x$–C (0.86 V) [112], $La_{0.7}(Ba_{0.5}Sr_{0.5})_{0.3}Co_{0.8}Fe_{0.2}O_3-\delta$/C (0.88 V) [113], $La_{0.58}Sr_{0.4}Co_{0.2}Fe_{0.8}O_3$/N-CNT (0.826 V) [111], $LaNi_{0.8}Fe_{0.2}O_3$ (1.02 V) [114], $LaNi_{0.85}Mg_{0.15}O_3$ (1.15 V) [115], and the state-of-art catalysts RuO_2 (0.8 V) and IrO_2 (1.32 V) [112].

Water splitting has been arising interest for its possibility to generate hydrogen in a carbon-neutral manner. This process can be achieved by using electrodes under a light source, converting H_2O to H_2 and O_2 through a photoelectrochemical (PEC) process [116]. Several transition metal oxides, and oxygen and halide perovskites have been used in the PEC evaluation of H_2 [117]. Kumar and co-workers [94] provide an example after having developed an anode of NaNbO3 nanorods decorated with plasmonic Ag nanoparticles applied to PEC H_2 evaluation. The results showed that both bare $NaNbO_3$ and the Ag-decorated nanorods exhibit low current density due to H_2 generation in dark conditions, but when the system was irradiated with a halogen lamp, the Ag-decorated nanorods anode exhibit 4-fold current density in relation to bare $NaNbO_3$ catalyst. The origins of such an enhancement were investigated using the EIS technique, which revealed that the decoration of $NaNbO_3$ increases its donor density, indicating the transfer of majority carriers from plasmonic Ag nanoparticles to perovskite matrix resulting in a higher concentration of charge carriers.

5. Final Considerations

Photocatalysis and electrocatalysis have been great routes to degrade persistent organic pollutants based on advanced oxidative processes. Because of their electrical and optical properties, perovskite structured materials represent a good alternative to replace TiO_2-based materials in environmental treatment using AOPs. The perovskite properties are improved with the presence of precious metals in its crystalline structure and/or the formation of a precious metal/perovskite structure hybrid material. The most used precious metals for this purpose are silver and gold. This occurs due to the surface plasmon resonance phenom presented by metals. These semiconductors can be processed as powder and/or films. Chemical syntheses are efficient routes to produce these catalysts, especially hydrothermal synthesis. Based on the processing route discussed here, it is possible to obtain a catalyst with appropriate characteristics to be applied in both the dye decomposition and water-splitting process.

Author Contributions: Bibliographical review about photocatalysis and the use of perovskites in this issue, chemical synthesis survey and eletrocathalysis approach were done by G.F.T.; E.S.J. and R.V. under supervision of M.A.Z. and F.C. All authors contributed to the manuscript preparation, and as the same way, all authors have given approval for the final version of the review.

Funding: This research was funded by Chemistry Postgraduate Program of Federal University of Goiás, CAPES (Process number 88882.306480/2018-1) and FAPESP-CEPID/CDMF 2013/07296-2

Acknowledgments: The authors would like to thank the Brazilian research agencies CAPES (Process: 88882.306480/2018-1), CNPq, (Process 554569/2010-8 and 475609/2008-5) and, FAPESP-CEPID/CDMF 2013/07296-2 for granting the support to research groups.

Conflicts of Interest: The authors declare no conflict of interest.

References

1. IUPAC. Compendium of chemical terminology. In *The Gold Book*, 2nd ed.; McNaught, A.D., Wilkinson, A., Eds.; Blackwell Scientific Publications: Oxford, UK, 1997.
2. Deng, Y.; Zhao, R. Advanced Oxidation Processes (AOPs) in Wastewater Treatment. *Curr. Pollut. Rep.* **2015**, *1*, 167–176. [CrossRef]
3. Yang, L.; Yang, L.; Ding, L.; Deng, F.; Luo, X.-B.; Luo, S.-L. Principle for the application of nanomaterials in environmental pollution control and resource reutilization. In *Nanomaterial for the Removal of Pollutants and Resource Reutilization*; Luo, X., Deng, F., Eds.; Elsevier: Amsterdam, The Netherlands, 2019; pp. 1–23.
4. Glaze, W.H.; Kang, J.-W.; Chapin, D.H. The chemistry of water treatment processes involving ozone, hydrogen peroxide and ultraviolet radiation. *Ozone Sci. Eng.* **1987**, *9*, 335–352. [CrossRef]
5. Parsons, S. *Advanced Oxidation Processes for Water and Wastewater Treatment*; IWA Publishing: London, UK, 2004; pp. 1–347.
6. Fan, X.; Hao, H.; Shen, X.; Chen, F.; Zhang, J. Removal and degradation pathway study of sulfasalazine with Fenton-like reaction. *J. Hazard. Mater.* **2001**, *190*, 493–500. [CrossRef] [PubMed]
7. Rehman, S.; Ullah, R.; Butt, A.M.; Gohar, N.D. Strategies of making TiO2 and ZnO visible light active. *J. Hazard. Mater.* **2009**, *170*, 560–569.
8. WHO. Fluoride in drinking-water. In *Guidelines for Drinking-Water Quality*, 3rd ed.; WHO Press: Geneva, Switzerland, 2006; Volume 1, pp. 375–377.
9. Ibrahim, M.; Asimrasheed, M.; Sumalatha, M.; Prabhakar, P. Effects of fluoride contents in ground water: A review. *Inter. J. Pharm. Appl.* **2011**, *2*, 128–134.
10. Singh, J.; Singh, P.; Singh, A. Fluoride ions vs. removal technologies: A study. *Arab. J. Chem.* **2016**, *9*, 815–824. [CrossRef]
11. Modi, S.; Soni, R. Merits and Demerits of different technologies of defluoridation for drinking water. *IOSR J. Environ. Sci. Toxicol. Food. Technol.* **2013**, *3*, 24–27. [CrossRef]
12. Gupta, V.K.; Ali, I.; Saleh, T.A.; Nayak, A.; Agarwal, S. Chemical treatment technologies for waste-water recycling—An overview. *RSC Adv.* **2012**, *2*, 6380–6388. [CrossRef]
13. Gogate, P.R.; Pandit, A.B. A review of imperative technologies for wastewater treatment II: Hybrid methods. *Adv. Environ. Res.* **2004**, *8*, 553–597. [CrossRef]
14. Oller, I.; Malato, S.; Sanchez-Perez, J.A. Combination of advanced oxidation processes and biological treatments for wastewater decontamination—A review. *Sci. Total Environ.* **2011**, *409*, 4141–4166. [CrossRef]
15. Esplugas, S.; Bila, D.M.; Krause, L.G.T.; Dezotti, M. Ozonation and advanced oxidation technologies to remove endocrine disrupting chemicals (EDCs) and pharmaceuticals and personal care products (PPCPs) in water effluents. *J. Hazard. Mater.* **2007**, *149*, 631–642. [CrossRef] [PubMed]
16. Saravanan, R.; Gracia, F.; Stephen, A. Basic principles, mechanism, and challenges of photocatalysis. In *Nanocomposites for Visible Light-Induced Photocatalysis*; Khan, M.M., Pradhan, D., Shon, Y., Eds.; Springer: Cham, Switzerland, 2018; pp. 19–40.
17. Serpone, N.; Horikoshi, S.; Emeline, A.V. Microwaves in advanced oxidation processes for environmental applications. A brief review. *J. Photochem. Photobiol. C* **2010**, *11*, 114–131. [CrossRef]

18. Legrini, O.; Oliveros, E.; Braun, A.M. Photochemical processes for water treatment. *Chem. Rev.* **1993**, *93*, 671–698. [CrossRef]
19. Huang, C.P.; Dong, C.; Tang, Z. Advanced chemical oxidation: Its present role and potential future in hazardous waste treatment. *Waste Manag.* **1993**, *13*, 361–377. [CrossRef]
20. Pignatello, J.J.; Oliveros, S.E.; Mackay, A. Advanced oxidation processes of organic contaminant destruction based of the Fenton reaction and related chemistry. *Crit. Rev. Environ. Sci. Technol.* **2006**, *36*, 1–84. [CrossRef]
21. Carraway, E.R.; Hoffman, A.J.; Hoffmann, M.R. Photocatalytic oxidation of organic acids on quantum-sized semiconductor colloids. *Environ. Sci. Technol.* **1994**, *28*, 786–793. [CrossRef] [PubMed]
22. Neppolian, B.; Choi, H.S.; Sakthivel, S.; Arabindoo, B.; Murugesan, V. Solar light induced and TiO_2 assisted degradation of textile dye reactive blue 4. *Chemosphere* **2002**, *46*, 1173–1181. [CrossRef]
23. Reza, K.M.; Kurny, A.S.; Gulshan, F. Parameters affecting the photocatalytic degradation of dyes using TiO_2: A review. *Appl. Water Sci.* **2015**, *4*, 1569–1578. [CrossRef]
24. Lim, M.; Son, Y.; Khim, J. Frequency effects on the sonochemical degradation of chlorinated compounds. *Ultrason. Sonochem.* **2011**, *8*, 460–465. [CrossRef]
25. Doodeve, C.F.; Kitchener, J.A. Photosensitisation by titanium dioxide. *Trans. Faraday Soc.* **1938**, *34*, 570–579.
26. Hashimoto, K.; Irie, H.; Fujishima, A. TiO_2 photocatalysis: A historical overview and future prospects. *Jpn. J. Appl. Phys.* **2005**, *44*, 8269–8285. [CrossRef]
27. Ameta, S.C.; Ameta, R. Introduction. In *Advanced Oxidation Processes for Wastewater Treatment*; Ameta, S.C., Ameta, R., Eds.; Academic Press: Oxford, UK, 2018; pp. 1–12.
28. Gaya, U.I. Principles of heterogeneous photocatalysis. In *Heterogeneous Photocatalysis Using Inorganic Semiconductor Solids*; Gaya, U.I., Ed.; Springer Science: Dordrecht, The Netherlands, 2014; pp. 1–34.
29. Gratzel, M. Photoelectrochemical cells. *Nature* **2001**, *414*, 338–344. [CrossRef]
30. Fujishima, A.; Honda, K. Electrochemical photolysis of water at a semiconductor electrode. *Nature* **1972**, *238*, 37–38. [CrossRef]
31. Cowan, A.J.; Durrant, J. Long-lived charge separated states in nanostructured semiconductor photoelectrodes for the production of solar fuels. *Chem. Soc. Rev.* **2013**, *42*, 2281–2293. [CrossRef]
32. Al-Ekabi, H.; Serpone, M. Kinetic studies in heterogeneous photocatalysis. 1. Photocatalytic degradation of chlorinated phenols in aerated aqueous solutions over TiO_2 supported on a glass matrix. *J. Phys. Chem.* **1988**, *92*, 5726–5731. [CrossRef]
33. Serpone, N. Brief introductory remarks on heterogeneous photocatalysis. *Sol. Energy Mater. Sol. Cells* **1995**, *38*, 369–379. [CrossRef]
34. Paramasivam, I.; Jha, H.; Liu, N.; Schmuki, P. A review of photocatalysis using self-organized TiO_2 nanotubes and other ordered oxide nanostructures. *Small* **2012**, *8*, 3073–3103. [CrossRef]
35. Shen, Y.; Guo, X.; Bo, X.; Wang, T.; Guo, X.; Xie, M.; Guo, X. Effect of template-induced surface species on electronic structure and photocatalytic activity of $g-C_3N_4$. *Appl. Surf. Sci.* **2017**, *396*, 933–938. [CrossRef]
36. Ravelli, D.; Dondi, D.; Fagnoni, M.; Albini, A. Photocatalysis. A multi-faceted concept for green chemistry. *Chem. Soc. Rev.* **2009**, *38*, 1999–2011. [CrossRef]
37. Liu, B.; Zhao, X.; Terashima, C.; Fujishima, A.; Nakata, K. Thermodynamic and kinetic analysis of heterogeneous photocatalysis for semiconductor systems. *Phys. Chem. Chem. Phys.* **2014**, *16*, 8751–8760. [CrossRef]
38. Henderson, M.A. A surface science perspective on TiO_2 photocatalysis. *Surf. Sci. Rep.* **2011**, *66*, 185–297. [CrossRef]
39. Grabowska, E. Selected perovskite oxides: Characterization, preparation and photocatalytic properties—A review. *Appl. Catal. B Environ.* **2016**, *186*, 97–126. [CrossRef]
40. Zhang, G.; Liu, G.; Wang, L.; Irvine, J.T.S. Inorganic perovskite photocatalysts for solar energy utilization. *Chem. Soc. Rev.* **2016**, *45*, 5951–5984. [CrossRef]
41. Rojas-Cervantes, M.L.; Castillejos, E. Perovskites as catalysts in advanced oxidation processes for wastewater treatment. *Catalysts* **2019**, *9*, 230. [CrossRef]
42. Shi, J.; Guo, L. ABO_3-based photocatalysts for water splitting. *Prog. Nat. Sci. Mater. Inter.* **2012**, *22*, 592–615. [CrossRef]
43. Mitchell, R.H. *Perovskites: Modern and Ancient*; Almaz Press: Thunder Bay, ON, Canada, 2002; pp. 1–318.
44. Glazer, A.M. The classification of tilted octahedra in perovskites. *Acta Crystallogr. Sect. B* **1972**, *28*, 3384–3392. [CrossRef]

45. Tilley, R.J.D. *Perovskites. Structure-Property Relationships*; John Wiley & Sons Ltd.: Chichester, UK, 2016; pp. 1–315.
46. Mishra, A.; Prasad, R. Preparation and application of perovskite catalyst for diesel soot emission control: An overview. *Catal. Rev.* **2014**, *56*, 57–81. [CrossRef]
47. Knight, K.S. Structural phase transitions, oxygen vacancy ordering and protonation in doped $BaCeO_3$: Results from time-of-flight neutron powder diffraction investigations. *Solid State Ion.* **2001**, *145*, 275–294. [CrossRef]
48. Woodward, P.M. Octahedral tilting in perovskites. I. geometrical considerations. *Acta Crystallogr. Sect. B Struct. Sci.* **1997**, *53*, 32–43. [CrossRef]
49. Goldschmidt, V.M. Die Gesetze der Krystallochemie. *Naturwissenschaften* **1926**, *14*, 477–485. [CrossRef]
50. Shannon, R.D. Revised effective ionic radii and systematic studies of interatomie distances in halides and chaleogenides. *Acta Crystallogr. Sect. A Cryst. Phys. Diffr. Theor. Gen. Crystallogr.* **1976**, *A32*, 751–767. [CrossRef]
51. Fan, Z.; Sun, K.; Wang, J. Perovskites for photovoltaics: A combined review of organic–inorganic halide perovskites and ferroelectric oxide perovskites. *J. Mater. Chem. A* **2015**, *3*, 18809–18828. [CrossRef]
52. Yamada, I.; Takamatsu, A.; Ikeno, H. Complementary evaluation of structure stability of perovskite oxides using bond-valence and density functional-theory calculations. *Sci. Technol. Adv. Mater.* **2018**, *19*, 102–107. [CrossRef]
53. Zhang, L.; Zhou, Y.; Guo, L.; Zhao, W.W.; Barnes, A.; Zhang, H.-T.; Eatom, C.; Zheng, Y.; Brahlek, M.; Haneef, H.F.; et al. Correlated metals as transparent conductor. *Nat. Mater.* **2016**, *15*, 204–210. [CrossRef]
54. Imran, Z.; Rafiq, M.A.; Hasan, M.M. Charge carrier transport mechanisms in perovskite $CdTiO_3$ fibers. *AIP Adv.* **2014**, *4*, 067137. [CrossRef]
55. Nikonov, A.V.; Kuterbekov, K.A.; Bekmyrza, K.Z.; Pavzderin, N.B. A brief review of conductivity and thermal expansion of perovskite-related oxides for SOFC cathode. *Eurasian J. Phys. Funct. Mater.* **2018**, *2*, 274–292. [CrossRef]
56. Yang, F.; Yang, L.; Ai, C.; Xie, P.; Lin, S.; Wang, C.-Z.; Lu, X. Tailoring band gap of perovskite $BaTiO_3$ by Transition metals co-doping for visible-light photoelectrical applications: A first-principles study. *Nanomaterials* **2018**, *8*, 455. [CrossRef]
57. Hwang, S.W.; Noh, T.H.; Cho, I.S. Optical properties, electronic structures, and photocatalytic performances of bandgap-tailored $SrBi_2Nb_{2-x}V_xO_9$ compounds. *Catalysts* **2019**, *9*, 393. [CrossRef]
58. Linic, S.; Christopher, P.; Ingram, D.B. Plasmonic-metal nanostructures for efficient conversion of solar to chemical energy. *Nat. Mater.* **2001**, *10*, 911–921. [CrossRef]
59. Linic, S.; Aslam, U.; Boerigter, C.; Morabito, M. Photochemical transformations on plasmonic metal nanoparticles. *Nat. Mater.* **2015**, *14*, 567–576. [CrossRef]
60. Nycenga, M.; Cobley, C.M.; Zeng, J.; Li, W.; Moran, C.H.; Zhang, Q.; Qin, D.; Xia, Y. Controlling the synthesis and assembly of silver nanostructures for plasmonic applications. *Chem. Rev.* **2011**, *111*, 3669–3712. [CrossRef]
61. Wang, M.; Ye, M.; Iocozzia, J.; Lin, C.; Lin, Z. Plasmon-mediated solar energy conversion via photocatalysis in noble metal/semiconductor composites. *Adv. Sci.* **2016**, *3*, 1600024. [CrossRef]
62. Hou, W.; Liu, Z.; Pavaskar, P.; Hung, W.H.; Cronin, S.B. Plasmonic enhancement of photocatalytic decomposition of methyl orange under visible light. *J. Catal.* **2011**, *277*, 149–153. [CrossRef]
63. Nie, J.; Scheneider, J.; Sieland, F.; Zhou, L.; Xia, S.; Bahnemann. New insights into the surface plasmon resonance (SPR) driven photocatalytic H_2 production of $Au-TiO_2$. *RSC Adv.* **2018**, *8*, 25881–25887. [CrossRef]
64. Liz-Marzán, L.M. Tailoring surface plasmons through the morphology and assembly of metal nanoparticles. *Langmuir* **2006**, *22*, 32–41. [CrossRef]
65. Zhang, X.; Zhu, Y.; Yang, X.; Wang, S.; Shen, J.; Lin, B.; Li, C. Enhanced visible light photocatalytic activity of interlayer-isolated triplex $Ag@SiO_2@TiO_2$ core-shell nanoparticles. *Nanoscale* **2013**, *5*, 3359–3366. [CrossRef]
66. Hou, W.; Cronin, S.B. A review of surface plasmon resonance-enhanced photocatalysis. *Adv. Funct. Mater.* **2013**, *23*, 1612–1619. [CrossRef]
67. Wang, H.; Tam, F.; Grady, N.K.; Halas, N.J. Cu nanoshells: Effects of interband transitions on the nanoparticle plasmon resonance. *J. Phys. Chem. B* **2005**, *109*, 18218–18222. [CrossRef]
68. Musialic-Piotrowwska, A.; Landmesser, H. Noble metal-doped perovskites for the oxidation of organic air pollutants. *Catal. Today* **2008**, *137*, 357–361. [CrossRef]

69. Smart, L.E.; Moore, E.A. *Solid State Chemistry*, 3rd ed.; Taylor & Francis: Abingdon, UK, 2005; pp. 148–177.
70. West, A.R. *Solid State Chemistry and Its Applications*, 2nd ed.; Wiley: Chichester, UK, 2014; pp. 187–228.
71. Iwashina, K.; Kudo, A. Rh-doped SrTiO$_3$ photocatalyst electrode showing cathodic photocurrent for water splitting under visible-light irradiation. *J. Am. Chem. Soc.* **2011**, *133*, 13272–13275. [CrossRef]
72. Irie, H.; Maruyama, Y.; Hashimoto, K. Ag$^+$ and Pb^{2+} doped SrTiO$_3$ Photocatalysts. A correlation between band structure and photocatalytic activity. *J. Phys. Chem. C* **2007**, *111*, 1847–1852. [CrossRef]
73. Saadetnejad, D.; Yıldırım, R. Photocatalytic hydrogen production by water splitting over Au/Al-SrTiO$_3$. *Int. J. Hydrog. Energy* **2018**, *43*, 1116–1122. [CrossRef]
74. Wang, F.; Wang, T.; Lang, J.; Su, Y.; Wang, X. Improved photocatalytic activity and durability of AgTaO$_3$/AgBr heterojunction: The relevance of phase and electronic structure. *J. Mol. Catal. A Chem.* **2017**, *426*, 52–59. [CrossRef]
75. Konta, R.; Ishii, T.; Kato, H.; Kudo, A. Photocatalytic activities of noble metal ion doped SrTiO$_3$ under visible light irradiation. *J. Phys. Chem. B* **2004**, *108*, 8992–8995. [CrossRef]
76. Sajjadi, S.P. Sol-gel process and its application in Nanotechnology. *J. Polym. Eng. Technol.* **2005**, *13*, 38–41.
77. Rao, B.G.; Mukherjee, D.; Reddy, B.M. Nanostructures for novel therapy novel approaches for preparation of nanoparticles. In *Nanostructures for Novel Therapy*; Fricai, D., Grumezescu, A., Eds.; Elsevier: Amsterdam, The Netherlands, 2017; pp. 1–36.
78. Nayak, A.K.; Das, B. Introduction to polymeric gels. In *Polymeric Gels. Characterization, Properties and Biomedical Applications*; Pal, K., Banerjee, I., Eds.; Elsevier: Duxford, UK, 2018; pp. 3–27.
79. Zhang, H.; Chen, G.; He, X.; Xu, J. Electronic structure and photocatalytic properties of Ag-La codoped CaTiO$_3$. *J. Alloy. Compd.* **2012**, *516*, 91–95. [CrossRef]
80. Reddy, K.H.; Martha, S.; Parida, K.M. Erratic charge transfer dynamics of Au/ZnTiO$_3$ nanocomposites under UV and visible light irradiation and their related photocatalytic activities. *Nanoscale* **2018**, *10*, 18540–18554. [CrossRef]
81. Li, Y.; Li, J.; Chen, L.; Sun, H.; Zhang, H.; Guo, H.; Feng, L. In situ synthesis of au-induced hierarchical nanofibers/nanoflakes structured BiFeO3 homojunction photocatalyst with enhanced photocatalytic activity. *Front. Chem.* **2019**, *6*, 649–657. [CrossRef]
82. Feng, S.; Li, G. Hydrothermal and solvothermal syntheses. In *Modern Inorganic Synthetic Chemistry*, 2nd ed.; Xu, R., Xu, Y., Eds.; Elsevier: Amsterdam, The Netherlands, 2017; pp. 73–104.
83. Yan, Y.; Yang, H.; Yi, Z.; Li, R.; Wang, X. Enhanced photocatalytic performance and mechanism of Au@CaTiO$_3$ composites with au nanoparticles assembled on CaTiO$_3$ nanocuboids. *Micromachines* **2019**, *10*, 254. [CrossRef]
84. Malkhasian, A.Y.S. Synthesis and characterization of Pt/AgVO$_3$ nanowires for degradation of atrazine using visible light irradiation. *J. Alloy. Compd.* **2015**, *649*, 394–399. [CrossRef]
85. Zielińska, B.; Borowiak-Palen, E.; Kalenczuk, R.J. Preparation, characterization and photocatalytic activity of metal-loaded NaNbO$_3$. *J. Phys. Chem. Solids* **2011**, *72*, 117–123. [CrossRef]
86. Chen, W.; Hu, Y.; Ba, M. Surface interaction between cubic phase NaNbO$_3$ nanoflowers and Ru nanoparticles for enhancing visible-light driven photosensitized photocatalysis. *Appl. Surf. Sci.* **2018**, *435*, 483–493. [CrossRef]
87. Zhang, T.; Lei, W.; Liu, P.; Rodriguez, J.A.; Yu, J.; Qi, Y.; Liu, G.; Liu, M. Organic Pollutant Photodecomposition by Ag/KNbO$_3$ Nanocomposites: A combined experimental and theoretical study. *J. Phys. Chem. C* **2016**, *120*, 2777–2786. [CrossRef]
88. Xing, P.; Wu, S.; Chen, Y.; Chen, P.; Hu, X.; Lin, H.; Zhao, L.; He, Y. New Application and Excellent Performance of Ag/KNbO$_3$ Nanocomposite in Photocatalytic NH$_3$ Synthesis. *ACS Sustain. Chem. Eng.* **2019**, *7*, 12408–124118. [CrossRef]
89. Xu, D.; Yang, S.; Jin, Y.; Chen, M.; Fan, W.; Luo, B.; Shi, W. Ag-decorated ATaO$_3$ (A = K, Na) nanocube plasmonic photocatalysts with enhanced photocatalytic water-splitting properties. *Langmuir* **2015**, *31*, 9694–9699. [CrossRef]
90. Jayabal, P.; Sasirekha, V.; Mayandi, J.; Jeganathan, K.; Ramakrishnan, V. A facile hydrothermal synthesis of SrTiO$_3$ for dye sensitized solar cell application. *J. Alloy. Compd.* **2014**, *586*, 456–461. [CrossRef]
91. Gao, H.; Yang, H.; Wang, S. Hydrothermal synthesis, growth mechanism, optical properties and photocatalytic activity of cubic SrTiO$_3$ particles for the degradation of cationic and anionic dyes. *Optik* **2018**, *175*, 237–249. [CrossRef]

92. Zhang, Q.; Huang, Y.; Xu, L.; Cao, J.; Ho, W.; Lee, S.C. Visible-light-active plasmonic Ag-SrTiO$_3$ nanocomposites for the degradation of NO in air with high selectivity. *ACS Appl. Mater. Interfaces* **2016**, *8*, 4165–4178. [CrossRef]
93. Kalyani, V.; Vasile, B.S.; Ianculescu, A.; Testino, A.; Carino, A.; Buscaglia, M.T.; Buscaglia, V.; Nanni, P. Hydrothermal synthesis of SrTiO$_3$: Role of interfaces. *Cryst. Growth Des.* **2015**, *15*, 5712–5725. [CrossRef]
94. Kumar, D.; Singh, S.; Khare, N. Plasmonic Ag nanoparticles decorated NaNbO$_3$ nanorods for efficient photoelectrochemical water splitting. *Int. J. Hydrog. Energy* **2018**, *43*, 1–8. [CrossRef]
95. Jeong, S.Y.; Shin, H.-M.; Jo, Y.-R.; Kin, Y.J.; Kim, S.; Lee, W.-J.; Lee, G.J.; Song, J.; Moon, B.J.; Seo, S.; et al. Plasmonic Silver nanoparticle-impregnated nanocomposite BiVO$_4$ photoanode for plasmon-enhanced photocatalytic water splitting. *J. Phys. Chem. C* **2018**, *122*, 7088–7093. [CrossRef]
96. Zhang, S.; Zhang, B.-P.; Li, S.; Huang, Z.; Yang, C.; Wang, H. Enhanced photocatalytic activity in Ag-nanoparticle-dispersed BaTiO$_3$ composite thin films: Role of charge transfer. *J. Adv. Ceram.* **2017**, *6*, 1–10. [CrossRef]
97. Gpel, W.; Schierbaum, K.D.; Vayenas, C.G.; Yentekakis, I.V.; Gerischer, H.; Vielstich, W.; Savage, P.E.; Moyes, R.B.; Bond, G.; Suslick, K.S. Special Catalytic Systems. In *Handbook of Heterogeneous Catalysis*; Ertl, G., Knözinger, H., Weitkamp, J., Eds.; Wiley-VCH Verlag GmbH: Weinheim, Germany, 1997; pp. 1283–1357.
98. Wang, Y.-J.; Zhao, N.; Fang, B.; Li, H.; Bi, X.T.; Wang, H. Carbon-Supported Pt-Based Alloy Electrocatalysts for the Oxygen Reduction Reaction in Polymer Electrolyte Membrane Fuel Cells: Particle Size, Shape, and Composition Manipulation and Their Impact to Activity. *Chem. Rev.* **2015**, *115*, 3433–3467. [CrossRef]
99. Shan, S.; Wu, J.; Kang, N.; Cronk, H.; Zhao, Y.; Zhao, W.; Skeete, Z.; Joseph, P.; Trimm, B.; Luo, J.; et al. Nanoscale Alloying in Electrocatalysts. *Catalysts* **2015**, *5*, 1465–1478. [CrossRef]
100. Zhu, H.; Zhang, P.; Dai, S. Recent Advances of Lanthanum-Based Perovskite Oxides for Catalysis. *ACS Catal.* **2015**, *5*, 6370–6385. [CrossRef]
101. Ghosh, S.; Basu, R.N. Multifunctional nanostructured electrocatalysts for energy conversion and storage: Current status and perspectives. *Nanoscale* **2018**, *10*, 11241–11280. [CrossRef]
102. Ghosh, S.K.; Rahaman, H. Noble metal-manganese oxide hybrid nanocatalysts. In *Noble Metal-Metal Oxide Hybrid Nanoparticles*; Elsevier: Amsterdam, The Netherlands, 2019; pp. 313–340.
103. Zhang, J.; Xia, Z.; Dai, L. Carbon-based electrocatalysts for advanced energy conversion and storage. *Sci. Adv.* **2015**, *1*, e1500564. [CrossRef]
104. Xu, Y.; Zhang, B. Recent advances in porous Pt-based nanostructures: Synthesis and electrochemical applications. *Chem. Soc. Rev.* **2014**, *43*, 2439–2444. [CrossRef]
105. Shao-Horn, Y.; Hwang, J.; Yu, Y.; Katayama, Y.; Rao, R.R.; Giordano, L. Perovskites in catalysis and electrocatalysis. *Science* **2017**, *358*, 751–756.
106. Zhang, Y.Q.; Tao, H.B.; Liu, J.; Sun, Y.F.; Chen, J.; Hua, B.; Thundat, T.; Luo, J.L. A rational design for enhanced oxygen reduction: Strongly coupled silver nanoparticles and engineered perovskite nanofibers. *Nano Energy* **2017**, *38*, 392–400. [CrossRef]
107. Zhu, Y.; Zhou, W.; Ran, R.; Chen, Y.; Shao, Z.; Liu, M. Promotion of Oxygen Reduction by Exsolved Silver Nanoparticles on a Perovskite Scaffold for Low-Temperature Solid Oxide Fuel Cells. *Nano Lett.* **2016**, *16*, 512–518. [CrossRef]
108. Huang, K. An emerging platform for electrocatalysis: Perovskite exsolution. *Sci. Bull.* **2016**, *61*, 1783–1784. [CrossRef]
109. Mitlitsky, F.; Myers, B.; Weisberg, A.H. Regenerative fuel cell systems. *Energy Fuels* **1998**, *12*, 56–71. [CrossRef]
110. Retuerto, M.; Calle-Vallejo, F.; Pascual, L.; Lumbeeck, G.; Fernandez-Diaz, M.T.; Croft, M.; Gopalakrishnan, J.; Peña, M.A.; Hadermann, J.; Greenblatt, M. La$_{1.5}$Sr$_{0.5}$NiMn$_{0.5}$Ru$_{0.5}$O$_6$ double perovskite with enhanced ORR/OER bifunctional catalytic activity. *ACS Appl. Mater. Interfaces* **2019**, *11*, 21454–21464. [CrossRef]
111. Elumeeva, K.; Masa, J.; Tietz, F.; Yang, F.; Xia, W.; Muhler, M.; Schuhmann, W. A Simple Approach towards High-Performance Perovskite-Based Bifunctional Oxygen Electrocatalysts. *ChemElectroChem* **2016**, *3*, 138–143. [CrossRef]
112. Rincón, R.A.; Masa, J.; Mehrpour, S.; Tietz, F.; Schuhmann, W. Activation of oxygen evolving perovskites for oxygen reduction by functionalization with Fe-Nx/C groups. *Chem. Commun.* **2014**, *50*, 14760–14762. [CrossRef]
113. Elumeeva, K.; Masa, J.; Sierau, J.; Tietz, F.; Muhler, M.; Schuhmann, W. Perovskite-based bifunctional electrocatalysts for oxygen evolution and oxygen reduction in alkaline electrolytes. *Electrochim. Acta* **2016**, *208*, 25–32. [CrossRef]

114. Zhang, D.; Song, Y.; Du, Z.; Wang, L.; Li, Y.; Goodenough, J.B. Active LaNi$_{1-x}$Fe$_x$O$_3$ bifunctional catalysts for air cathodes in alkaline media. *J. Mater. Chem. A* **2015**, *3*, 9421–9426. [CrossRef]
115. Du, Z.; Yang, P.; Wang, L.; Lu, Y.; Goodenough, J.B.; Zhang, J.; Zhang, D. Electrocatalytic performances of LaNi$_{1-x}$Mg$_x$O$_3$ perovskite oxides as bi-functional catalysts for lithium air batteries. *J. Power Sources* **2014**, *265*, 91–96. [CrossRef]
116. Warwick, M.E.A.; Kaunisto, K.; Barreca, D.; Carraro, G.; Gasparotto, A.; Maccato, C.; Bontempi, E.; Sada, C.; Ruoko, T.-P.; Turner, S.; et al. Vapor Phase Processing of α-Fe$_2$O$_3$ Photoelectrodes for Water Splitting: An Insight into the Structure/Property Interplay. *ACS Appl. Mater. Interfaces* **2015**, *7*, 8667–8676. [CrossRef]
117. Guerrero, A.; Bisquert, J. Perovskite semiconductors for photoelectrochemical water splitting applications. *Curr. Opin. Electrochem.* **2017**, *2*, 144–147. [CrossRef]

 © 2019 by the authors. Licensee MDPI, Basel, Switzerland. This article is an open access article distributed under the terms and conditions of the Creative Commons Attribution (CC BY) license (http://creativecommons.org/licenses/by/4.0/).

MDPI
St. Alban-Anlage 66
4052 Basel
Switzerland
Tel. +41 61 683 77 34
Fax +41 61 302 89 18
www.mdpi.com

Catalysts Editorial Office
E-mail: catalysts@mdpi.com
www.mdpi.com/journal/catalysts

www.ingramcontent.com/pod-product-compliance
Lightning Source LLC
LaVergne TN
LVHW071946080526
838202LV00064B/6690